Introduction to Ion Beam Biotechnology

Yu Zengliang

Introduction to Ion Beam Biotechnology

Translated by Yu Liangdeng   Thiraphat Vilaithong
Ian Brown

With 132 illustrations, including 1 color plate.

 Springer

Yu Zengliang
Key Laboratory of Ion Beam Bioengineering of Chinese Academy of Sciences
Institute of Plasma Physics
Chinese Academy of Sciences
P.O. Box 1126
Hefei 230031
People's Republic of China
zlyu@ipp.ac.cn

Cover illustration: Instantaneous expression of the GUS gene entering rice embryo cells treated by 30-keV N ions at a dose of 2×10^{15} N$^+$/cm^2. Rice embryo cell on the left is the control; the embryo cell on the right is due to GUS, the gene transferred cells show blue color.

Library of Congress Control Number: 2005923442

ISBN-10: 0-387-25531-1   e-ISBN 0-387-25586-9
ISBN-13: 978-0387-25531-6

Printed on acid-free paper.

Printed in the United States of America. (EB)

9 8 7 6 5 4 3 2 1

springeronline.com

To all the researchers who have concerned themselves with this new and fascinating field, and with thanks to 37 unnamed graduate students

PREFACE

(to the English edition)

Although low-energy ion beam biotechnology has been the subject of active research in a few countries, particularly China, for over a decade, and impressive accomplishments in biology and agriculture have been achieved there, the field is new and has not yet been well recognized by most of the world. At Chiang Mai University we have known for quite some time that Prof. Yu Zengliang, Director of The Key Laboratory of Ion Beam Bioengineering, Institute of Plasma Physics, Chinese Academy of Sciences, was a founder of this new, highly interdisciplinary branch of science, and that he has carried out research in this field since the late 1980s. Stimulated by Prof. Yu's achievements, Thai scientists, joined by Chinese and American colleagues, organized research in this area at Chiang Mai University for serving economic development in Thailand. Only in a few short years, we have made great progress. In the laboratory of Fast Neutron Research Facility, a specialized bioengineering ion beam line has been installed. Research on both fundamental interactions between ions and biological organisms and applications to agricultural and horticultural crop mutation breeding and gene transfer has been carried out. A number of important results have been achieved and published. Besides having worked independently, the Thai researchers have had the pleasure of frequent and close academic exchange with Prof. Yu Zengliang and his colleagues. As work progressed, the need for a text elaborating the fundamentals of low-energy ion beam biotechnology grew increasingly clear. We recently learned that Prof. Yu Zengliang had written and published a book in Chinese on the fundamentals of this fascinating technology – precisely the kind of monograph needed. The idea of translating Prof. Yu's book into English was immediately suggested by Prof. Thiraphat Vilaithong, President of the Thai Institute of Physics, and warmly supported by Prof. Yu Zengliang himself. The original Chinese edition of the book consists of twelve chapters and an appendix. In order to reflect more recent developments in research on the technology, Prof. Yu has added two chapters (Chapter 13 and Chapter 14 of this English edition) as well as some parts in Section 2.4. of Chapter 2, Section 6.6. of Chapter 6 and Section 11.3. of Chapter 11. Translating Prof. Yu's book has been a great challenge, as the job involves aspects of physics, biology, agriculture, chemistry, and more. We needed assistance and consultation. We would like to sincerely thank those who have made important contributions to the creation of this English edition of Prof. Yu's book. They are

- Prof. Yu Zengliang, for his full support and suggestions for the translation and kind cooperation and concern;

- Dr. Gordon L. Hiebert, for his complete reading and commenting of the book manuscript, two times;
- Dr. Xu An and Dr. Hu Zhiwen, for their enthusiastic assistance in coordinating the work between the Thai and Chinese sides and effective management of the work in China;
- Fifty-two graduate students, most of whom are Ph.D. candidates, of The Key Laboratory of Ion Beam Bioengineering, Institute of Plasma Physics, Chinese Academy of Sciences, for their invaluable assistance in checking the translations and re-preparing many figures and references;
- Prof. Pimchai Apavajrut, Prof. Somboon Anuntalabhochai, Prof. Adisorn Krasaechai, Dr. Somjai Sangyuenyongpipat, and Mr. Boonrak Phanchaisri of Chiang Mai University, for their reading of and commenting on various chapters of the book manuscript;
- Dr. Liu Ming, for her help in managing relevant material for translation in the fields of biology and chemistry;
- Personnel at the Department of Physics, Faculty of Science, Chiang Mai University, for their support and concern;
- The Thailand Research Fund for its support;
- Mr. Aaron Johnson of Springer Science & Business Media for his efforts in coordinating the publication;
- Our wives for their long-suffering support and patience.

Yu Liangdeng
Thiraphat Vilaithong
Ian Brown

June 2004
Chiang Mai, Thailand

PREFACE

(from the Chinese edition)

In the mid-1980's, encouraged by the successful development of ion implantation modification of material surfaces, the author made attempts to apply ion implantation techniques to the improvement of agricultural crop varieties. At that time the author was completely a layman in the area of genetic modification and thus early experiments met with only partial success. Subsequently, Prof. Wang Xuedong and Prof. Wu Yuejin of the Anhui Academy of Agriculture were invited to join the research work, and in 1986, biological effects of ion implantation on rice were discovered. Since then ion implantation as a new tool for genetic modification has been applied to the breeding of crops and microbes. In 1988, the author discovered the etching effect of ion beams on cells and proposed the idea of ion beam processing of cells for gene transfer. Through the independent research of three Ph.D. students, new varieties of gene-transferred rice were developed.

Energetic ion implantation of complex biological materials exists in nature. There are always some energetic ions in the environment. Some of these are implanted into biological cells, and affect the evolution of life and the health of human beings. In the past century in particular, mankind's conquest of space has included long journeys to the moon, and a trip to Mars may not remain just a scientific fantasy. It will become a long-term research field for scientists to simulate the behavior of energetic charged particles and to study their effects on human health. The ion implantation technique itself is relatively simple. We can closely examine radiation damage to organisms due to implanted ion deposition in cells, as well as the deposited ions themselves. Thus, studies of the interaction between implanted ions and complex biological systems have attracted significant interest.

The practice of ion beam genetic modification necessitates study of mechanisms. This is indeed a difficult subject. Starting from the moment the ions are implanted to the final biological effects, the timescales involved span from 10^{-19} to 10^9 seconds and the relevant spatial dimensions scale from microbiological damage to macro-property changes; hence the effects vary tremendously. To understand all the processes in such broad time and space extremes is not yet possible with our present knowledge and technological abilities. Even with studies of the primary physical processes, the extreme complexity of the ion-organism system makes attempts to establish physical models for the ion-organism interaction very difficult.

In 1989, the author published papers proposing that three factors affected the interaction between implanted ions and organisms. These are an energy deposition effect, a mass deposition effect, and a charging effect. This proposal appealed to both physicists and biologists. In the 2nd National Conference on Ion Implantation Effects on Biology in 1993, three American experts were invited to join Chinese experts. It was at this conference where the participants and researchers from the Anhui Science and Technology Press conceived of creating a book that would reflect current research achievements and future development directions in this new field. The book was finally completed in three years through an intermittent writing effort. Some of the opinions put forward in the book have been supported by experiments, but some still remain speculative. The author's purpose was to "attract jade by tossing a brick", aiming at opening discussions on the subject by physicists and biologists working in the field so that mistakes in the book could be corrected and an improved theoretical system established. Thus the title of this book is headed "Introduction".

Research on low-energy ion interaction with biological materials has been significantly supported by the National Planning Committee, National Foundation of Natural Sciences, Chinese Academy of Sciences, and the Anhui Provincial Planning Committee. This work has become a new growing discipline with an independent research basis and wide applications to genetic modification in both agriculture and fermentation industries. During the period of the 7th National Five Year Plan for Economic Development, the initiation of Director's funds from the Mathematics-Physics Division of the National Foundation of Natural Sciences as well as critical support from the National Planning Committee and the Anhui Provincial Committee of Sciences played a catalytic role in promoting the discipline.

In the development of this field, more than thirty of my graduate students have made significant contributions. Research of Ph.D. students Shao Chunlin, Huang Weidong, Han Jianwei, Wu Yuejin, Yang Jianbo, Cui Hairei, Cheng Beijiu, and Deng Jianguo have provided this book with rich and solid data. The researchers in the Institute of Plasma Physics, Chinese Academy of Sciences, have offered much assistance in the writing of the book. Ms. Feng Huiyun and Ms. Liu Junhong worked hard on the language. I would like to express my sincere appreciation to them all.

Last, but not least, I wish to thank Mrs. Kang Ming. She introduced Prof. Liang Yude, a famous senior biologist, for later collaborations with biologists. My 10-year study, from physics into biology, has been supported by Prof. Xu Guanren, Prof. Huo Yuping and many other biologists and physicists. If it is said that the ten-year work has achieved something, this success should also be attributed to these people.

Yu Zengliang

CONTENTS

NOTATIONS OFTEN USED

A	Element mass number, area
a($\mathbf{a}$)	Acceleration
$B(\boldsymbol{B})$	Magnetic field strength
bp	Base pair
c	Light speed in vacuum
C	Coulomb, capacitance, specific heat
CaMV	Cauliflower mosaic virus
CVg	Genetic coefficient of variation
d	Molecular weight (Dalton), number of times of hitting the target
D	Dose, diffusion coefficient
DMSO	Dimethyl Sulfoxide
$e(e_0)$	Charge (basic charge)
$E(\boldsymbol{E})$	Electric field strength
ΔE	Absorption energy
E_1	Incident ion energy
ECR	Electron cyclotron resonance
E_d	Critical energy for target atom displacement
eV	Electron volt (energy unit)
$F(\boldsymbol{F})$	Electromagnetic force acting on the charged particle
F_0	Neutralization efficiency
$F_{1,2,\ldots n}$	Hybrid generations
$\Delta G(\Delta G')$	Genetic progress (relative genetic progress)
GUS	GUS gene
H	Chemical change coefficient of a volume unit absorbing per 100 implanted ions; also the Hamiltonian operator
h	Plank's constant, number of times of hitting the target
ΔH	Molar heat of vaporization
h_2	Heritability
H_m	Hygromycin
hpt	Hygromycin phosphotransferase gene
I^+	Implanted positive ion
J	Ion flux density
K	Reaction rate constant, various constants, radiation coefficient
kb	Kilo base pair
kn	Radiation quality
Krens'F	One kind of the media for exogenous DNA transfer
KT	Kinetin (one kind of cytokinins)

l	Length, distance
LET	Linear energy transfer
$m(m_0)$	Mass (rest mass)
Δm	Mass of the irradiated volume
m_1	Incident ion mass
m_2	Target atom mass
$M_{1,2,\ldots n}$	Mutation generations
M_{13}	One kind of plasmid DNA (7.2 kb)
n	Denisty
N	Number of particles, extrapolation number, Newton
NAA	Naphthalene acitic acid
N_6	One kind of plant culture medium
OER	Oxygen enhancement ratio
P	Pressure
P_2	Portion of long-life radicals
pBI	One kind of plasmid DNA (4.7 kb)
pCR	Polymerase Chain Reaction
pUC	One kind of plasmid DNA (2.7 kb)
Q	Electric charge, heat, quality factor
q	Electric charge
$-Q_b$	Cellular electric characteristic value
ΔQ_t	Change in the cellular electric characteristic value
R, r	Ion radius, distance, gas constant
RBE	Relative biological effectiveness
R_t	Incident ion total range
R_p	Incident ion projected range
S	Survival rate, sputtering yield
S_e	Electronic stopping power
S_n	Nuclear stopping power
SSC	Mixture solution of NaCl and citric acid
T	Temperature
t	Time
TE	Tris-EDTA (mixture solution of trismethylaminomethane and ethylene-diamine-tetra-acetic acid)
U	Electric potential
V	Voltage, volume, target volume
$v(\mathbf{v})$	Velocity, target volume used in the Target Theory
W	Ion energy, ion flux acceleration energy
$W_{k,l}$	Emission probability of X-luminescence
Y_i	Emission coefficient of secondary electrons
Z	Charge number
Z_1	Charge number of incident ion
Z_2	Charge number of target atom
α	Adaptation coefficient, helium nucleus
β	Coefficient
γ	Proportional coefficient, protection coefficient
ε	Permittivity

η	Radiation effect, proportional coefficient, portion
θ	Incident angle, angle
λ	Free path, wave length, inactivation constant
μ	Permeability
ν	Photon frequency
ρ	Density, probability
σ_{ij}	Charge exchange cross section

1

GENERAL IDEAS ABOUT IONS

Ions exist everywhere in nature – in cosmic rays, air, water, and living bodies. Whether they originate in matter or are implanted from external radiation sources into biological systems, ions affect life importantly.

1.1. PHYSICAL PROPERTIES OF IONS

About a half century before electrons were discovered, Faraday called the particles that flow between two electrodes in electrochemical experiments 'ions'. An atom or molecule becomes a positive ion when it loses one or more electrons due to ionization, and it becomes a negative ion when it captures one or more electrons.

Ions and electrons are charged particles. The mass of an electron is about 1/1840 that of a proton, and the charge of a proton or electron is 1.60×10^{-19} Coulomb. The charge carried by an electron is the smallest unit of charge, and the total charge carried by any charged object is an integral multiple of this unit charge. Ions are different from electrons in that they can be of many different species. Ions can be defined by a charge number Z and a mass number A. If the quantity of charge carried by an ion is e, then

$$e = Ze_0, \tag{1.1.1}$$

where e_0 is the elementary charge, i.e. the charge of an electron (1.60×10^{-19} Coulomb). If the rest mass of an ion is m_0, then

$$m_0 \approx Am_P, \tag{1.1.2}$$

where m_P is the proton rest mass (1.67×10^{-27} kg). This is approximate because the binding energy of the nucleons and the mass of the orbital electrons are neglected.

The relativistic relationship between mass m and the rest mass is

$$m = m_0\gamma, \tag{1.1.3}$$

where

1

$$\gamma = 1/\sqrt{1-\beta^2}, \tag{1.1.4}$$

and

$$\beta = v/c, \tag{1.1.5}$$

and v is the particle velocity and c the speed of light in vacuum.

The diameter* of an electron is about 10^{-15} m, whereas the diameter of an ion is much greater, of order 10^{-10} m, about the size of an atom. Generally, the diameter of a negative ion is somewhat greater than that of a positive ion. The radius of almost all negative ions is around $(1.3–2.5) \times 10^{-10}$ m, while the radius of positive ions is about $(0.1–1.7) \times 10^{-10}$ m. For example, the radius of a potassium ion, K^+, is 0.133 nm (1 nm = 1 nanometer = 1 $\times 10^{-9}$ m), while the radius of Cl^- is 0.181 nm. Ions of the same element in different valence states have different sizes. The more positively charged ions are smaller than less positively charged ions; this follows since electrons are removed to increase the ion valence state. If all of the orbital electrons are removed, the diameter of the nucleus is only $10^{-14}–10^{-16}$ m. The mass of an ion is almost totally in the nucleus.

An ion produces an electrostatic field surrounding itself. If e is the ion charge, the electrostatic field produced by the ion at a distance r is given by

$$E = e/4\pi\varepsilon r^2, \tag{1.1.6}$$

where $\varepsilon = \varepsilon_r\varepsilon_0$, ε_r is the relative permittivity, and $\varepsilon_0 = 8.85 \times 10^{-12}$ F/m (the permittivity of free space). In vacuum, $\varepsilon_r = 1$, whereas in media other than vacuum $\varepsilon_r > 1$. The direction of the electric field is defined as the direction of the force on a test positive charge at the same point. The electrostatic energy of the ion (i.e., the total energy of the electrostatic field due to the ion) is given by

$$W = e^2/2\varepsilon R, \tag{1.1.7}$$

where R is the ion radius. The ion energy can thus be estimated. Since the ion diameter is $\sim 2 \times 10^{-10}$ m, the ion energy in vacuum is about 5 eV. In media other than vacuum the ion electrostatic energy is $1/\varepsilon_r$ of that in vacuum.

A moving ion produces a magnetic field in the surrounding space. If the ion velocity is v, the magnetic field at a distance $\boldsymbol{r}$ from the ion is given by

$$B = \frac{\mu e v}{4\pi r^2} \sin\alpha, \tag{1.1.8}$$

---

* In the quantum-mechanical treatment of the electron, its size depends on the size of the potential energy well that confines it. However, treating the electron as a hard sphere, it should be noted that the electron is very small compared to the diameter of an atom or an ion.

where μ is the permeability, α is the angle between the velocity vector v and the position vector r. Expressed vectorially the magnetic field is given by

$$B = \frac{\mu e}{4\pi r^3} v \times r. \qquad (1.1.9)$$

The force experienced by an ion in electric and magnetic fields is given by the Lorentz expression

$$F = eE + ev \times B. \qquad (1.1.10)$$

Here, eE is the force due to the electric field, having a direction (for a positive ion) that is the same as the E direction; and $ev \times B$ is the force due to the magnetic field, with direction perpendicular to the v-B plane. It follows that the kinetic equation for an ion moving in electric and magnetic fields is given by

$$eE + ev \times B = ma, \qquad (1.1.11)$$

where m is the ion mass and a is the ion acceleration.

In the absence of externally applied electric and magnetic fields, the interaction between ions, or between ions and other charged particles, is determined by the fields produced by the moving ions themselves. Let us consider a very simple example of two ions with the same charge, separated by a distance of d, moving parallel to each other with the same velocity v. The electric and magnetic forces experienced by each ion are given by

$$F_e = e^2/4\pi\varepsilon_0 d^2, \qquad (1.1.12)$$

and

$$F_m = \mu_0 e^2 v^2/4\pi d^2, \qquad (1.1.13)$$

respectively. Thus the relative magnitudes of the magnetic force and the electric force are given by

$$F_m/F_e = \varepsilon_0\mu_0 v^2 = v^2/c^2. \qquad (1.1.14)$$

For non-relativistic velocities, $v \ll c$, we have that $F_m \ll F_e$. That is, the magnetic force between ions is much smaller than the electric force. Thus, for the case of high current ion beams, we may consider only the electric field produced by the ions and neglect the magnetic field. This electric field – the field of an ion beam formed by the ions in the beam – is called the space-charge field of the beam, and the force experienced by the ions due to this field is called the space charge force. These concepts will be used in next Chapter.

4

CHAPTER 1

1.2. FORMATION AND RECOMBINATION OF IONS

Let us represent ground state atoms of type A and B by the symbols A and B, and the ground state molecule formed by their combination by AB. Let: A*, B* and AB*, and A**, B** and AB** represent these particles in the first and second excitation states, respectively; A^+, B^+ and AB^+, and A^{++}, B^{++} and AB^{++} represent singly-charged and doubly-charged positive ion species, respectively; and A^-, B^- and AB^- represent negative ions; and finally, let e represent an electron, and hv a photon. Then we can write the various ion formation and recombination processes as:

Ionization processes:

Collision with an electron:	$A + e \rightarrow A^+ + 2e$	(1.2.1)
Collision between two neutral particles:	$A + A \rightarrow A^+ + A + e$	(1.2.2)
Photon ionization:	$A + hv \rightarrow A^+ + e$	(1.2.3)
Charge transfer:	$A^- + B \rightarrow B^- + A$	(1.2.4)
	$A^+ + B \rightarrow B^+ + A$	(1.2.5)
Stepwise ionization:	$A^* + e \rightarrow A^+ + e + e$	(1.2.6)
	$A^* + hv \rightarrow A^+ + e$	(1.2.7)
Dissociative ionization:	$AB \rightarrow A^+ + B^-$	(1.2.8)
Negative ion formation:	$A + e \rightarrow A^-$	(1.2.9)
Superexcited preionization:	$AB^* \rightarrow AB^- + e.$	(1.2.10)

Recombination processes:

Radiative recombination:	$A^+ + B^- \rightarrow AB + hv$	(1.2.11)
Mutual recombination:	$A^+ + B^- \rightarrow AB$	(1.2.12)
Stepwise recombination:	$A^+ + e \rightarrow A^{**} \rightarrow A + hv$	(1.2.13)
Three-body collision recombination (where M is the third particle):		
	$A^+ + B^- + M \rightarrow AB + M$	(1.2.14)
Dissociative recombination:	$AB^+ + e \rightarrow A + B.$	(1.2.15)

These ion reactions are rather complicated. Each kind of reaction is usually accompanied by other kinds also. All of these reactions can take place within the scope of the discussion in this book. For example, in the production of ion beams, reactions (1.2.1), (1.2.3), (1.2.6), and (1.2.10) can occur; in the interaction between the energetic ions and biological materials, the possibility of simultaneous occurrence of (1.2.2)–(1.2.15) is very high. There are also some special reaction processes, such as ion glomeration, which is very significant to ion beam induced biological damage.

1.3. ION POLARIZATION AND GLOMERATION

An ion is originally not polarized. However, in an external electric field, the positive nucleus and the surrounding negative electrons of an ion are displaced so as to form an induced dipole (Figure 1.1). This process is called ion polarization, which causes the ion to be deformed. For any given electric field, the ion polarization deformation depends on the structure of the ions.

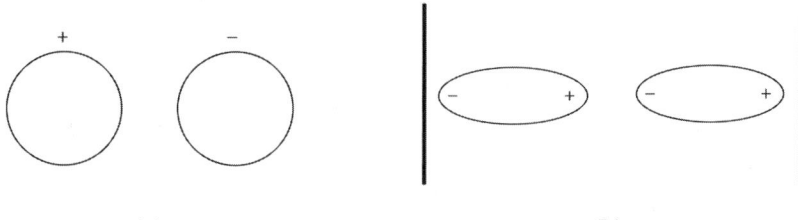

(a)   (b)

Figure 1.1. Polarization of ions in an electric field. (a) Ions without an electric field, and (b) polarization of ions in an electric field.

(1) For ions with the same electron shell structure, ion deformation increases with decreasing positive charge and increasing negative charge:

$$O^{--} > F^{-} > Na^{+} > Mg^{++} > Al^{+++} > Si^{4+}.$$

(2) For ions with similar structure and with the same charge number Z (i.e., ions of elements in the same group in the Periodic Table), the deformation increases with increasing electron shell number:

$$I^{-} > Br^{-} > Cl^{-} > F^{-}, \qquad Cs^{+} > Rb^{+} > K^{+} > Na^{+} > Li^{+}.$$

(3) Ions with 18 outermost electrons have a greater deformation than ions of similar radii but with inert-gas type structure. For example, the deformation of Hg^{++} (18 outermost electrons) is greater than that of Ca^{++} (inert-gas type), though they have very close to the same ion radii.

An ion is a charged particle with an electric field surrounding it. Therefore, when an ion is close to another ion or molecule, the electric field of the ion can cause deformation of the nearby ion or molecule, thus forming an induced dipole.

A molecular dipole can also be induced in an externally applied electric field (Figure 1.2a). A polarized molecule reorients its direction according to the electric field direction; this process is called orientation (Figure 1.2b). The interactions between the electric field and the molecule, and the poles of the nearby molecules and the molecule (attraction between unlike poles and repulsion between like poles) tend to separate the positive and negative charges of polarized molecules, so producing induced dipoles, and consequently causing changes in the molecular properties.

A gas can be ionized by radiation processes. Because the electric field surrounding the ions polarizes nearby neutral molecules, polarized molecules may attach to the ion, which now acts as a core to form a glomeration (Figure 1.3). This process can be expressed as

$$A^{+} + nB \rightarrow A^{+}B_n \quad (n = 1, 2, \ldots), \tag{1.3.1}$$

where A and B can be either like or unlike molecules. For example,

$$H^{+} + n(H_2O) \rightarrow H^{+}(H_2O)_n \tag{1.3.2}$$

$$H^{+} + n(CH_3OH) \rightarrow H^{+}(CH_3OH)_n. \tag{1.3.3}$$

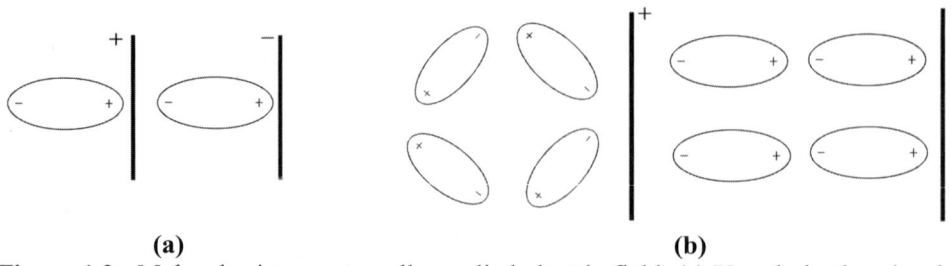

(a)   **(b)**

Figure 1.2. Molecules in an externally applied electric field. (a) Unpolarized molecules showing induced polarization, and (b) polarized molecules showing orientation to the applied field.

Figure 1.3. Ion glomeration.

This process is called ion glomeration. Ion glomerations exist in reacting systems, with ions located inside liquid cage-like structures. This can affect the reaction mechanisms and yields.

In energetic ion implantation into biological cells, all of the above processes can take place, and they can affect the biological activity of the cells.

1.4. BIOLOGICAL EFFECTS OF IONS

About 30 elements are used to form biological systems, including 11 abundant elements (C, H, O, N, P, S, Cl, K, Na, Ca, Mg), 3 rare elements (Fe, Zn, Cu), and 16 trace elements, of which 6 are nonmetallic elements (F, I, Se, Si, As, B) and 10 are metals (Mn, Mo, Co, Cr, V, Ni, Cf, Sn, Pb, Li). Some of these elements are present in biological organisms in ionic form. They participate in various complicated biochemical processes such as material transport, energy transformation, information exchange, and control of metabolism so as to maintain normal activity of the living system.

The totality of any biosystem can be broken down conceptually into individual biological cells. The cell itself is partitioned internally by membranes that limit material diffusion between smaller compartments inside the cell. Inside an actively living cell, water, ions and neutral solute, metabolite, and dissolved gas are all continuously exchanged across the membrane. These various component materials pass through the cell membrane in different ways. Some automatically diffuse inward in the cell in the direction of decreasing potential gradient. This is the biological equivalent of the thermodynamic transport of heat from a warmer body to a cooler body. Other substances, however, can be accumulated by the cell in a direction opposite to the potential gradient. Their motion into the cell interior, being against the potential gradient, must consume energy. If a potential difference is established across the membrane, this can be very important for producing an "uphill" driving force for the ions. The production and maintenance of the potential difference across the membrane is closely connected to the concentrations of positive and negative ions within the cell.

Let us consider a simple KCl cellular system consisting of two small compartments partitioned by a membrane. Since the K^+ ion has a smaller radius than the Cl^- ion, the membrane has a much higher permeability for K^+ transport than for Cl^- (Figure 1.4). If the two compartments in the cell contain KCl solution of different concentrations, then initially K^+ is transported out of the higher concentration compartment. Because the membrane has a higher permeability for K^+ than Cl^-, in a relatively short time period, negative charges accumulate in the higher concentration compartment, and positive charges accumulate in the lower concentration compartment. This results in the formation of a potential difference across the membrane. This potential difference attracts K^+ ions back to the higher concentration compartment, reducing and eventually preventing further loss of K^+ from the compartment. In this way a substantial potential difference can be maintained between the compartments. The permeability of the membrane for Cl^- is low but not zero, and thus after a relatively long time period the potential difference and the concentration difference will disappear as Cl^- also crosses the membrane. If the membrane transported no Cl^- ions at all, the potential, once established, could be maintained indefinitely. In reality the permeability of the cell membrane for negative ions is 1/10 to 1/100 of that for singly-ionized positive ions. Consequently, maintenance of the membrane potential requires that the rate at which negative ions enter the cell compartment be greater than the rate at which ions are lost. Thus a kind of ion-pump mechanism is called for.

Not only do ions lead to the formation of an electrostatic field across the membrane; they also give rise to an electron-proton field, oxidation-reduction gradient field, mechanical stress field, and more. These fields and the associated material fluxes create and maintain the life system. The particle flux causes the physical field, and the formation of the field across the membrane then plays a critical role in the feedback control of the flux.

Ions play an important role in life activities. In a sense, without ions there is no life. The questions are: where are the various ion species, and why are they concentrated where they are by the cell? For the case of normal concentration, what are the biological functions of ions? If the concentration is higher or lower than normal, what are the effects on the biological system? As the cell is an ever-changing living body, studying the biological functions inside the cell environment is still a challenge to present technology. A primary difficulty has to do with techniques for detecting the ion concentration regions, and another with the operations inside the cell, particularly the deviations from equilibrium ion concentrations on either side of the membrane. Actually, ever since life originated on earth there has always been some high energy ion irradiation of biological organisms, from cosmic ray air showers, leading to ion implantation into all organisms and deposition of these ions in certain locations within the cell. This ion implantation modification of the ion cellular concentrations, or viewed another way, of the implanted ions acting as dopant "impurities" in the cell composition, plays an important role in biological evolution. Up to now, research has considered in some detail the biological effects of the energy deposition accompanying the implanted ions, but little consideration has been given to the biological effects of the deposited ions consequent to the implantation.

1.5. ION IMPLANTATION AND GENETIC MODIFICATION

The energy of the primary cosmic radiation is very high, say about 10^9–10^{16} eV and up to a maximum detected energy of order $\sim 10^{20}$ eV. This primary radiation contains mostly protons together with electrons, α particles and a small fraction of heavier ions. Because of the shielding effect of the earth's magnetic field and atmosphere, the secondary radiation that arrives at the earth's surface is diffused, with annual total energy area density of only about 6.3×10^{-3} J/cm^2, approximately equal to the energy flux on the earth from starlight. We can reasonably conclude that cosmic radiation did not play a significant role in chemical evolutionary processes in the early earth. However, subsequent to the creation of life, this radiation might have played a significant role in the biological evolutionary stage.

The effects of the less-energetic secondary particles from cosmic radiation (cosmic ray air showers) implanted into biological cells over the long eons of cell evolution remain unknown. An actual experimental demonstration of mutation brought about by high energy heavy ions from the cosmic radiation was carried out in the 1975 Apollo-Soyuz flight. In this experiment, solid state detectors were installed around maize (corn) seed specimens to examine the tracks of the heavy ion cosmic rays that bombarded the seed embryos. It was found that of 150 seeds, one seed was bombarded by two heavy ions which had a LET (linear energy transfer) of about 1.0–1.5 MeV/μm, and the plant that developed from this seed had leaves with large areas of yellow stripes.

The effects of ion irradiation of biological samples have been investigated over the past several decades using a wide range of ion species produced by high energy particle accelerators. The results have led to an appreciation of the risks associated with radiation in space flight, and have laid the foundations of a novel interdisciplinary subject – heavy ion biology. In recent years heavy ion biology has emerged as a tool for medical applications as well as for genetic improvement of plant varieties, and has attracted much interest.

The starting point in the study of the interaction between high energy ions and biological organisms lies in understanding the energy transfer events along the ion path, without consideration of whether or not the ion comes to rest after losing all its energy within the cell. This is because cosmic ray heavy ions have very high energies and long ranges such that they completely penetrate biological materials. Ions produced using accelerators also have very high energies. Although the ion energy can be adjusted to a level at which the ion comes to rest inside the biological material, in most experiments the role played by the implanted ions is ignored.

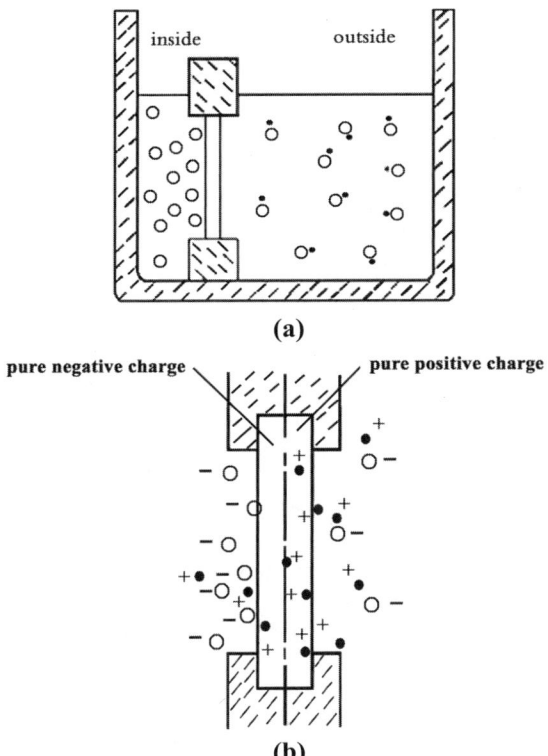

Figure 1.4. A simulated system for a membrane that has selectivity for positive ions. (a) Initial state, and (b) equilibrium state of the system.

In the mid-1980's, the biological effects of ion implantation were recognized and demonstrated experimentally [1,2]. In these experiments, the ion energies were 3 to 5 orders of magnitude lower than the energies typical of heavy ion biology. It had generally been assumed that these low energy ions had a range in water that was far too small to induce biological aberrations. In some pioneering work carried out by the author, 30 keV nitrogen ions were implanted into dry rice seeds. When the implantation dose (number of ions implanted) was sufficiently high, yellow stripes were seen on the leaves of rice plants grown from the seeds, just as was found for the maize leaves in the Apollo-Soyuz space flight experiment, and these yellow characteristics could be stably inherited to later generations.

Present research shows that low energy ion implantation into solid biological materials such as dry crop seeds has a unique mode of interaction. Ions with energies less than about 50 keV/amu transfer energy primarily through atomic collisions with target molecules. Atomic collisions lead to atomic displacements and vacancies in the solid. A small fraction of the incident ion energy is transferred by collisions within the solid, leading to rapid ionization of atoms in the neighborhood of the ion trajectory and the formation of unstable charged areas where ions repel each other and are displaced. After a chain of physical and chemical processes, the displaced atoms recombine with atoms or molecules nearby. If the implanted ion species are of a kind required for life or viability, they are retained inside the biological organism and recombine with other atoms or molecules. On the other hand, due to the poor conductivity of biological material and the existence of a large number of biological channels, when the incident ion velocity is less than the velocities of the orbital electrons, the orbital electrons are captured and the ions neutralized. Accumulation of surface charge in the biological material can cause Coulomb explosions, leading to shedding of biological molecules and fragments. This phenomenon is associated with surface sputtering, in which the low energy ion beam act like a surgical tool to remove the biological material surface. This process can cause the channels in some layers of the biological material to be linked, thus increasing the permeability.

It can be seen that low energy ion beams have wide applications in life science. Accurate and quantitative implantation of particular ion species into cells can be used for studying biological effects of the ions. Radiation emitted during the implantation process can provide a tool for measuring detailed cellular properties at the atomic or molecular scale. Fast ions provide an analytical analysis technique for biological macromolecular mass spectra. Ion beam bombardment of the cell, the cell nucleus, and the organelle, as well as ion beam cutting of genes, can be applied to observe damage and repair. Ion-implantation-induced displacement and recombination in genetic material can provide a new approach to mutation breeding. Application of ion cutting and etching can modify DNA (deoxyribonucleic acid) macromolecules and provide a means for gene transfer. Finally, following the development of micro-beam technology, ion beams can be applied for cell surgery.

REFERENCES

1. Yu, Z.L., Deng, J.G., He, J.J., *et al.*, Mutation Breeding by Ion Implantation, *Nucl. Instr. Meth.*, B**59/60** (1991)705.
2. Yu, Z.L., Yang, J.B., Wu, Y.J., *et al.*, Transferring Gus Gene into Intact Rice Cells by Low Energy Ion Beam, *Nucl. Instr. Meth.*, B**80/81** (1993)1328.

FURTHER READING

1. Hall, E.J., *Radiobiology for the Radiologist* (5th ed., Lippincott, Williams and Wilkins, Philadelphia, 2000).
2. Liu, P.H., *Biological Effects of Physical Factors* (in Chinese, Science Press, Beijing, 1992).
3. Qiu, G.Y. and Feng, S.Y., *Radiobiophysics* (in Chinese, Wuhan University Press, Wuhan, 1990).
4. Smith, H., ed. *The Molecular Biology of Plant Cells* (University of California Press, Berkeley, 1977).
5. Xia, S.X., *Radiobiology* (in Chinese, Military Medical Science Press, Beijing, 1998).
6. Xu, Y.K., Gu, G.W., Wu, X.K, *Mutation Breeding of Crops* (in Chinese, Shanghai Science and Technology Press, Shanghai, 1985).
7. Yang, T.C. and Mei, M.T., Space Radiation Biology (in Chinese, Zhongshan University Press, Guangzhou, 1995).
8. Smirnov, B.M., Physics of Atoms and Ions (Springer Verlag, Heidelberg, 2003).

2

PRODUCTION AND ACCELERATION OF IONS

Since the 1950's, growth in our understanding of the interaction of energetic ions with matter has led to some large scale applications of low energy ion beams, for example for materials analysis, semiconductor industry, and materials science. Compact ion accelerators, being the basic technology of the applications, have been developed in a number of embodiments. An accelerator is a device in which charged particles are formed and accelerated to substantial energy. Generally, three primary components are employed: an ion (or electron) source, a beam line, and a target chamber. Ions are produced in an ion source, accelerated to the required energy, transported from the source region to the application region, and finally allowed to bombard the experimental samples in a target chamber. This chapter begins with a description of the basic features of ion implantation into biological samples. Then the working principles of the components of a low-energy accelerator are described as well as the requirements for genetic modification by ion implantation.

2.1. EQUIPMENT REQUIREMENTS FOR ION IMPLANTATION OF BIOLOGICAL SAMPLES

Successful applications of low energy ion beams for genetic modification include ion beam induced mutation for breeding [1,2] and ion beam induced gene transfer [3,4]. In the former case, the ions are required to slow down and finally stop in the genetic substance so that they can cause displacements and recombination of the molecules and atoms of the genetic substance. In the latter case, only ion sputtering and desorption are considered but not the ion penetration range. In other words, ion sputtering or etching thins out the surface of the biological material so that the existing channels can be linked to form the pathways for gene transfer. In fact, ion implantation and etching occur almost simultaneously. An increase in the permeability of the biological material surface enhances deeper penetration of the later coming ions which pass through the channels. Therefore, the ion energy required in ion beam biotechnology is not very high.

For mutation breeding, generally a large number of samples are to be treated. This is not a problem for high-penetration mutation sources such as γ-rays. However, in the case of ion implantation, because of the limited ion range, the ion beam must be caused to bombard the embryo of each seed. If the beam has a diameter of 30 cm, a maximum of about 3,000 seeds of rice or wheat can be put in this irradiation area even for a very precisely designed sample holder. If the beam size is small, mechanical scanning for large-area homogeneous implantation may cause perturbation of the sample alignment and orientation. Electric or magnetic scanning of the beam is not practical for large implantation areas either. Therefore, the ion beam used for mutation breeding purposes is usually as large and homogeneous as possible.

For studies of ion-beam-induced biological effects, ions are required to be implanted accurately and quantitatively into specific parts of the organisms. Mass spectra of fast ion desorption or photon emission induced by ion impact can be used for measurement of the cell fine structure at the atomic and molecular levels. Particular kinds of ion beams may be used for cell surgery or crafting. All of these applications require the ion beam to have high spatial resolution. Therefore, ion beams applied in the life science should be developed in two extreme directions, namely, high-current broad beams, and micro beams.

In order to avoid loss or scattering of ions due to collisions with gas molecules, the production and acceleration of ions should be carried out in high vacuum. Biological samples, except dry crop seeds, contain a large amount of water. This gives rise to a primary problem. Evaporation of water usually leads to poor vacuum quality, and the biological material can be inactivated by rapid cooling or loss of water. In order to overcome this damage to the bio-samples, careful design of the vacuum system is needed as well as some samples pre-treatment. Additionally, possible contamination from organic materials in the vacuum system from the ion source cathode can occur and should be taken into consideration when designing the ion source.

One possible solution to these dual problems of beam line vacuum and bio-sample survival is to make use of ion beams that are coupled out of the vacuum system. This requires beam energy of at least 2 MeV.

The bio-samples are directly exposed to the ion beam during ion implantation. Thus the design of the sample chamber should guarantee sterile conditions for the plant cells and microbial samples as they are positioned in and removed from the chamber.

2.2. BASIC BEAM LINE STRUCTURE

Although processing of bio-samples calls for some special requirements of the ion beam facility [5,6], the basic structure is not greatly different from that of an industrial ion implanter. Figure 2.1 shows a schematic diagram of a typical accelerator beam line. It consists of the following basic subsystems: (1) ion source; (2) beam focusing system; (3) mass analyzer; (4) accelerating system; (5) measurement system; and (6) target chamber and vacuum system. The components and their arrangement may vary somewhat depending on the specific setup. For example some configurations use pre-acceleration and post-analysis, and the focusing system may be installed at different positions along the beam line according to the ion optics requirements to reduce beam loss.

The working principle is as follows. In the ion source, an electrical discharge in gas or vapor produces plasma. Macroscopically, the plasma is electrically neutral, containing approximately the same amount of positive ion charge as negative electron charge. Ions are extracted from the plasma by an extraction electrode and allowed to drift to a mass

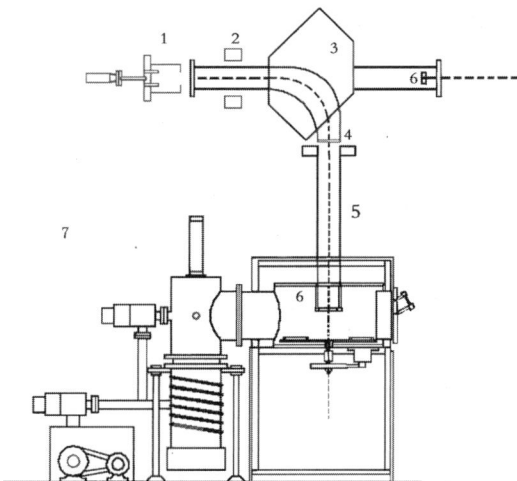

Figure 2.1. Schematic diagram of an ion beam line. (1) Ion source, (2) focusing lens, (3) mass analyzer, (4) diaphragm, (5) accelerating tube, (6) measurement devices, and (7) target chamber and vacuum system.

analyzer. Ions of the same species and with the same charge, selected by the mass analyzer from ions of different masses or charges, go to the acceleration section.

The extraction system accelerates the ions to the selected energy and implants them into the bio-sample inside the target chamber. For monitoring the ion beam parameters, various measurement systems are installed, such as a beam profile monitor and an implantation dose measurement device.

2.2.1. Ion Source

An ion source is a device that produces a beam of ions. It consists of two basic parts – the plasma generator and the ion extraction system. The extraction system is usually composed of two, three or four electrodes. Each electrode has one or several or even several hundred small holes or slits, which are well aligned. The first electrode is at the same potential as the plasma in the discharge chamber, and the last electrode is at ground potential. Figure 2.2 shows a schematic of a two-electrode extraction system. Between the two electrodes a second or third electrode might possibly be added, to form a three- or four- electrode extraction system. Each electrode is biased to the appropriate potential, and ions in the plasma at the boundary of the discharge chamber are accelerated and formed into an ion beam. It is possible for three- or four-electrode extraction systems to form beams of ions with energy up to several hundred keV.

The extraction system forms an ion beam with more-or-less small divergence. A good extractor design can generate an ion beam with divergence less than $0.5°$, under optimal conditions. The beam divergence is dependent not only on the arrangement of the extractor optical elements, but also, and usually more importantly, on the density and noise of the plasma near the extraction region. Thus an important factor in reducing the beam divergence is that the plasma density in the discharge chamber is adjusted to match

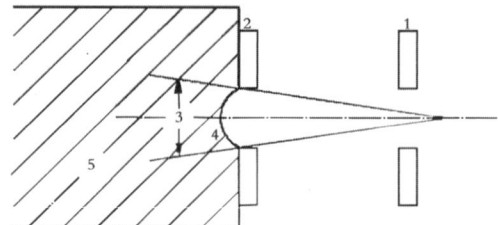

Figure 2.2. Schematic diagram of a dipole extraction system (the shaded part is the plasma). (1) Ground electrode, (2) extraction electrode, (3) effective angle, (4) plasma boundary, and (5) discharge chamber.

that required for the particular the extraction voltage. In first order approximation, the ion beam current density J and the extraction voltage V are related by a 3/2-power-law relationship:

$$J = kV^{3/2}, \tag{2.2.1}$$

where k is a constant dependent on the geometrical parameters of the electrodes and the extracted ion species. Although the value of k can be calculated, it is better determined by experiment. In such an experiment, fixing the extraction voltage V and adjusting the plasma density in the discharge chamber, or fixing the discharge parameters and measuring of the value of J when the beam divergence is the minimum, allows k to be determined from the above equation. In normal operation of an ion source, if the 3/2-power law is satisfied, the divergence of the extracted beam is usually small. In this case the ion beam suffers no significant loss when transported through the beam line.

Ion sources can be classified into two kinds: solid surface ion sources and gas/vapor sources. No matter which type of ion sources is used, formation of a stable discharge is pivotal for good source operation. Many different kinds of ion sources have been developed, such as for example, high-frequency sources, electron cyclotron resonance sources, arc discharge sources, and many more.

2.2.1.1. RF Ion Sources

In an RF (radio frequency) discharge ion source, a high frequency electric field is used to accelerate free electrons to energies that are high enough to lead to ionization of atoms or molecules with which they collide. The plasma chamber may be glass or ceramic, and is surrounded by an RF induction coil fed by high frequency RF power. The RF frequency might be a few MHz up to a few tens of MHz. Gas pressure is typically in the range $0.13 - 1.33$ Pa, and a stable plasma can be formed in the discharge chamber. The extraction system forms a more-or-less energetic ion beam from the plasma. The RF ion source has a simple structure and long life. Ions in the beam are mostly (say, ~80%) "atomic ions", meaning that they are ionized atoms as opposed to ionized molecules ("molecular ions"). The basic feed material can be a gas or liquid or solid; liquid or solid feed materials are evaporated in an oven and then enter the discharge chamber as a gas.

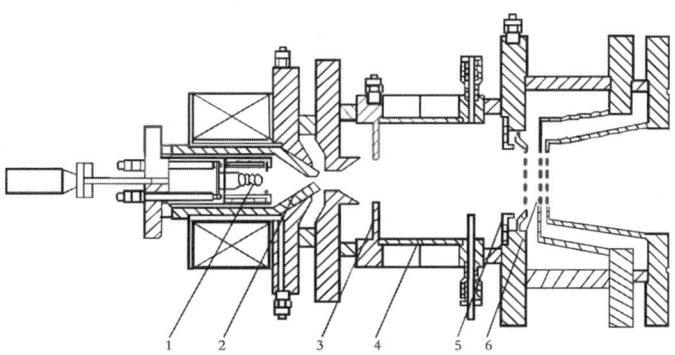

Figure 2.3. Schematic of a duo-Penning ion source. (1) Cathode, (2) intermediate electrode, (3) the first anode, (4) the second anode, (5) target cathode, and (6) extraction system.

2.2.1.2. Duo-Penning Ion Sources

The combination of a duoplasmatron ion source with a Penning discharge is called a duo-Penning ion source (Figure 2.3) [7,8]. The cathode (1) emits electrons which ionize gas in the mid-electrode area (2) to form a plasma. In the mid-electrode area, compression of the plasma by both geometric effects and by a magnetic field causes the plasma to form a double sheath in the throat region. The sheath accelerates the electrons, some of which then have enough energy to pass through the expander of the first anode (3) to enter the anodic discharge chamber (4) in a helical motion along the magnetic field lines. In this way the electron trajectory is lengthened and the collision probability between the electrons and the gas molecules is increased. Since the potential of the target-cathode (5) is lower than that of the anode, the primary electrons are reflected before reaching the cathode and oscillate in the anodic chamber so that the collision probability between the electrons and the gas molecules is yet further increased. The higher ionization efficiency of the anodic chamber gas leads to higher density of the anodic plasma. Ions in the uniform and stable anodic plasma are extracted through the extraction system (6).

The duo-Penning ion source can produce multiply-charged ions. Characteristic features of this type of ion source are high plasma density, stability, and homogeneity. When a multi-aperture extraction system is used, high current and large area ion beams can be formed.

2.2.1.3. ECR Ion Sources

The ECR ion source employs microwave power to supply energy to the electrons in the discharge chamber. When the magnetic field strength is adjusted appropriately, a resonance between the microwave frequency and the electron cyclotron frequency occurs; this is the electron cyclotron resonance, or ECR, condition. Electrons are heated resonantly to high energy, allowing ionization of the background gas by electron collisions. Basic components of the ECR ion source are (1) microwave source (for

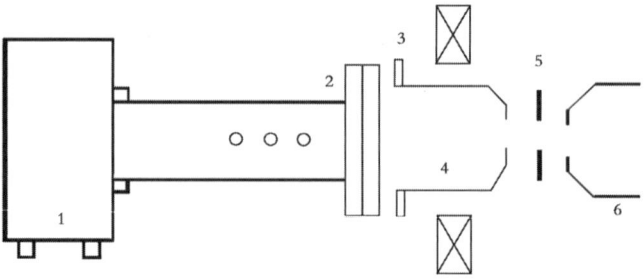

Figure 2.4. Schematic of an ECR ion source. (1) Microwave power supply, (2) microwave input window, (3) magnet, (4) discharge chamber, (5) anode, and (6) extraction electrode.

example, 2.45 GHz), (2) input window, (3) magnetic field, (4) discharge chamber, (5) anode, and (6) extraction electrode (Figure 2.4). This kind of ion source usually operates at low pressures (< 1 Pa) with high ionization efficiency. Because there is no cathode, cathodic sputtering and contamination are avoided and long source lifetime is achieved. Because of the absence of a hot cathode, strongly oxidizing ions can be formed, making this kind of ion source particularly attractive for ion beam bioengineering.

2.2.2. Analysis System

Investigations of biological effects due to different ion species require high purity ion beams [9]. Ions extracted from an ion source are usually a mixture of various components including impurity ions and multiply-charged ions. To purify the beam, mass analysis is employed. The most common kind of mass analysis system employs a magnetic field, which we describe here.

The magnetic analyzer consists of a sector electromagnet and a flat vacuum chamber (Figure 2.5). Ions with mass m, charge q and a velocity v moving in a uniform magnetic field B in a direction perpendicular to the field experience a force, called the Lorentz force, given by

$$\boldsymbol{F} = q\boldsymbol{v} \times \boldsymbol{B}. \tag{2.2.2}$$

The magnitude of the Lorentz force is $F = qvB$, and its direction is perpendicular to both the direction of motion and the magnetic field – hence the ion trajectory is the arc of a circle. The radius of the circle can be obtained from the force balance, $qvB = mv^2/R$, whence

$$R = mv/qB. \tag{2.2.3}$$

If the energy of the ion is W, then $v = (2W)^{1/2}/m$ and the radius of the circular arc can be expressed as

$$R = (2W)^{1/2}/qB. \tag{2.2.4}$$

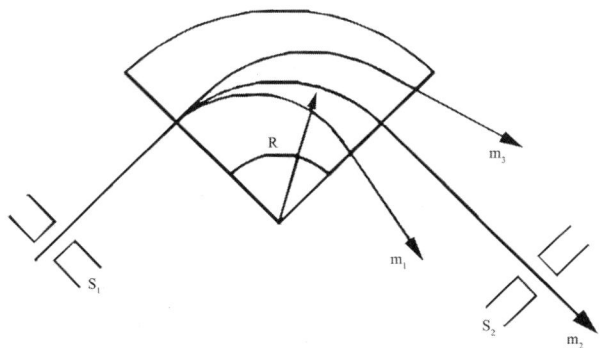

Figure 2.5. Conceptual diagram of a magnetic analyzer.

Since the incident ion energy is $W = qV$, where V is the extraction voltage of the ion source, the beam bending radius can also be expressed as

$$R^2 = 2mV/qB^2. \qquad (2.2.5)$$

We see that if the ion charge q, the magnetic field B, and ion energy W are fixed, the ion bending radius depends on its mass m; and when W, B and m are fixed, the radius depends on the charge number q. Therefore, for a given location of the magnetic analyzer exit (such as the slit S_2 in Figure 2.5), ions of selected mass number or charge number can be extracted. It is usual that magnetic analyzers have a fixed bending radius, thus allowing selection of chosen ion species by adjusting the magnetic field strength B.

2.2.3. Accelerating and Focusing System

The energy at which ions can be extracted from an ion source is limited by high voltage breakdown considerations. For example, for a three-electrode extraction system the upper limit to extraction voltage is about 100 kV. In order to achieve higher ion energies, the ions must be accelerated by an ion accelerating system. For ion bioengineering applications the ion energy typically required is below a few hundred keV, and for this energy range it is convenient to employ an electrostatic accelerating method. The 250 keV (singly charged) ion beam facility designed by the Institute of Plasma Physics, Chinese Academy of Sciences, for ion beam biotechnology uses a double-cylinder accelerating technique, as shown in Figure 2.6. The curves in the figure represent equi-potentials. The electric field E at point A on the left side of the figure, which is perpendicular to the equal-potentials, can be split into a component E_z parallel to the z-axis and a radial component E_r perpendicular to z. When an ion enters this field, it is acted upon by both field components. The field E_z accelerates the ion in the axial direction, while E_r focuses the ion along the axis z. On the right hand side of the central plane M-N, a positive ion diverged from the z-axis due to the E_r component. The ion is accelerated over the entire region. In the field region on the right hand side, the ion speed is higher and the time for which the force acts is shorter. The divergence is small, and hence the overall effect of the field is to form a convergent beam. It should be noted that the second slit is held at negative potential mainly to impede the backflow of electrons.

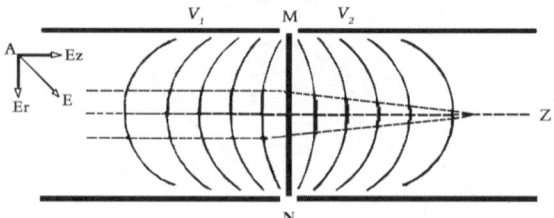

Figure 2.6. Conceptual diagram of a double-cylinder accelerating tube with electrostatic focusing.

This slit plays a divergent role to the positive ions. But, since the potential of this slit is much lower than that of the first slit, the entire system still focuses the ion beam.

If the charge carried by an ion is $q = Ze$, the ion acceleration energy is given by

$$W = q[(V_+ - V_-) + (V_- - V_0)] = q(V_+ - V_0), \tag{2.2.6}$$

where V_+ is the plasma electrode potential, V_- is the accelerating electrode potential and V_0 is the ground potential, or zero,. Thus the ion acceleration energy is given simply by $W = qV_+$.

If a higher ion energy is required, say, greater than several hundred keV, a double-cylinder, single-gap acceleration system is not suitable. Generally, MeV ion energy is produced by a tandem accelerator. The tandem accelerator is one kind of DC high voltage accelerator. The high voltage is applied between two accelerating tubes and both extreme ends of the tubes are grounded. Positive ions are produced in an ion source and converted into negative ions in a charge exchange chamber. The negative ions are accelerated through the first accelerating tube and then stripped of electrons to become positive ions, at the midplane. The positive ions are accelerated again through the second accelerating tube. Thus the ion energy is doubled. A Van der Graaff accelerator (such as that installed at the Institute of Plasma Physics, Chinese Academy of Sciences, and originally from the University of Texas), is an electrostatic accelerator capable of achieving an acceleration energy of 5.5 MeV.

An isolated metal sphere with charge Q and capacitance C will reach a potential given by $U = Q/C$. Addition of charge to the sphere increases the potential U. When the limiting charge is exceeded, a high voltage discharge to the surrounding air will occur. Thus the maximum electric field E_{max} at the sphere surface should be lower than the discharge voltage E_{disch}, namely $E_{max} \leq V/r = E_{disch}$, where r is the sphere radius and V is its voltage relative to ground. The maximum electric field strength that can be supported without breakdown in dry atmospheric air is about 3×10^6 V/m. Thus an acceleration voltage of 3 MV requires a sphere radius of at least 1.0 m. It can be seen that increasing the radius of the metal sphere to increase the acceleration voltage has its limitations. In order to improve the high voltage performance of the accelerator, normally the accelerator is installed inside a steel tank that is filled with a good quality insulating gas. The 5.5 MeV accelerator tank is filled with a mixed gas of 70% N_2 and 30% CO_2 at a pressure of 10–20 atm. The high pressure increases the maximum voltage holdoff, and thus the working voltage of the accelerator. To prevent breakdown, electrostatic accelerators commonly use multiple-stage tubes to distribute the high voltage uniformly along each stage (Figure 2.7). In this way, although the total voltage of the accelerator is very high the voltage safety coefficient at each stage is sufficient to ensure that breakdown through the air does not occur.

Ions travel a fairly long distance from the ion source to the target chamber. Because of space-charge repulsive forces (Chapter 1), the size of the ion beam increases during beam transport, and there may be a loss of particles from the beam. Scattering due to collisions between ions and residual gas molecules in the beam line can also lead to increased beam size. In order to reduce beam loss as well as well as to have an appropriately small beam size at the target, the ion beam must be focused. An ion beam can be focused by electric and magnetic fields. The double-tube single-gap accelerator shown in Figure 2.5 is actually a typical electrostatic lens. Ion beam focusing systems that use electrostatic forces are called electrostatic lenses, whereas systems that us magnetic fields are called magnetic lenses. The electro-quadrupole lens and the magnetic quadrupole lens, both of which are strongly focusing lenses, are often used. Lens systems are frequently installed at locations along the beam line for optimal beam focusing according to the ion optics requirements.

2.2.4. Target Chamber and Vacuum System

The target chamber is where the bio-sample is ion-implanted. As opposed to nonliving materials, biological samples are viable and in various shapes, and this can result spatially inhomogeneous ion implantation. In some experiments delivery of the sample into and out of the chamber requires sterile conditions. Thus the design of the chamber and its associated vacuum system must take into account these constraints.

At the present time, ion implantation for mutation breeding requires a large number of samples to be treated. Each individual sample to be implanted should be exposed directly to the ion beam. A vertical ion beam may be preferred for implantation with a beam spot as large and homogeneous as possible. In this way the bio-samples can be ion bombarded with the samples completely motionless. So long as every individual sample is installed in the sample holder in the same orientation, it is guaranteed that the sensitive location of each individual sample is exposed to the ion beam and receives a similar implantation dose.

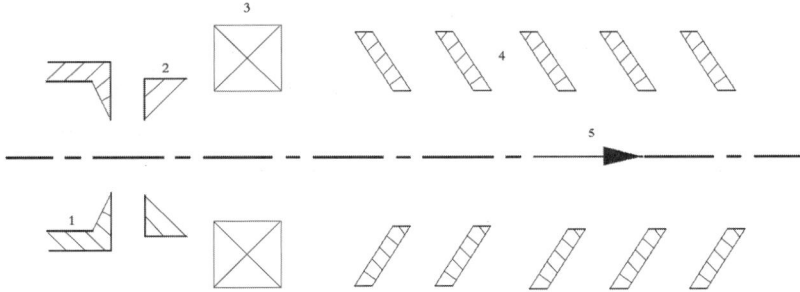

Figure 2.7.   Schematic of a multiple-stage accelerating tube. (1) Ion source, (2) extraction electrode, (3) lens, (4) accelerating section, and (5) ion beam.

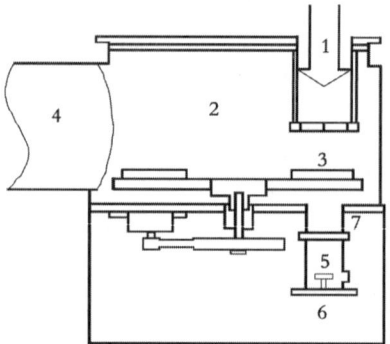

Figure 2.8. Schematic diagram of a target chamber. (1) Ion beam, (2) big target chamber, (3) rotating plate, (4) vacuum pump, (5) small target chamber, (6) sterile chamber, and (7) valve.

Figure 2.8 shows a schematic diagram of a typical target chamber for ion implantation of bio-samples. The ion beam extracted from the ion source and accelerated is incident vertically onto the chamber. The chamber is 90 cm in diameter and 1.0 m in height. A rotating plate 80 cm in diameter is located on the bottom of the chamber. On the plate there are six 20-cm-diameter concave holes, only one of which is completely through. Six 20-cm-diameter sample dishes can be fixed in the concave holes. On the sample dishes there are holes with different diameters for holding various bio-samples. Samples of the same variety, such as rice seeds, are placed in the sample dishes, which are then fixed onto the large rotating plate. The angular position of the plate is adjusted to align one of the dishes with the incident ion beam. After implantation of the first dish, another five sample dishes can automatically be aligned in sequence to the ion beam by computer control, which can also set different implantation parameters. Thus six different sample groups with different implantation parameters can be obtained in a single equipment operation cycle.

Ion implantation of microbes or plant cells is normally carried out in a small target chamber under the main chamber connected by a gate valve. The samples are installed under sterile conditions. The hole on the rotating plate is aligned to the entrance of the small chamber and the valve is then opened to allow ion implantation.

Ion implantation of bio-samples does not require a maximally clean vacuum, and the vacuum produced by a normal oil diffusion pump is adequate. Calculation of the vacuum system should take into account the volume ratio of the large and small chambers so that the small chamber can be pumped down to the desired pressure in a short time. Note that the living bio-samples in the small chamber, that contain a great deal of water, will be rapidly frozen due to pumping of water from the bio-surface, subsequently leading to less water evaporation (see Chapter 4).

2.3. MEASUREMENT OF BEAM CURRENT

Measurement of the ion beam current can be done both electrically and by heat measurements [10]. The electrical method measures total ion charge delivered and then converts this to a beam current; the heat method measures the heating due to the ion beam and then converts this to a beam current. Often the electrical method is used as a relative measurement, for example to measure the beam current density distribution in the radial direction. If a part of the ion beam is neutralized by background gas collisions during beam transport, the heat measurement may be preferable. The heat power method provides an absolute measurement, and can measure both ions and neutral particles by their energies.

Figure 2.9 shows the principle of the electrical measurement method. Ions pass through a suppression slit and enter a Faraday cup. The ion charge delivered to the cup is measured by the current flow through resistor R, producing a voltage drop V. Thus the ion current can be calculated by

$$I^+ = V/R. \tag{2.3.1}$$

Bombardment of ions on the Faraday cup surface can cause secondary electron emission. This secondary electron current is suppressed by application of a negative voltage.

The thermal measurement method, also called calorimetry, includes adiabatic and non-adiabatic types. Figure 2.10 shows a schematic diagram of an adiabatic calorimeter. When the ion beam bombards the target surface, the kinetic ion energy is transferred to the target internal energy which increases the target temperature. If the metal target has a specific heat C, mass m, initial temperature T_1, and temperature after one beam pulse T, the energy that the target gains from the ion beam is given by

$$Q = Cm(T - T_1), \tag{2.3.2}$$

and the beam current is

$$I^+ = Q/Vt, \tag{2.3.3}$$

where V is a sum of the extraction voltage and the acceleration voltage (i.e., total applied voltage), and t is the beam pulse length. Thus the target initial temperature T_1 and the temperature after one pulse T are measured, and the beam current can be calculated from (2.3.2) and (2.3.3).

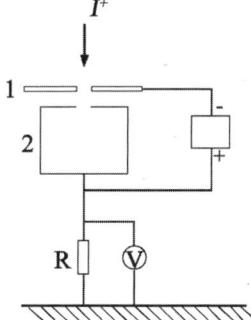

Figure 2.9. Conceptual diagram of electrical measurement. (1) Secondary electron suppression electrode, and (2) Faraday cup.

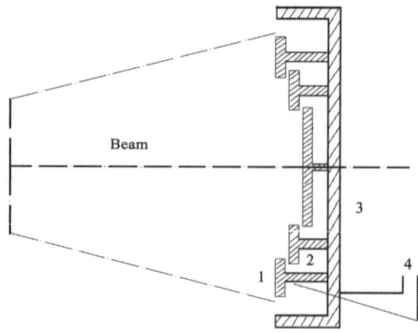

Figure 2.10. Schematic of an adiabatic calorimeter. (1) Partial target, (2) ceramic insulator, (3) screen, and (4) thermocouple.

A metal target has good thermal conductivity. In a long pulse interval, the temperature everywhere on the target surface may be taken to be uniform. Thus the position of the thermocouple that measures the initial target temperature T_1 is not important. Shortly after the target is irradiated with one ion beam pulse, the temperatures at various locations of the target surface will vary. Using of the temperature at one location on the target surface to represent the mean temperature is obviously not reasonable. However, no matter how different the temperatures at different locations of the target surface are, after a relatively long time they will eventually be the same.

If only radiation loss is considered, the temperature change as a function of time after the target reaches an equilibrium temperature is given by

$$ln[(T - T_0)/(T + T_0)] - 2 \tan^{-1}(T/T_0) = A_1 t + A_0, \qquad (2.3.4)$$

where T_0 is the ambient temperature, and A_1 and A_0 are constants to be determined. In the experiment, firstly a number of the pairs (T_i, t_i) are measured based on a certain time interval length, and then the constants A_1 and A_0 are determined using the least-squares method. Eq. (2.3.4) expresses the temperature change after the target reaches an equilibrium temperature. The equivalent mean temperature T instantly after the pulse ends can be obtained from Eq. (2.3.4) with t = 0:

$$ln[(T - T_0)/(T + T_0)] - 2 \tan^{-1}(T/T_0) = A_0. \qquad (2.3.5)$$

Substitution the solution for T from Eq. (2.3.5) into Eq. (2.3.2) then provides the beam energy due to one pulse received by the calorimeter.

In ion beam biotechnology experiments, periodic measurement of the ion beam current is necessary. If the ion beam bombardment area is A, the measured beam current I^+, the implantation time t, and the ions are singly charged, then the implantation dose can be calculated from

$$D = 6.25 \times 10^{18} (I^+/A) t, \qquad (2.3.6)$$

where I^+ is in Amperes, A is in cm^2, and D is in number of ions implanted per cm^2.

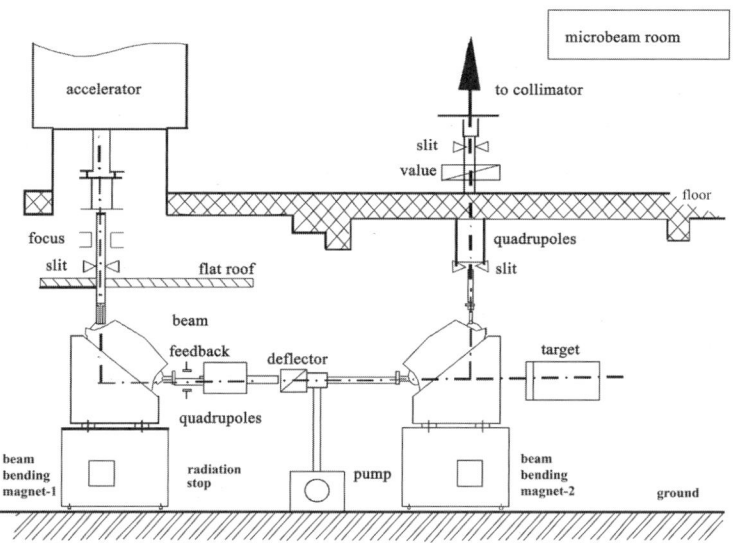

Figure 2.11.   Schematic diagram of the ASIPP microbeam line.

2.4 SINGLE-ION MICROBEAM FACILITY

Since the late 1980's, there has been a resurgence of interest in developing and applying single-ion microbeam (SIM) technique [11] to problems in radiation biology [12-17]. One of the studies is on cell and tissue damage caused by ionizing radiations. In this section, SIM facilities will be generally introduced.

A SIM facility mainly consists of an electrostatic accelerator, deflection magnets, beam regulators, quadrupoles, electrostatic deflectors, diaphragms, collimators and so on. A general view of the installation of the ASIPP SIM at the Key Laboratory of Ion Beam Bioengineering, Chinese Academy of Sciences is depicted in Figure 2.11. The microbeam utilizes a CN-5.5 Van de Graaff accelerator, which is able to accelerate singly- or doubly-charged particles, generally produced from a radio-frequency ion source, with the maximum high voltage of 5.5 MV. Accelerated particles of the desired mass and energy are selected using an analyzing magnet which bends the vertical particle beam into a horizontal section of the beam line. Along the beam line is an arrangement of slits, beam feedback, quadrupole magnets and electrostatic steerers, used to define the beam profile and trajectory. The horizontal beam line transports the particles into the second analyzing magnet which bends the beam vertically up to the microbeam experimental room through slits.

SIM, as a new mutation source, is uniquely capable of delivering precisely a predefined number of charged particles (precise doses of radiation) to individual cells or subcelluar targets *in situ* normally at an area with a diameter of a few micrometers or smaller. However, even if the beam profile has already been optimized, it is still difficult to confine the beam profile within a desired range such as a diameter of smaller than 1 mm. In order to further reduce the beam size down to an order of micrometer, two methods can be adopted, namely, magnetic focusing and collimation. These two types of arrangement have some specific technological advantages and some disadvantages respectively. It is clear that a focused arrangement can offer the finest beam and a better defined linear energy transfer (LET) because no particles are scattered inside a collimator [11,18,19]. As an additional bonus, there is more space to install a detector before the target and a higher potential throughput can be gained as the beam line can be deflected to the position of the cells instead of positioning the cells into the beam exit. But a focused arrangement faces more challenges because the focused beam will have to pass through a vacuum window and the cost of installation is considerably greater than a collimated counterpart. On the other hand, there are several advantages by a collimated arrangement. Firstly, there is greater flexibility with regard to the overall configuration of the beam line. This is important, bearing in mind that it is desirable to operate a wet-cell irradiation system vertically. Secondly, collimation offers a straightforward method of reducing the dose-rate to radiobiological levels. Regardless of which method adopted, three crucial factors in general are the high accuracy of targeting, the control of a precise number of charged particles delivered, and high throughput of cells.

A glass capillary collimator is adopted by the ASIPP microbeam to gain a fine beam. A 1 mm long fine bore capillaries with internal diameters as small as 1 μm fits into a stepped hole machined by spark erosion in a 2.1 mm thick stainless steel disk. A 3 μm thick Mylar film is used as a vacuum window. A 7 μm thick aluminum foil and an 18 μm thick scintillator are sandwiched by another 3 μm thick mounting Mylar film. The arrangement of the collimator is illustrated in Figure 2.12. The number of photons emitted and energy consumption will depend on the type and energy of the traversing particle and the total thickness of the thin films used. The number of near-monoenergetic photons emitted is 10^4/s when a 2 ~ 3 MeV proton pass through the collimator with the above-described arrangement. A typical energy spectrum from the aligned SIM is shown in Figure 2.13. In practice, the particle number detected will depend on geometry and coupling arrangement of the detector, the quantum efficiency of the detector and other factors, such as losses due to scattering.

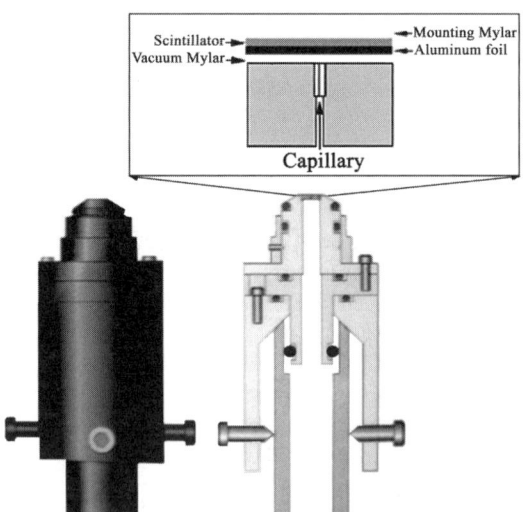

Figure 2.12.   The arrangement of the collimator at the ASIPP SIM.

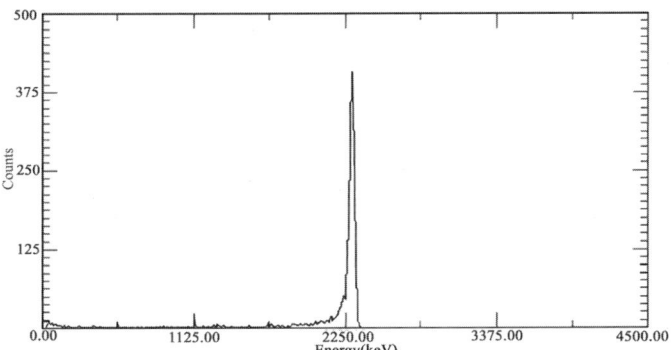

Figure 2.13. The spectrum of 2.2 MeV monoenergetic protons (after passage through the vacuum window and scintillator), collimated by a glass capillary collimator with a 1.0-μm diameter and a 0.98-mm length.

The precision of quantitative radiation and the veracity of localization radiation are the most important technical specifications. To achieve high precision and high veracity of delivery the predefined number of particles to the desired targets, the preconditions that control the key components of the system should be set up. At the same time, diagnosing the active status of the key instruments and promptly evaluating the feedback are necessary to ensure that the experiment is smoothly carried out.

The integrated control system can be divided into three modules logically: imaging acquisition and processing, microscope stage control, and particle counter and beam shutter, as shown in Figure 2.14. The cells to be irradiated are stained and attached to a specific dish. The light spot, which is produced when emitted particles excite the scintillator, can be used as an optical reference to localize the position of the exit aperture. Firstly, the nucleus of each cell is identified and located by a computer/microscope-based image analysis system, which detects the fluorescent staining pattern of the cells by UV light. Secondly, the dish is moved under computer control to the position where the first cell nucleus is positioned over a highly collimated particle beam. The ability to deliver a single or preset number of particles to each cell depends critically on collaborative work of a detector and a beam shutter. During irradiation, the charged particles are detected after passing through the cell. The beam shutter is a fast electrostatic deflection system allowing each irradiated nucleus to be quickly removed from the beam when enough particles have been detected. The beam shutter is opened until the required number of particles has passed though the nucleus. The shutter is then closed, and the next cell is positioned over the beam.

So far, there has been a rapid growth in the number of institutions which are developing SIMs or planning to do so. There are currently more than 14 SIMs worldwide [18,20-37], several SIMs in routine use for radiobiology, and their current major specifications are shown in Table 2.1, and they have been in a constant state of evolution. Much of the recent research using SIMs has been to study low-dose effects and cell signaling, genomic instability, bystander effects and adaptive responses. And the list of possible applications for SIM in radiobiology continues to grow and diversify day by day. Without doubt, scientists don't settle for only targeting nucleus or cytoplasm, and in so doing, cell organelles and tissue becomes the next targets naturally.

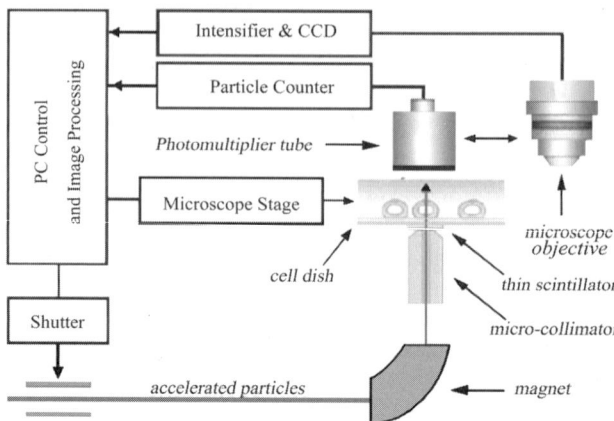

Figure 2.14. Schematic diagram of the overall layout of the ASIPP microbeam facility and the integrated control system.

SIMs have been in constant development and much progress has been made over the past few years. For example, a laser ion source (LIS) enables us to use ions of sufficient range from hydrogen to iron and provide a wider range of LET [38,39]. However, a number of aspects of SIM performance should continue to be scrutinized and if possible, improvements should be implemented, such as increasing the cell throughput, increasing the target accuracy, increasing the penetration of the irradiation and applying non-UV methods for target visualization. The throughput of the SIM needs to further increase because statistical uncertainty and the low incidence of some biological endpoints (e.g. the expression of genes in terms of a cell population in general comes from samples involving millions of cells). And the combination of pre-cell detection and post-cell detection may be the only way to eliminate the possibility of overcount or undercount. Only in this way, SIM could meet the need of mountainously growing applications in biology.

SIM, as a powerful tool, opens up the possibility of probing the answers to many enigmas in radiobiology, but some new technologies should be grafted in or related by marriage. Provided that visual observation of the cells and particle detection occur simultaneously (lucite light guide technique provides an elegant solution for this intent), different technologies could bring out the best in each other. Many optical screening technologies would be expected to be contributors, notwithstanding cell-based applications in high throughput screening (HTS) form a considerable challenge to both instrumental design and fully automated data handling and evaluation, especially the miniaturized HTS and ultra-HTS (uHTS) at ambient temperature. Most importantly, they open up the possibility to online *in situ* physiological monitoring the cells or tissues in a noninvasive way [40]. In doing so, scientists could fix attention on the original process of the interaction between low-energy ions and complicated organisms rather than endpoints, and gain deeper insight into the true nature. Fortunately, biologists and biophysicists are now joining forces to bring major advances in the art of SIM. It is believed that this is not the end of the saga.

Table 2.1. SIMs worldwide in routine use and their current main specifications.

Institution	SIM	Accelerator	Particles	Energy /LET	Mode of detection	Targeting accuracy	Through-put
RARAF, Columbia University, USA	RARAF micro-beam	4.2MeV Van de Graaff accelerator	From hydrogen to iron	10 - 4500 keV/μm	Gas-filled ionization chamber and a SSD detector	~ ±2μm	~11,000 cells per hour
CRC Gray Laboratory, Northwood, UK	GCI micro-beam	4MeV Van de Graaff accelerator	Protons, Helium ions	5.7MeV He^3	Scintillator + PMT	< ±2μm	~36,000 cells per hour
JAERI-Takasaki, Japan	TIARA micro-beam	AVF-cyclotron accelerator	From C ions to Fe ions	8-1800 keV/μm	a CaF2 (Eu) scintillator + PMT	~ ±1μm	>60,000 cells per hour
CENBG, Bordeaux, France	CENBG micro-beam	4MeV Van de Graaff accelerator	Protons, Helium ions	3.5MeV alpha or proton	a thin scintillating foil + Gas transmission detector	±2μm	>2,000 cells per hour
PTB, Braunsch-weig, Germany	PTB micro-beam	3.75 MeV Van de Graaff and 20~35MeV cyclotron accelerators	Protons, Helium ions	3 - 200 keV/μm	Scintillator + PMT	±1.5μm	15,000 cells per day
NIRS, Chiba, Japan	SPICE micro-beam	HVEE Tandem accelerator	Protons, Helium ions	3.4 MeV H^+, 5.1 MeV He^{2+}	Scintillator + PMT	<±2μm	2,000 cells per hour
University of Melbourne, Parkville, Australia	MARC Micro-beam	5U NEC Pelletron accelerator	Protons, Helium ions, C, N, O	4.6~7.7 MeV alpha particles	Boron doped thin diamond film + SSD + CEM detector	<= ±300nm	>3,000 cells per hour
IPP, CAS, Hefei, China	CAS-LIBB micro-beam	5.5 MV Van de Graaff accelerator	$^1H^+$, $^2H^+$	3.5 MeV H^+, 3.5 MeV $^2H^+$	Scintillator + PMT	~ ±1μm	-
MIT LABA, Boston, USA	MIT LABA micro-beam	1.5 MV single stage electrostatic accelerator	He^{2+}, H^+, He^+	15-30 keV/μm	Light guide + plastic scintillator + PMTs	~ ±3μm	-
GSI, Darmstadt, Germany	SIHF micro-beam	GSI linear accelerator (UNILAC)	From Carbon to Uranium	1.4MeV/μm – 11.4 MeV/μm	Si_3N_4 foil + channeltron	< ±1μm	-
Technische Universität München, Germany	SNAKE micro-beam	14MV Munich Tandem accelerator	From protons to uranium	2 KeV/μm – several MeV/μm	Position-sensitive detector (PSD)	~ ±700nm	-
University of Leipzig, Leipzig, Germany	LIPSION micro-beam	3.5 MV Singletron™ accelerator	Protons, Helium ions	10-20 keV/μm	Multi-dectector	~ ±130nm	-
WERC, Japan	W-MAST micro-beam	5 MV tandem accelerator	Protons	10 MeV proton	Si surface-barrier detector (SSD)	~ ±10 μm	-
INFN-LNL, Padova, Italy	INFN micro-beam	7MeV Van de Graaff CN accelerator	$^1H^+$, $^2H^+$, $^3He^{2+}$, $^4He^{2+}$	7-150 keV/μm	Silicon detector	-	-

REFERENCES

1   Yu, Z.L., Deng, J.G. and He, J.J., Mutation Breeding by Ion Implantation. *Nucl. Instr. Meth.*, B**59/60** (1991)705-708.

2   Wu, L.F. and Yu, Z.L., Radiobiological Effects of a Low-energy Ion Beam on Wheat, *Radiat Environ Biophys*, **40** (2001)53–57.

3   Yu, Z.L., Ion Beam Application in Genetic Modification. *IEEE Trans. on Plasma Sci.*, **28** (2000)128-135.

4   Yu, Z.L., Yang, J.B., Wu, Y.J., *et al.*, Transferring Gus Gene into Intact Rice Cells by Low Energy Ion Beam, *Nucl. Instr. Meth.*, B**80/81** (1993)1328-1331.

5   Yu, Z.L., Ion Beam and Biology Science, *Physics* (Chinese), **26** (1997)333–338.

6   Yu, Z.L., Interaction between Low Energy Ions and the Complicated Organism, *Plasma Sci. & Tech.*, **1** (1999) 79-85.

7   Yu, Z.L., He, J.J., Zhou, J. and Deng, J.G., High-Current DC Ion Source with Large Radiation Area, *Vacuum Science and Technology*, **9**(6) (1989)379–382.

8   Yu, Z.L., Xu, P., He, J.J., Gao, S.Q. and Zhou, J., Studies on IS-A 10cm-Duopigatron Ion Source, *Chinese Journal of Nuclear Science and Engineering*, **7**(3,4) (1987) 318–321.

9   Yu, Z.L., Xu, P., He, J.J. and Zhou, J., Experiments and Applications of IS-A Type Duopigatron, *Nuclear Technology*, **12**(10) (1989)614–617.

10 Yu, Z.L., Power Measurements of Heavy Current Pulsed Ion Beam, *Nuclear Technology*, **12**(4) (1989)205–208.

11 Watt, F. and Grime, G.W., *Principles and Applications of High Energy Ion Microbeams* (Adam Hilger, Brostol., 1987).

12 Zirkle, R.E and Bloom, W., Irradiation of Parts of Cells, *Science*, **117**(1953)487–493.

13 Bloom, W., Cellular Responses, *Rev. Modern Phys.*, **31**(1959)66–71.

14 Legge, G.J.F., A History of Ion Microbeams, *Nucl. Instr. Meth.*, B**130** (1997)9–19.

15 Brenner, D.J. and Hall, E.J., Microbeams: A Potent Mix of Physics and Biology, *Radiat. Prot. Dosim*, **99**(2002)283–286.

16 Hei, T.K., Wu, L.J., Liu, S.X., *et al.*, Mutagenic Effects of a Single and an Exact Number of α Particles in Mammalian Cells, *Proc. Natl. Acad. Sci. USA*, **94** (1997)3765–3770.

17 Wu, L.J., Randers-Pehrson, G., Xu, A., *et al.*, Targeted Cytoplasmic Irradiation with Alpha Particles Induces Mutations in Mammalian Cells, *Proc. Natl. Acad. Sci. USA*, **96**(1999)4959 ~ 4964.

18 Cholewa, M., Saint, A., Legge, G.J.F., *et al.*, Design of a Single Ion Hit Facility, *Nucl. Instr. Meth.*, B**130**(1997)275–279.

19 Sakai, T., Naitoh, Y., Kamiya, T., *et al.*, Single Ion Hitting to Living Samples. *Nucl. Instr. Meth.*, B**158**(1999)250–254.

20 Folkard, M., Vojnovic, B., Prise, K.M., *et al.*, A Charged-Particle Microbeam: I. Development of an Experimental System for Targeting Cells Individually with Counted Particles, *Int. J. Radiat. Biol.*, **72** (1997)375–385.

21 Folkard, M., Vojnovic, B., Hollis, K.J., *et al.*, A Charged-Particle Microbeam: II. A Single-Particle Micro-Collimation and Detection System, *Int. J. Radiat. Biol.*, **72** (1997)387–395.

22 Dymnikov, A.D., Brenner, D.J., Johnson, G., *et al.*, Theoretical Study of Short Electrostatic Lens for the Columbia Ion Microprobe, *Rev. Sci. Instr.*, **71**

(2000)1646-1650.

23 Butz, T., Flagmeyer, R-H., Heitmann, J., *et al.*, The Leipzig High-Energy Ion Nanoprobe: A Report on First Results, *Nucl. Instr. Meth.,* B**161-163**(2000)323- 327.

24 Fischer, B.E., Cholewa, M. and Hitoshi, N., Some Experiences on the Way to Biological Single Ion Experiments, *Nucl. Instr. Meth.*, B**181**(2001)60–65.

25 Randers-Pehrson, G., Geard, C.R., Johnson, G., *et al.*, The Columbia University Single-Ion Microbeam, *Radiat. Res.*, **156**(2001)210–214.

26 Michelet, C., Moretto, Ph., Barberet, Ph., *et al.*, A Focused Microbeam for Targeting Cells with Counted Multiple Particles, *Radiat. Res.*, **158**(2002)370–371.

27 Dollinger, G., Datzmann, G., Hauptner, A., *et al.*, The Munich Ion Microprobe: Characteristics and Prospect, *Nucl. Instr. Meth.*, B**210** (2003)6–13.

28 Greif, K. D., Brede, H. J., Giesen, U., *et al.*, The PTB Focused Microbeam for High and Low LET Radiation, *Radiat. Res.*, **161**(2004)89–90.

29 Oikawa, M., Kamiya, T., Fukuda, M., *et al.*, Design of a Focusing High-Energy Heavy Ion Microbeam System at the JAERI AVF Cyclotron, *Nucl. Instr. Meth.*, B**210**(2003)54-58.

30 Cholewa, M., Fischer, B. E. and Heiß, M., Preparatory Experiments for a Single Ion Hit Facility at GSI, *Nucl. Instr. Meth.*, B**210**(2003)296–301.

31 Folkard, M., Vojnovic, B., Gilchrist, S., *et al.*, The Design and Application of Ion Microbeams for Irradiating Living Cells and Tissues, *Nucl. Instr. Meth.,* B**210** (2003)302-307.

32 The Radiological Research Accelerator Facility, *RARAF Annual Report* 2002, Columbia University, USA.

33 Wu, L.J., Hei, T.K., Randers-Pehrson, G., Wang, S.H., and Yu, Z.L., Columbia University Microgeam: Development of an Experimental System for Targeting Cells Indiviually with Counted Particles, *Nucl. Sci. and Tech.*, **10**(3)(1999).

34 Hu, Z.W., Yu, Z.L. and Wu, L.J., An Optimization Control Program for the ASIPP Microbeam, *Nucl. Instr. Meth.*, A**507**(2003)617–621.

35 Wang, X.F., Chen, L.Y., Hu, Z.W., Wang, X.H., Zhang, J., Li, J., Wu, L.J., Wang, S.H., Yu, Z.L., *et al.*, Quantitative Single-Ion Irradiation by ASIPP Microbeam, *Chin. Phys. Lett.*, **21**(2004)821–824.

36 Moretto, Ph., Michelet, C., Balana, A., *et al.*, Development of a Single Ion Irradiation System at CENBG for Applications in Radiation Biology, *Nucl. Instr. Meth.*, B**181**(2001)104–109.

37 Cherubini, R., Conzato, M., Galeazzi, G., *et al.*, Light-Ion Microcollimated Beam Facility for Single-Ion, Single Mammalian Cell Irradiation Studies at LNL-INFN, *Radiat. Res.*, **158**(2002)371– 372.

38 Bigelow, A.W., Randers-Pehrson, G. and Brenner, D.J., Laser Ion Source Development for the Columbia University Microbeam, *Rev. Sci. Instr.*, **73** (2002)770-772.

39 Bigelow, A.W., Randers-Pehrson, G. and Brenner, D.J., Proposed Laser Ion Source for the Columbia University Microbeam, *Nucl. Instr. Meth.*, B**210**(2003)65-69.

40 Nelms, B.E., Maser, R.S., MacKay, J.F., *et al.*, *In Situ* Visualization of DNA Double-Strand Break Repair in Human Fibroblasts, *Science*, **280**(1998)590-592.

FURTHER READING

1. Brown, I.G., ed. *The Physics and Technology of Ion Sources*, 2nd Edition, (Wiley-VCH, Berlin, 2004).
2. Ivanov, В.И. and Razov, В.И., *Fundamentals of Micro-Dosimetry*, transl.: Hua, M.C. (in Chinese, Atomic Energy Press, Beijing, 1987).
3. Nastasi, M., Mayer, J.W. and Hirvonen, J., *Ion-Solid Interactions: Fundamentals and Applications* (Cambridge University Press, Cambridge, 1996).
4. Wang, Y.H., *et al.*, *Fundamentals of Ion Implantation and Analysis* (in Chinese, Aviation Industry Press, Beijing, 1992).
5. Zhang, G.H. and Zhong, S.L., *Ion Implantation Technology* (in Chinese, Engineering Industry Press, Beijing, 1982).
6. Zhu, R.S., *Principles and Applications of Solid Nuclear Trace Detectors* (in Chinese, Science Press, Beijing, 1987).
7. Ziegler, J.F., ed. *Ion Implantation: Science and Technology* (Academic Press, New York, 1984).

3

FUNDAMENTALS OF ION IMPLANTATION

Energetic ion bombardment of materials is like target practice. The ion can be thought of as the "bullet" and the bombarded material as the "target". Understanding the bullet-target interaction process involves many fields of modern physics. We need to know the detailed structure of the target material, including its atoms and the nuclei of those atoms, and also details about the ion. We need to invoke not only macroscopic phenomena such as thermodynamic and chemical reaction processes, but also the microscopic processes of atomic collisions and nuclear reactions. In this chapter we discuss the fundamentals of interactions between energetic ions and metal or semiconductor solid materials, establishing a base for later studies of the interaction between implanted ions and biological materials.

3.1. INTERACTIONS OF ENERGETIC IONS WITH SOLID SURFACES

When energetic ion beams bombard solid surfaces, charged and neutral particles are emitted from the surface, as illustrated in Figure 3.1 [1-5]. The charged particles include rebounding primary ions, emitted secondary ions, sputtered ions, backscattered ions, etc. The neutral particles are usually gas molecules that were originally absorbed and attached to the solid surface and later knocked off the surface by the incident energetic ions, or ions neutralized by electrons escaping from the surface, or individual atoms or atom clusters sputtered from the surface. In addition, high-energy ion bombardment of target atoms can also produce X-ray and photon emission.

3.1.1. Secondary Electron Emission

As described in Chapter 2, when an electrical method is used to measure ion beam current, secondary electron emission from the collector can affect measurement accuracy. Biological materials are usually very poor electrical conductors. The emission of secondary electrons from the interaction of energetic ions with biological materials can affect the electric characteristic of the material. Changes in the biological electrical characteristic can disturb normal life activity of the biological organisms, and can cause various physiological reactions, even possibly damaging genetic substances.

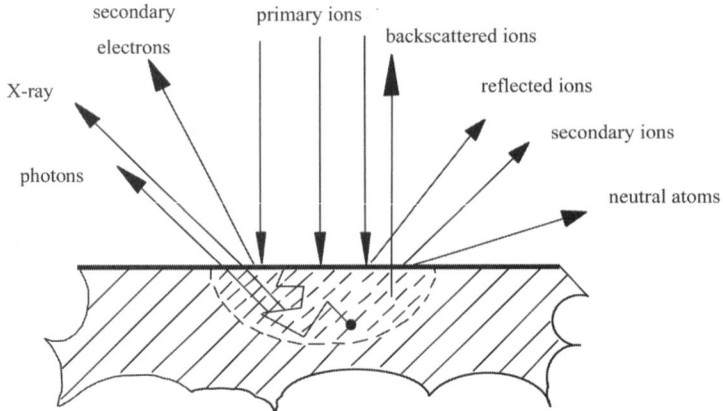

Figure 3.1. Principle of interaction between ions and solid.

Electrons emitted from a solid surface that is bombarded with energetic incident ions are called secondary electrons, and the phenomenon is called secondary electron emission [6-8]. The secondary electron flux includes electrons that are both elastically and inelastically scattered.

The kinetic energy of secondary electrons emitted from a metal surface has a minimum value equal to the work function of the material surface. The incident ions transfer their energy firstly to the crystal lattice and then secondarily to the electrons in the solid, thus providing some electrons with enough energy to overcome the surface potential energy (the work function) and allowing them to escape from the surface. This is called kinetic emission. The process may consist of multiple steps. When an incident energetic ion collides with many lattice atoms, or lattice atoms collide with neighboring atoms, secondary electrons may be emitted. On the other hand, if the energy of the incident ion is greater than the metal work function, then when the incident ion is close to the metal surface (a distance of just a few atomic diameters), the interaction between the ion and conduction electrons in the metal lowers the surface binding potential energy and leads to secondary electron emission. This is called potential emission.

The parameter characterizing secondary electron emission in any given situation is the secondary electron emission coefficient Y_i. This is defined as the mean number of secondary electrons produced by each incident ion:

$$Y_i = n_e/n_i, \tag{3.1.1}$$

where n_e is the number of secondary electrons emitted and n_i is the number of incident ions. The secondary electron emission coefficient is also sometimes called the secondary electron emission yield.

The secondary electron yield, Y_i, is determined by the incident ion and the properties of the solid surface. When the bombarding ion energy is several tens to hundreds of eV, Y_i generally lies in the range of one to several percent. But when the incident ion energy is several keV, Y_i can assume quite high values, in excess of unity. The ion energy plays a secondary role, compared to the ion excitation energy, in determining Y_i. For the same solid target material, the higher the charge state of the incident ion, the higher the secondary electron emission yield. Secondary electrons originate in the near-surface region of the solid, of thickness say about 50 – 100 Å, and the secondary electron energy is typically less than 30 – 50 eV. For higher ion energy, secondary electrons originate from somewhat greater depths, but they lose energy as they move to the surface and only a fraction of them escape. Thus the secondary electron yield Y_i decreases for sufficiently high ion energy. There are many free electrons in the conduction band in metals, and secondary electrons lose significant energy due to collisions with free electrons. Thus when they arrive at the surface they have a low escape probability. In semiconductors and insulators, there are fewer free electrons and hence the secondary electron escape probability is greater. A rough or convoluted surface has a greater secondary electron emission yield than a smooth surface. Biological organisms have structurally complex surfaces and are electrically inert as well. Therefore secondary electron emission yields from biological organism surfaces are greater than from metallic or semiconductor surfaces when they are ion bombarded under similar implantation conditions.

3.1.2. Ion Sputtering

When a beam of accelerated ions bombards a solid surface, the ions exchange momenta with atoms at the surface and some atoms are ejected. This phenomenon is called sputtering [9-11]. It might more strictly be called bombardment evaporation, and it is intrinsically different from thermal evaporation. Thermal evaporation is due to energy transfer whereas sputtering is due to momentum transfer. Thus the sputtered atom flux has directional properties.

The sputtering ratio (or sputtering yield as it is sometimes called) is a characteristic parameter describing sputtering and is defined as

$$S = \frac{\text{number of sputtered target atoms}}{\text{number of incident ions}} \qquad (3.1.2)$$

The sputtering ratio (yield) is a complicated parameter that is determined by a number of factors [12-14].

For bombardment of different target materials by ions of fixed species and energy, the sputtering yield varies according to the group in the Periodic Table to which the target atom belongs. On the other hand, for bombardment of a fixed target material with ions of different species and energy, the sputtering yield varies according to the group in the Periodic Table to which the bombarding ions belong. Figure 3.2 shows the relationship between the relative sputtering yield and the atomic number of the incident ion.

S varies with the energy of the incident ion. The incident ion energy has a critical value below which no sputtering takes place. For most metal target materials this critical value is between ten and several tens of eV. If the ion energy is lower than this value, the sputtering yield is zero. As the incident ion energy increases, S reaches a maximum and then decreases as the incident ion energy is further increased. This is because the incident ions are implanted to a greater depth and displaced atoms cannot easily reach the surface. Figure 3.3 shows the sputtering ratio for bombardment of a Cu surface with Kr ions in the energy range 1 keV – 1 MeV. Maximum sputtering yield occurs for an ion energy near 100 keV.

For a single-crystal target material, S varies with orientation of the crystalline plane. In this case the direction of the ejected sputtered atom does not follow a cosine law but instead tends to be directed along the denser crystal planes.

The sputtered atom energy is 1 – 2 orders of magnitude greater than the energy of thermally evaporated atoms (~ 0.1 eV).

The production of sputtered atoms and molecules from a target material surface is a very complicated process. For single-element materials, fewer single atoms are sputtered and more multi-atom particles are sputtered as the ion energy increases. For example, for a polycrystalline copper target, more Cu_2 particles are sputtered with increasing Ar^+ energy: for 100 eV ion energy the Cu_2 fraction is about 5% of the total sputtered particle flux, while for 12 keV ion energy Cu_n^+ (n = 1 – 11) multi-atom clusters are observed. For Ar^+ bombarded Al, Al_n (n = 1 – 7) are observed; and for Xe^+ bombarded Al, Al_n (n = 1 – 8).

If the target material is a compound such as GaAs, 99% of the sputtered atoms and molecules are neutral Ga or As single atoms, and the remainder are neutral GaAs molecules. When an ion beam interacts with a biological organism, formation of sputtered atoms and molecules is more complicated.

Sputtering is a particular case of radiation damage created by energetic ions interacting with a solid surface. Just as strong acids can strip the metal surface layer by layer, so also ion sputtering can strip away the metal surface atoms. This is called etching or engraving. Ion beam engraving or etching has many industrial applications, such as metal surface polishing, atomic thinning, and production of semiconductor component circuits. Sputtering can also be used for thin film deposition on other solid surfaces. In studying ion sputtering of biological organisms, one can use this phenomenon to process the cell surface to cause porosity so as to enhance cell permeability.

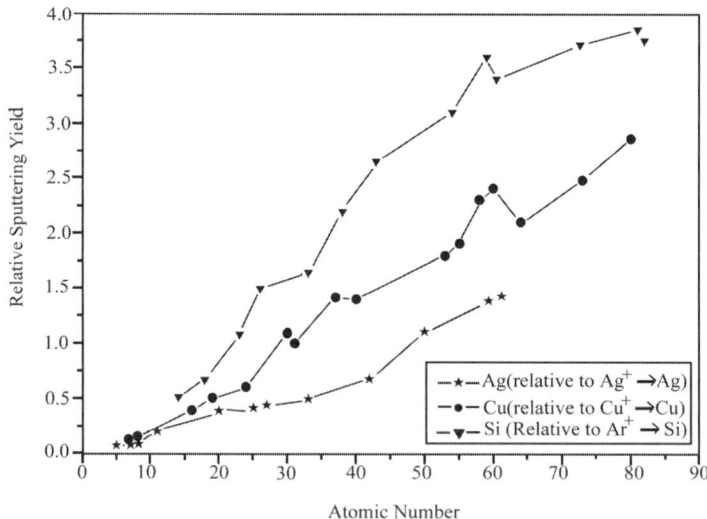

Figure 3.2. Relative sputtering yields for bombarding ion species of Si, Cu and Ag as a function of target atomic number.

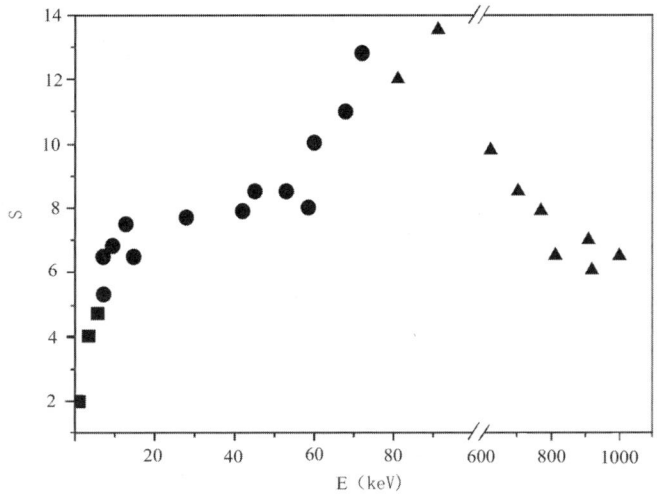

Figure 3.3. Sputtering yield of Kr^+ for Cu target as a function of ion energy. Symbols of square, circle and triangle indicate data in low ($\sim < 10$ keV), medium ($\sim 10 - 80$ keV) and high ($\sim > 80$ keV) energy regions, respectively.

3.1.3. X-Ray Emission

When the incident ion energy is sufficiently high, ion bombardment can cause ejection of inner electrons from the target atoms, leaving vacancies in the electron shell structure. When, subsequently, outer electrons fall into these inner vacancies, fluorescence radiation with wavelength in the X-ray range or Auger electrons are emitted, releasing the extra energy in the atom [1,2]. Which way the energy is released depends on relative probabilities. Generally the fluorescence yield is used to express the probability of producing X-ray fluorescence:

ω_K = number of K X-ray photons / number of vacancies in K shell,

ω_L = number of L X-ray photons / number of vacancies in L shell.

According to this definition, the K-group fluorescence yield, ω_K, is the probability of producing a K X-ray when a vacancy forms in the K shell. Similarly ω_L represents the probability of producing an L X-ray when a vacancy forms in the L shell. It has been shown from experiments that the value of ω_K varies strongly with different elements. The ω_K value for low-atomic-number elements is close to zero, whereas that for high atomic number elements is nearly 1.

The X-rays emitted from different elements have different characteristic energies. From the energies of these characteristic spectral lines, the elemental species in the sample can be determined; and from the intensities of the spectral lines, the concentration of the elements can be calculated. In low atomic number elements, the electron shells are bound by the nucleus relatively loosely, and thus Auger electrons are more easily emitted and the emitted X-ray energy is low, in the ultra-soft X-ray regime. Ultra-soft X-rays have a short penetration depth and strong self-absorption, and thus they are not easy to detect.

3.2. ION IMPLANTATION RANGE

3.2.1. Energy Loss and Range

An ion implanted with an energy E_1 experiences a series of collisions with the target atoms and loses its energy in the process. If the energy loss of the implanted ion per unit distance in the depth direction is dE_1/dx, the energy loss of the ion in a distance Δx is

$$\Delta E_1 = |dE_1/dx| \Delta x. \tag{3.2.1}$$

Once (dE_1/dx) is known, the ion range in the target material can be calculated:

$$R_t(E_0) = - \int_{E_0}^{0} \frac{dE_1}{-(dE_1 / dx)}, \tag{3.2.2}$$

where E_0 is the original implantation energy of the ion.

If the volume density of target atoms is N, thickness of the target Δx, and bombarded area A, then $N\Delta x A$ is the total number of bombarded target atoms. The number of bombarded target atoms per unit area is $N \Delta x$, which increases linearly with Δx, as does also the energy loss ΔE_1 linearly increase with Δx. Let the energy loss $(dE_1/dx)\Delta x$ be proportional to the target atom number $N\Delta x$ and the proportionality coefficient $S(E_1)$ be defined as the stopping power. Then

$$S(E_1) \equiv - \frac{1}{N} \frac{dE_1}{dx}. \tag{3.2.3}$$

As dE_1/dx is negative, the right hand side of the above equation is given a minus sign so that $S(E_1)$ is positive [1,2,15-20].

In a solid target there are three energy loss mechanisms for interactions between the energetic ion and the target atoms: (i) Nuclear collisions: in this kind of collision the incident ion transfers its energy to the target atom nucleus, the ion is scattered through a large angle, and the lost energy displaces lattice atoms. (ii) Electronic collisions: the incident ion excites or ionizes the electrons of the target atom, or causes the atom to capture electrons; less energy is lost in each collision, the ion is scattered through a smaller angle, and lattice disordering can be neglected. (iii) Energy loss due to charge exchange between the ion and the target atoms: this process is a relativistic effect which normally accounts for only a few percent of the total ion implantation energy and can be neglected.

We now consider energy loss processes via nuclear and electronic collisions. The stopping power $S(E_1)$ of the target atoms to the incident ion is the sum of the nuclear stopping power $S_n(E_1)$ and the electronic stopping power $S_e(E_1)$. In a small distance dx, the energy lost by ion collisions with nuclei and with electrons of atoms in the target, is on average $-dE_n$ and $-dE_e$, respectively. Thus, the stopping powers are

$$S_n(E_1) = - \frac{1}{N} (\frac{dE}{dx})_n, \tag{3.2.4}$$

and

$$S_e(E_1) = -\frac{1}{N}\left(\frac{dE}{dx}\right)_e.$$ (3.2.5)

For a single incident ion the total energy loss per unit distance can be expressed as

$$\left(-\frac{dE}{dx}\right)_n + \left(-\frac{dE}{dx}\right)_e = N[S_n(E_1) + S_e(E_1)].$$

E_1 is the energy of the incident ion at position x in the target, and thus

$$-\frac{dE_1}{dx} = N[S_n(E_1) + S_e(E_1)].$$ (3.2.6)

Substitution of Eq. (3.2.6) into Eq. (3.2.2) gives the total mean range R_t of the incident ion:

$$R_t = -\frac{1}{N}\int_{E_0}^{0} \frac{dE_1}{S_n(E_1) + S_e(E_1)},$$

or

$$R_t = \frac{1}{N}\int_{0}^{E_0} \frac{dE_1}{S_n(E_1) + S_e(E_1)}.$$ (3.2.7)

The range projected onto the incident direction is called the *projected range*, expressed by R_p (Figure 3.4). The total range is

$$R_t = l_1 + l_2 + \ldots \ldots = \sum_i l_i,$$

and thus the projected range is

$$R_p = l_1\cos\theta_1 + l_2\cos\theta_2 + \ldots \ldots = \sum_i l_i \cos\theta_i.$$

The use of analytical expressions to express the nuclear stopping power $S_n(E_1)$ and the electronic stopping power $S_e(E_1)$ is not straightforward. To a first order approximation the nuclear stopping power is roughly independent of the incident ion energy:

$$S_n^0 = 2.8 \times 10^{-15} \frac{Z_1 Z_2}{(Z_1^{2/3} + Z_2^{2/3})^{1/2}} \frac{m_1}{m_1 + m_2} \quad (eV \cdot cm^2).$$ (3.2.8)

This formula is useful for estimating the range. Here Z_1, m_1 and Z_2, m_2 are the atomic numbers and mass numbers of the incident ion and the target atom, respectively. For instance, for 95 keV B^+ ions implanted into a Cu target we have that $Z_1 = 5, m_1 = 11$ amu, $Z_2 = 29, m_2 = 64$ amu, and from the above equation,

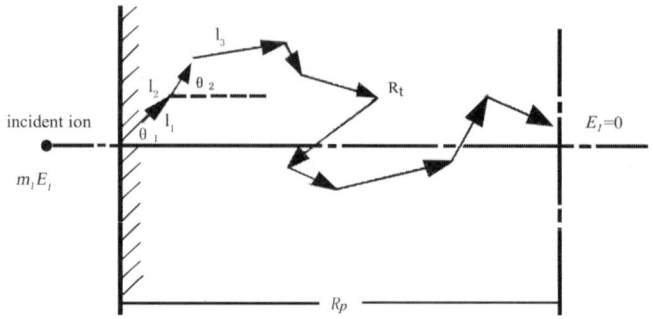

Figure 3.4. Total range R_t and projected range R_p.

$$S_n^0 = 2.8 \times 10^{-15} \frac{5 \times 29}{(5^{2/3} + 29^{2/3})^{1/2}} \frac{11}{11 + 64} \approx 1.7 \times 10^{-14} \ (eV \cdot cm^2).$$

According to the LSS (Lindhard-Scharff-Schiøtt) theory, electrons in a solid can be thought of as forming a free electron gas, and the resistance by these electrons to an energetic incoming ion is similar to the resistance of gas molecules to a projectile passing through the gas. At low velocities, the resistance is proportional to the velocity. Lindhard et al. found the electronic stopping power using the Thomas-Fermi model to be [9-11]

$$S_e(E_1) = \xi_e \cdot 8\pi e^2 a_0 \frac{Z_1 Z_2}{(Z_1^{2/3} + Z_2^{2/3})^{2/3}} \frac{v_1}{v_b}, \qquad (3.2.9)$$

where $\xi_e \propto Z_1^{1/6}$, $a_0 = 0.53$ Å is the Bohr radius, and the incident ion velocity $v_1 < Z_1^{2/3} v_b$ ($v_b = e^2/h$, the Bohr velocity). When practical units are used, then

$$S_e(E_1) = 1.22 \times 10^{-16} \frac{Z_1^{7/6} Z_2}{(Z_1^{2/3} + Z_2^{2/3})^{2/3}} \sqrt{\frac{E_1}{m_1}} \ (eV \cdot cm^2). \qquad (3.2.10)$$

This also indicates that the electronic stopping power is proportional to the square root of the ion energy:

$$S_e(E_1) = K E_1^{1/2}. \qquad (3.2.11)$$

Note that the above formula is correct only for $v_1 \ll v_b$. For high velocity ions, $S_e(E_1)$ decreases with increasing ion velocity and reaches a maximum at $v_1 = Z_1^{2/3} v_b$.

3.2.2. Estimation of the Range

Figure 3.5 shows theoretical curves of S_n and S_e as a function of the incident ion energy. The following can be seen from the figure.

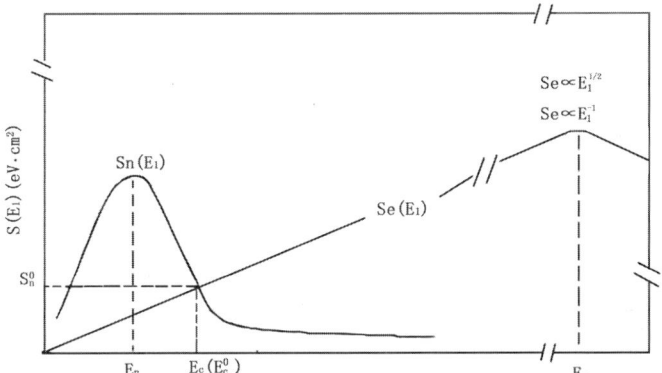

Figure 3.5. Theoretical curve of stopping cross section $S(E_1)$ as a function of energy E_1.

- $S_n(E_1)$ and $S_e(E_1)$ have similar shape, which have maxima. The maximum of $S_n(E_1)$ occurs in the low energy region ($E_n = 10^3 – 10^5$ eV); the maximum of $S_e(E_1)$ occurs in the high energy region ($E_e = 10^6 – 10^8$ eV).
- At the intersection of the two curves there exists a critical energy E_c (or E_c^0). In the low energy region, $E_1 < E_c$, the nuclear stopping power dominates; in the medium energy region, $E_1 \approx E_c$, the electronic stopping power and the nuclear stopping are equally important; and in the high energy region, $E_1 > E_c$, the electronic stopping power dominates and the nuclear stopping power can be neglected. However, the high energy region extends beyond the ion implantation regime and lies in the nuclear physics energy range.
- The critical energy E_c^0 can be estimated. Let $S_e = S_n$. From Eq. (3.2.8) and Eq. (3.2.1) we have

$$(E_c^0)^{1/2} = \frac{S_n^0}{K} = \frac{2.8 \times 10^{-15}}{K} \frac{Z_1 Z_2}{(Z_1^{2/3} + Z_2^{2/3})^{1/2}} \frac{m_1}{m_1 + m_2} \ (eV)^{1/2}. \quad (3.2.12)$$

If the target is non-crystalline Si, $K_{Si} \approx 0.2 \times 10^{-15}$ [$(eV)^{1/2} \cdot cm^2$]. If the implanted ion is B^+ ($Z_1 = 5$, $m_1 = 11$), $E_c^0 \approx 10$ keV, and if the implanted ion is P ($Z_1 = 15$, $m_1 = 31$), the critical energy is $E_c^0 \approx 200$ keV.
- The range in the low energy regime can be estimated. In the low energy regime, the electronic stopping power can be neglected, $S(E_1) \approx S_n$, and from Eq. (3.2.7) we have [15-20]

$$R_t = \frac{1}{N} \int_0^{E_0} \frac{dE_1}{S_n(E_1)} \approx \frac{1}{N} \int_0^{E_0} \frac{dE_1}{S_n}. \quad (3.2.13a)$$

If a more exact S_n value is used, it can be shown that when $E_1 \ll E_c$, R_t is approximately proportional to the incident ion energy, i.e. $R_t \propto E_1$. Thus S_n is independent of energy and can be moved out of the integration in Eq. (3.2.13a):

$$R_\mathrm{t} = \frac{1}{NS_n} \int_0^{E_0} dE_1 = \frac{1}{NS_n} E_0 . \qquad (3.2.13\mathrm{b})$$

For the cases when small angle scattering dominates, the range estimated using S_n^0 to replace S_n is an excellent approximation. Substitution of Eq. (3.2.8) into Eq. (3.2.13b) yields

$$R_\mathrm{t} = \frac{1}{NS_n} E_0 = \frac{1}{NS_n^{\,0}} E_0$$

$$= \frac{1}{2.8 \times 10^{-15} N} \frac{(Z_1^{2/3} + Z_2^{2/3})^{1/2}}{Z_1 Z_2} \frac{m_1 + m_2}{m_1} E_0 . \qquad (3.2.13\mathrm{c})$$

- The range in the high energy regime can be estimated also. In the high energy regime, nuclear stopping can be neglected, $S(E_1) \approx S_e(E_1) \approx KE_1^{1/2}$, and thus

$$R_\mathrm{t} = \frac{1}{N} \int_0^{E_0} \frac{dE_1}{S_e(E_1)} \approx \frac{1}{NK} \int_0^{E_0} E_1^{-1/2} dE_1 = \frac{2}{NK} E_0^{1/2} . \qquad (3.2.14)$$

For non-crystalline targets, $R_\mathrm{t} \approx 20E_0^{1/2}$ (Å). R_t expresses the range. For the case when nuclear stopping dominates, the projected range R_p can be expressed by the following empirical formula:

$$\frac{R_t}{R_p} = 1 + \frac{m_2}{3m_1} . \qquad (3.2.15)$$

3.2.3. Standard Deviation of the Range

The standard deviation of the range is an important parameter describing the range distribution, and is defined as

$$\overline{\Delta R^2} = \overline{(R - \overline{R})^2} = \overline{R^2} - \overline{R}^2 .$$

ΔR^2 is normally used to express $\overline{\Delta R^2}$ and thus

$$\Delta R^2 = \overline{R^2} - \overline{R}^2 . \qquad (3.2.16)$$

The standard deviation of the mean projected range R_p, ΔR_p^2, is another important parameter. It depends on the distribution of the ion concentration in the target. Similar to Eq. (3.2.16), the standard deviation of the mean projected range R_p can be written as

$$\Delta R_p^{\,2} \;=\; \overline{R_p^{\,2}} - \overline{R_p}^{\,2}. \tag{3.2.17}$$

When the angle of incidence of the ion beam is close to normal, we have $\overline{\Delta R} = \Delta R_p$. For an off-normal angle of incidence,

$$\overline{\Delta R_p^{\,2}} \;=\; \overline{\Delta R_p^{\,2}} \cos^2 \theta + \tfrac{1}{2}(R_\perp \sin \theta)^2, \tag{3.2.18}$$

where θ is the angle between the incident ion beam and the normal to the target surface, and $R_\perp$ is the transversal deviation, i.e. the displacement of the ion in the plane perpendicular to the incident ion beam.

The calculation of the mean deviation is fairly complicated and is helped by tabulated data. When the energy is very low, the mean deviation is a simple function of m_2/m_1.

When the target is composed of more than two species of atoms, the ion range and deviation can be written as the following forms:

$$\frac{1}{R} = \sum_i \frac{X_i}{R_i}, \tag{3.2.19}$$

$$\frac{R^2}{\Delta R^2} = \left(\sum_i \frac{r_i X_i R_i^2}{R_i \Delta R_i^2} \right) \left(\sum_i r_i \frac{X_i}{R_i} \right)^{-1}, \tag{3.2.20}$$

where $r_i = \dfrac{4 m_1 m_i}{(m_1 + m_i)^2}$, X_i is the fraction of the ith atomic species in the total target mass, and R_i and ΔR_i are the ion range and deviation in a target composed of the ith atom species, respectively. Estimation of the range and deviation using this method normally has an error of less than 10%. Since ion implantation can alter the target density and the implanted ions move due to diffusion, the error in this estimate is actually less than errors due to target density changes and diffusion.

3.2.4. Distribution of Implanted Ions in Non-Crystalline Targets

In the absence of channeling effects (see next section), the implanted ion concentration is distributed approximately as a Gaussian distribution along the direction of ion incidence [1,2]:

$$N(x) = N_{\max} \exp(-\tfrac{1}{2} X^2), \tag{3.2.21}$$

where

$$X = \frac{x - R_p}{\Delta R_p}. \tag{3.2.22}$$

Let us assumed that the number of ions implanted per unit area is N; this is the dose D. Since a Gaussian distribution is symmetric about the X axis, D is the integral of the implanted ion distribution function from $-\infty$ to $+\infty$ along the X direction:

$$D = \int_{-\infty}^{+\infty} N(x)dx .$$

(3.2.23)

From Eq. (3.2.22), $dx = \Delta R_p \, dX$, and the Gaussian distribution is symmetric, thus,

$$D = \Delta R_p \, N_{max} \int_{-\infty}^{+\infty} e^{-\frac{1}{2}x^2} dx = 2\Delta R_p N_{max} \int_{0}^{+\infty} e^{-\frac{1}{2}x^2} dx .$$

Since $\int_{0}^{+\infty} e^{-\frac{1}{2}x^2} dx = \sqrt{\dfrac{\pi}{2}}$, we have that

$$D = \sqrt{2\pi} \Delta R_p N_{max} ,$$

(3.2.24)

or

$$N_{max} = \frac{D}{\sqrt{2\pi} \Delta R_p} .$$

(3.2.25)

Substitution of Eq. (3.2.25) and Eq. (3.2.22) into Eq. (3.2.21) yields an expression for the implanted ion distribution in the target:

$$N(x) = -\frac{D}{\sqrt{2\pi} \Delta R_p} \exp\left[-\frac{1}{2}\left(\frac{x - R_p}{\Delta R_p} \right)^2 \right].$$

(3.2.26)

3.2.5. Channeling Effects

In the theory outlines in the preceding, the target is taken to be non-crystalline, or amorphous. This means that the atoms in the solid target are arranged randomly, or disordered, while the atom density distribution in the target is homogeneous. Thus collisions of incident ions with target atoms are random. If the target has a crystalline structure, the arrangement of the atoms in space is ordered. When ions are implanted along the principal crystalline axes, they may collide with lattice atoms in a similar way. At each collision, the ion suffers only small-angle scattering and can penetrate the solid to deeper distances. This phenomenon is called *channeling* [1,2]. As shown in Figure 3.6, since the diameter of a positive ion is 0.02 – 0.34 nm, if a channel presents an opening greater than the ion diameter then the ion can be implanted in the open crystalline direction to a greater depth. For instance, for the (110) and (111) crystalline directions of Si, the channels are of size 0.12 nm and so easily accept ions with diameter smaller than this; the (110) channel direction of Si an opening of 0.18 nm is presented and thus almost all the elemental positive ions can be accepted.

Not all of the ions in a beam that is exactly parallel to a crystalline axis are accepted into the channels. A channeling ion must have three basic prerequisites: first, the ion must penetrate the open channel between the atomic arrays; second, there must be a force acting on the ion causing it to move to the channel center; and third, the channeling ion must be stable. Ions that do not meet all of these three conditions will quickly be scattered by the target atoms and will be distributed similarly to the case of a non-crystalline target, forming a non-crystalline peak (Gaussian peak) in the target. Channeling ions form a channeling peak at a deeper location in the target. The distribution is different from that for non-crystalline implantation, decreasing rapidly to zero near the maximum range R_{max}. The random beam and the channeling beam are not independent each other. A part of the ions in the random beam may enter the channels at a later time, while some channeling ions may be scattered due to lattice defects and thermal motion and thus dechanneled. The distribution of ions implanted along the channeling direction is shown in Figure 3.7. When the number of the dechanneled ions increases, the channeling peak becomes lower but the maximum range is unchanged.

The range distribution of channeled ions in the target is dependent on many factors:

- Dose. The incident ions cause damage to the crystalline lattice and thus the number of channeling ions decreases as the implanted ion dose increases. When the dose exceeds a certain value, the lattice is completely destroyed and channels no longer exist, and thus the ions penetrate as in a disordered region.

- Crystalline directions. The widths of the crystalline direction opening vary and the distance that ions travel along wider channels is greater than for along narrower channels.

- Incident angle. The more the angle of incidence deviates from the critical angle of the crystalline direction, the smaller the penetration distance.

- Temperature. The greater the amplitude of lattice thermal motion, the stronger the scattering effect of channels on the ions. As the temperature increases, the number of dechanneled ions increases and the channeling peak is reduced.

- Energy. In channeling ion implantation, "completely channeled" ions penetrate the deepest. These ions lose their energy during their motion in the channel mainly due to electronic stopping.

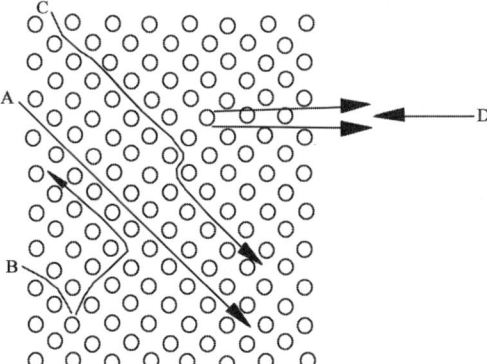

Figure 3.6. Examples of ion implantation in crystal. A: channeling direction, B: random direction, C: partial channeling direction, D: blocked direction.

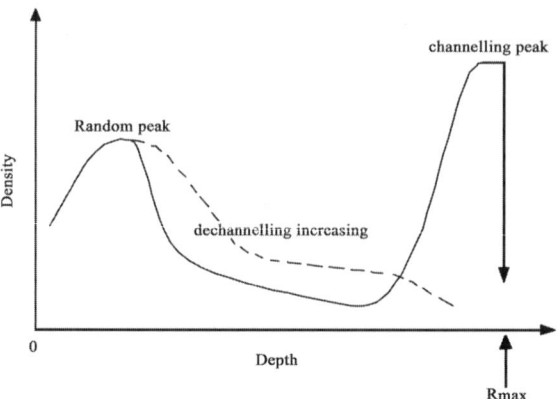

Figure 3.7. Distribution of ions implanted along channeling direction.

- Ion species. The penetration depths of different ions along the channel depend on the atomic number Z_1. For example, as the electronic stopping powers of ^{24}Na and ^{64}Cu in Si are very small, 40 keV ^{24}Na implanted in the (110) direction in Si can penetrate to a depth of 5 μm at a low temperature of 77 K.
- Target material. The penetration depth of ions in different target material channels relates to the target atomic number Z_2. As measured in experiments, the maximum range of ^{35}S in Ge (110) is about twice that in Si (110).

3.3. RADIATION DAMAGE

Radiation is energy propagation in space. Materials absorbing radiation will undergo structural damage. This kind of damage is called radiation damage [1,2]. In the process of ion implantation, not only do the foreign implanted ions alter the solid properties after they enter the target, but also the energy deposition of the energetic ions in the solid can result in radiation damage.

3.3.1. Energy Deposition of Ions in Solids

Energetic ions penetrating a solid target interact with atoms and electrons in the target, gradually transferring their kinetic energy to the target atoms and electrons until the ion completely loses its kinetic energy and comes to rest in the target. This process is called the *stopping* of the ion in the solid. The stopping process is a process in which the incident ion transfers and deposits its energy to the target. The energy deposition rate can be 10 – 100 eV/Å for lighter ions and of order keV/Å for heavier ions at the end of their range.

When an incident ion transfers its energy to target atoms in such a way that the mechanical (kinetic) energies of the particles participating in the collision are conserved, this is called an *elastic collision*. If the ion transfers energy to electrons, causing excitation or ionization, the mechanical energy of the particles participating in the collision are not conserved, and this is called an *inelastic collision*. In general, when an ion passes through a solid these two kinds of collision processes occur simultaneously. When the ion energy is higher, inelastic collision dominates, and when the ion energy is lower, elastic collision dominates.

If a lattice atom gains enough energy from a collision to exceed the potential that binds the atom to a fixed location in the lattice, and the bombarded atom then leaves its original position to enter the space between lattice atoms, this phenomenon is called *atomic displacement*. The minimum energy required for atomic displacement is called the *displacement critical energy*, indicated by E_d. E_d consists of two parts: one part is the bond-breaking energy that is equal to the atomic bonding energy; and the other part is the work done to overcome the potential. The work done to overcome the potential is related to the direction of the atomic recoil. If the energy gained by the atom from the incident ion is far more than the displacement critical energy, the atom can further collide with its neighboring atoms (in a *recoil collision*) to displace other atoms and form a sort of cascade process, called a *collision cascade* (Figure 3.8).

In insulators and semiconductors having very low electrical conductivities, the interaction between implanted ions and electrons in the solid can produce a kind of ion-explosion atomic displacement (Figure 3.9). The charged particle suddenly explosively ionizes the target atoms in the neighborhood of its path to form an unstable electrostatic region in which ions repel each other and many leave their original locations to enter interstitial locations. After the initial ionization, Coulomb repulsion action between the ions causes them to form matrices of interstitials and lattice vacancies. Subsequently, through an elastic relaxation process, this effect spreads over a larger region.

The condition for Coulomb explosion to cause atomic displacements is that the local electrostatic force greatly exceeds the local mechanical strength or binding strength. Thus this kind of phenomenon readily occurs in materials of lower mechanical strength, smaller permittivity, and smaller atomic spacing. In terms of the energy to cause an ion explosion, the energy to displace atoms is normally about 25 eV. In the case of ion implantation, this is clearly possible. Furthermore, the time for which exploding pieces with charges q_1 and q_2 maintain their high charge should be long enough to facilitate their movement through at least two lattice intervals. A theoretical calculation shows that this time about 10^{-14} seconds. But note that the typical time to capture an electron in the solid is greater than 10^{-13} seconds. This implies that the capture of an electron does not occur before the exploding ion moves through two lattice intervals.

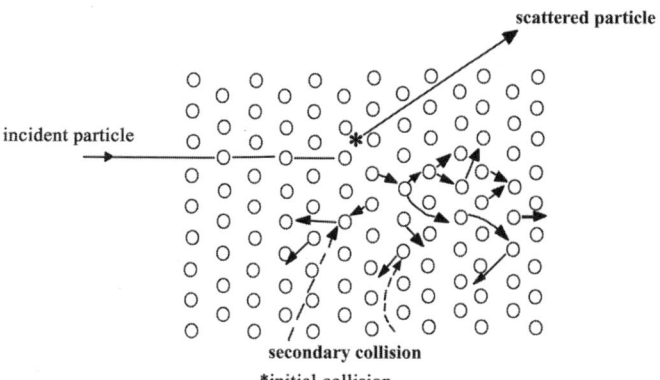

Figure 3.8. Collision cascade process.

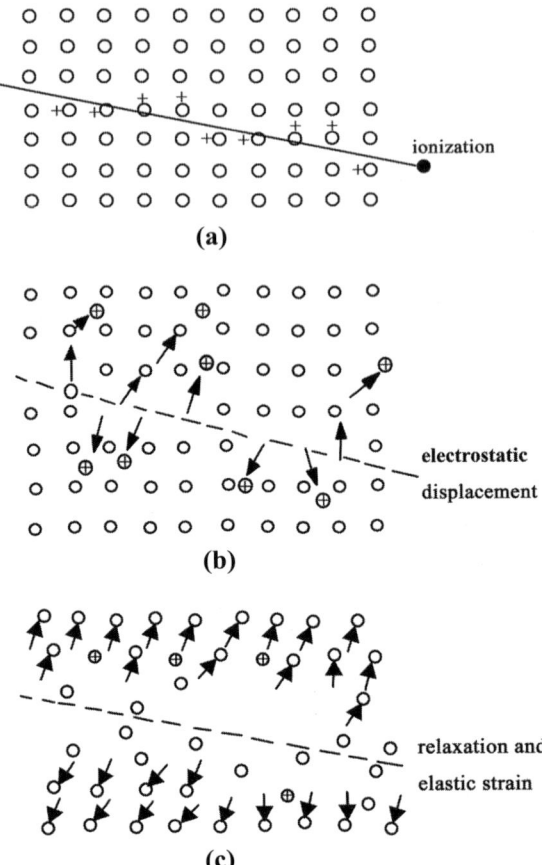

Figure 3.9. Schematic of atomic displacement due to explosive ionization nearby the ion trajectory. (a) Primary ionization induced by the incident charged particle, (b) induced mutual repulsion of ions resulting in atomic vacancies and interstitial ions, and (c) strain occurring in the undamaged area through elastic relaxation of the stressed area.

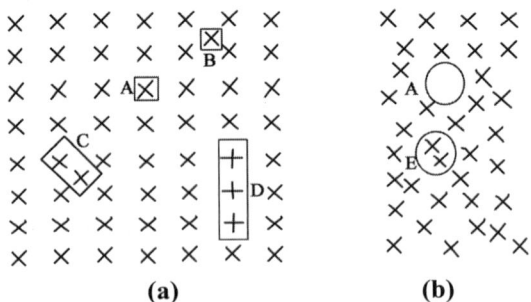

Figure 3.10. Simple defects in ordered (a) and disordered (b) solids. A: vacancy, B and C: interstitial, D: substitution, and E: interstitialoid.

3.3.2. Radiation Damage due to Ion Implantation

Radiation damage changes the physical and chemical properties of solid materials. The basic reason is production of various defects in the solid. The defects in crystals can be classified into two categories: structural defects and impurities. The structural defects can be divided into the following several kinds according to the defect geometry: point defects – vacancy and interstitial atoms; line defects – dislocations; and plane defects – layer dislocations and grain boundary dislocations. The impurities include substitutional impurities and interstitial impurities. From a geometric point of view, impurity defects belongs to the point defect class (Figure 3.10).

In the collision cascade process, target atoms are displaced. The displaced atom can occupy an interstitial position or a substitutional position, leaving a vacancy at its original position. Additionally, the implanted ion may stop at an interstitial position or vacancy to become a substitutional impurity. Therefore, ion implantation forms three types of defects in the solid: vacancy, interstitial impurity atom, and substitutional impurity atom. In the cases of low dose ion implantation, the defect concentration is low, dominated by simple defects (such as vacancies and interstitial atoms). With increasing dose, defect concentration increases and the defects become complicated. Single vacancies can combine to form double vacancies, triple vacancies and complexes of other multi vacancies. In the same way, interstitial atoms can also be combined into similar complexes. Finally, when the implantation dose is high, severe damage occurs and a large number of defects overlap each other to disorder completely the ordered arrangement of lattice atoms, and the implanted region becomes a disordered amorphous state. Note also that the cases of heavy ion implantation and light ion implantation are different – as the energy loss rate for a heavy ion is greater than for a light ion, in the collision cascade process, even though the dose may not be very high, serious defects can be produced by heavy ion implantation even to form an amorphous region.

In ion-bonded and covalent-bonded solids, the energy required to form a vacancy equals the sum of the energies needed to break the bonds binding the atom and to move the atom from its lattice position to the crystal surface. For metals and semiconductors, this energy is about $1 - 2$ eV. The energy required to form an interstitial is greater than that to form a vacancy, normally about $4 - 5$ eV. Therefore, in thermal equilibrium the density of interstitials in the solid is lower by several orders of magnitude than the density of vacancies. Deformations near interstitials facilitate interstitial motion. The energy required to move an interstitial is about a few tenths of an eV. Once an interstitial forms in the solid, it quickly moves around. If the interstitial meets a vacancy in its movement, they combine. If the interstitial atom is the same as a solid matrix atom, the solid recovers from the lattice damage; if they are different, the interstitial atom and the matrix atom may form a new compound molecule.

If vacancies and interstitials in the crystal aggregate into clusters to form a planar array embedded in the crystal but not matched with the crystalline planes, *dislocations* occur in the crystal. The presence of dislocations breaks the periodicity of the lattice array in the crystal. If other defects migrate to the dislocation region, the dislocation can expand or contract.

There is in general an inhomogeneous distribution of defects, and the defects diffuse to regions of low concentration. In the case of thermal equilibrium the defect diffusion motion is described by diffusion equations. However in the ion implantation process abnormal diffusion occurs, and the diffusion coefficient can be several orders of magnitude greater than that for normal thermal diffusion. Hence in some cases of implanted ion distributions, very long tails can occur, indicating the diffusion of ions to greater depths in the solid.

3.4. MONTE-CARLO CALCULATION OF ION IMPLANTATION

It can be seen from the discussion above that the interaction process between energetic ions and a solid is very complicated. The ion collision cascade effect, radiation damage, and damage propagation combine to make an exact analytical description difficult. Thus a Monte-Carlo approach for simulating the process is often used. The most commonly used program for this purpose is TRIM. The basic approach of this program is to trace the developing details of the collision cascade for each incident ion in the target. The program assumes the target to be a homogeneous medium where target atoms distributed randomly. At each random target atom location the program calculates the energy loss due to collisions between the energetic implanted ion and the target atom, as well as the angular distribution, defect distribution, collision cascade processes, atomic displacement, vacancy production, and ionization and excitation, so as to determine the trajectories of all atoms in the collision system. Successive calculation for a large number of implanted ions yields a statistical average of the implanted ion action and a physical picture of the collision cascade process can be displayed [21-26].

Let us suppose that the implanted ion is N^+ with an energy of 30 keV, and the target material is a compound composed of four elements, N, H, O, and C (with atomic ratios 1:5:2:2) that are homogeneously distributed with a density of $d = 1.6$ g/cm^3. The critical displacement energy for the four elements is $E_d = 20$ eV, the binding energy of the lattice is 1 eV, and the incident ion is implanted at normal incidence to the target surface [27].

3.4.1. Ion Trajectory

Figure 3.11 shows the collision cascades for 19 incident ions. We see that the implanted ion trajectories are very complicated, not at all simple straight lines, exhibiting a typical cascade structure.

Because the target material contains some atomic species that are the same as the incident ion species (i.e., N) as well as other atomic species that are different from the implantation ions (i.e., C,H,O), according to Eq. (3.2.8) and Eq. (3.2.10) their stopping powers are different and the energy losses are different also. When the energy transferred by the incident ion is comparatively high, the total energy of the target atom that gains the transferred energy exceeds the displacement critical energy so as to cause the atom to be displaced, and a cascade effect is produced. Based on the TRIM-calculated result for 454 incident ions, the mean range of the incident ions is 87.6 nm and the transverse diffusion distance is 24.1 nm. Each incident ion produces 210 vacancies. The energy loss is distributed as follows.

- The energy gained by ionized molecules that are directly produced by the incident ion: 51.1%;
- The energy gained by ionized molecules that are produced by recoil atoms: 30.8%;
- The energy loss of the incident ion in producing vacancies: 4.2%;

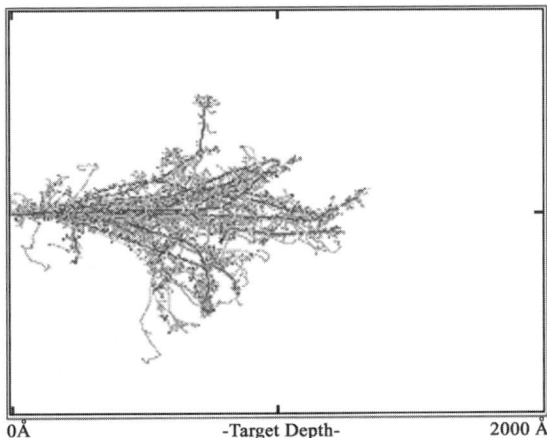

0Å   -Target Depth-   2000 Å

Figure 3.11. Ion implantation produced trajectories of ions simulated by TRIM.

- The energy loss for recoil atoms in producing vacancies: 11.3%;
- The energy gained by phonons produced by the incident ion: 2.3%;
- The energy gained by phonons produced by recoil atoms: 0.3%.

3.4.2. Distribution of Implanted Ions

The normally implanted N^+ ion exhausts all its energy and comes to rest in the solid. Figure 3.12 shows the depth distribution of the implanted species. It can be seen that the concentration distribution has a peak and a Gaussian-like shape. Although the mean ion implantation range is 87.6 nm, the 50 nm depth region from 70 to 120 nm contains 73% of the total number of implanted atoms. For an implantation dose of 1×10^{16} ions/cm^2, the total number of implanted ions in this 50 nm interval is 0.73×10^{16} ions/cm^2. If the target molecule is assumed to have a molecular weight of 75, one can calculate that the total number of target molecules in this interval is 6.42×10^{16} ions/cm^2. Thus the ratio of the number of implanted ions to target molecules is 0.73/6.42 = 11.4%. If the implanted ions combine with the target molecules, even though energy deposition effects (damage) are not taken into account, the addition of implanted ions alone can significantly change the properties of the target molecules.

3.4.3. Displaced Atoms

Each incident ion can lead to displacement of each of the four kinds of target atoms, C, N, O, and H. Figure 3.13 shows the distribution of displacement collisions for the four kinds of atoms. Statistical results show that each 30 keV N^+ ion can displace 72.12 carbon atoms, 99.64 hydrogen atoms, 74.02 oxygen atoms and 37.02 nitrogen atoms (including cascade effects), or each incident ion can displace 282.8 target atoms in total. Since the content ratio of four elements is C:H:O:N = 2:5:2:1, the relative probabilities of displacements for the elements are 72.12/2 : 99.64/5 : 74.02/2 : 37.02/1 = 36.06 : 19.93 : 37.01 : 37.02. This indicates that the displacement probabilities from high to low fall in the order N > O > C > H.

In the case of implantation of a large number of ions, the damage to the target molecules by displaced atoms is many times greater than that due to implanted ions alone. For our example with an implantation dose of 1×10^{16} ions/cm^2, the total number of displaced atoms is 2.828×10^{18} /cm^2. The calculation yields a maximum depth of the displaced atom distribution of 160 nm, and the number of target molecules in this depth is about 1.93×10^{17} /cm^2. If each molecule possesses 10 atoms, the number of target atom to this depth is 1.93×10^{18} /cm^2. Thus each target atom moves on average about 2.828×10^{18} / $1.93 \times 10^{18} \approx 1.46$ times through a distance of 160 nm. Actually, the depth distribution of displacement collisions shows a peak (Figure 3.13). In the region of this maximum all of the target molecules are rearranged to form a new molecular structure.

After coming to rest, the implanted ions may suffer replacement reactions with target atoms (Figure 3.14). The replacement collision distribution also shows a peak with depth. Comparing Figure 3.12 and Figure 3.14, it can be seen that not all of the implanted ions result in a replacement. The replacement reactions mostly take place near the target surface. Because the replacement reaction needs energy, and the implanted ions have a thermal energy that is lower than the replacement reaction energy, the implanted ions are deposited further from the surface while the replacement reactions occur near the surface. Rearrangement and combination of vacancies, displaced atoms, stopping ions, and target atoms cause significant damage to the target molecules. This damage has a peak in the depth direction, and the ion implantation energy determines the peak position. If the implanting ion beam is focused to a spot of size less than a micrometer (about the dimension of genes or chromosomes), the damaged region can be identified in 3-dimensional space.

3.5. CHEMICAL REACTIONS AND NUCLEAR REACTIONS

Radiation can destroy the chemical bonds in the solid and cause chemical changes [27]. Organic polymers, organic solids, nitrates, chlorates and so on can be dissociated by radiation. Radiation dissociates nitrates to produce nitrites and oxygen. If the radiated nitrate is dissolved in water, these two products can be observed. The ratio between nitrite and oxygen is 2:1,

$$\overline{NO_3} \rightarrow \overline{NO_2} + \frac{1}{2}O_2 .$$

Radiation can also be used for organic synthesis. A typical example is the radiation synthesis of high purity anhydrous bromethyl using γ-radiation of hydrogen bromide HBr and ethylene C_2H_4. This method is based on the radical chain reaction mechanism:

$$HBr \rightarrow \dot{H} + \dot{Br} ,$$
$$\dot{H} + HBr \rightarrow H_2 + \dot{Br} ,$$
$$\dot{Br} + CH_2 = \dot{CH}_2 \rightarrow CH_2 - CH_2Br ,$$
$$\dot{CH}_2 - CH_2Br + HBr \rightarrow C_2H_5Br + \dot{Br} ,$$
$$\dot{Br} + \dot{Br} \rightarrow Br_2 .$$

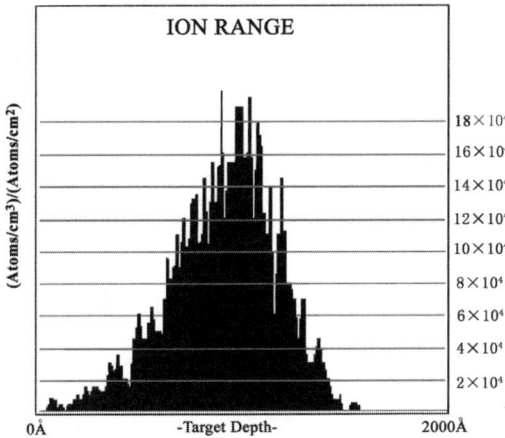

Figure 3.12. TRIM simulated distribution of implanted ions.

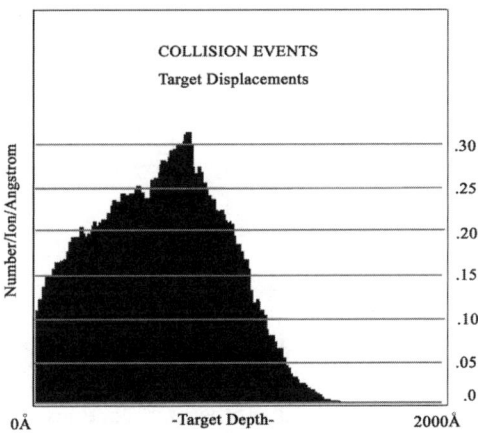

Figure 3.13. Displacement collision distribution simulated by TRIM.

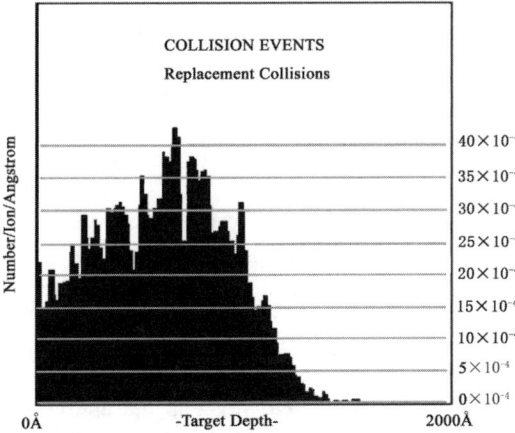

Figure 3.14. Replacement collision distribution simulated by TRIM.

For the case of low energy ion implantation, the ion energy deposition rate is $10 - 10^3$ eV/Å. Generally this energy is greater than the atomic displacement critical energy. In the collision cascade process, a large number of interstitials and substitutional atoms are produced. When ions are implanted into organic compounds, the compounds are usually dissociated.

After coming to rest the implanted ion becomes either an interstitial or a substitutional atom. Under some conditions, when the implanted ion energy drops down to the energy range of chemical reactions, the implanted atom may take part in chemical reactions with target atoms to form a new compound. Consider for example oxygen ions that are implanted into Si with an implantation energy of 60 keV with a beam flux of 20 μA/cm² for 200 minutes at a temperature range from room temperature to 300°C maintained in the target chamber. An infrared transmission microscope is used to observe the implanted target surface and an absorption peak is found to appear at 9.4 μm. This demonstrates that the implanted surface layer has been transformed into SiO_2 rather than SiO (which has its peak absorption at 10.0 μm). When N^+ ions are implanted at high dose into Si, the implanted surface layer forms Si_3N_4. N^+ ion implantation into Ti films can form TiN_x (where x = 0.6 – 1.0).

The above are some examples of low energy ion interactions with target materials which undergo chemical reactions. The chemical reaction refers only to changes in the atomic or molecular electron shell structure but not to changes in the nucleus. Collisions between light ions with energy of about 10^5 eV and light target atoms can result in nuclear reactions; and for heavy ions, they occur only when the collision energy is greater than 10^6 eV.

A nuclear reaction can be expressed in general as A + a → B + b, written in simplified form as A(a,b)B, where A and a are the target nucleus and the incident particle, respectively, and B and b are the remaining nucleus and the emitted particle, respectively. In an experimental system, if the target nucleus is static, the energy for a specific nuclear reaction to occur is provided by the incident particle energy.

Nuclear reactions are divided into endothermic reactions described by A + a → B + b + Q, where the reaction energy Q < 0, and exothermic reactions described by A + a → B + b + Q, where Q > 0. These two types of nuclear reactions have different requirements for the incident ion (particle a) energy E_a. For endothermic reactions, the threshold energy for the incident particle to cause a nuclear reaction is $E_{th} = \dfrac{m_a + m_A}{m_A}|Q|$, and it can be seen that E_a should be greater than $|Q|$. For exothermic reactions, the threshold energy is $E_a > \dfrac{m_a + m_A}{m_A}V_c$, where V_c is the Coulomb potential between the bullet (incident particle) and the target nucleus. Threshold energies for nuclear reactions to occur when light ions such as α-particles (He nucleus) and N ions are implanted in biological samples are shown in Table 3.1.

It can be seen that for ion implantation of biological samples, the energy of the incident particle needs to be greater than 6 MeV for an α particle and 10 MeV for a N^+ ion for any nuclear reactions occurs. In the scope of this book, the upper limit of ion energy will not cause nuclear reactions. However in the analysis of biological samples, nuclear reaction analysis is a good method for measuring the concentration of certain elements (impurity or implanted ions) and their depth distribution.

Table 3.1. Threshold energies for nuclear reactions to occur between α-particles and N ions and some biological molecule elements.

Reaction	Q (MeV)	E_{th} or V_c (MeV)
$^{12}C(\alpha,P)N^{15}$	-4.966	6.62
$C(\alpha,n)O$	-8.501	11.33
$O(\alpha,p)F$	-8.114	10.14
$O(\alpha,n)Ne$	-12.134	15.17
$N(\alpha,P)O$	-1.193	1.53
$N(\alpha,n)F$	-4.735	6.09
$P(\alpha,p)S$	0.627	6.57
$P(\alpha,n)Cl$	-5.648	6.38
$C(N,\gamma)Al$	14.842	9.19
$C(N,P)Mg$	8.765	
$C(N,n)Al$	3.705	
$H(N,\gamma)O$	7.293	2.2
$O(N,\gamma)P$	18.33	11.7
$O(N,n)P$	15.078	
$O(N,\alpha)Al$	7.608	
$N(N,n)Si$	10.04	10.05
$P(N,P)Sc$	8.945	19.5

REFERENCES

1. Wang, Y.H., *et al.*, *Fundamentals of Ion Implantation and Analysis* (in Chinese, Aeronautical Industry Press, Beijing, 1992).
2. Zhang, G.H. and Zhong, S.L., *Ion Implantation Technology* (in Chinese, Engineering Industry Press, Beijing, 1982).
3. Nastasi, M., Mayer, J.W. and Hirvonen, J., *Ion-Solid Interactions: Fundamentals and Applications* (Cambridge University Press, Cambridge, 1996).
4. Benninghoven, A., Developments in Secondary Ion Mass Spectroscopy and Applications to Surface Studies, *Surf. Sci.*, 53(1975)596.
5. Niehus, H., *et al.*, Low-energy Ion Scattering at Surfaces, *Surf. Sci. Rep*, 7(1993)213.
6. Gerlach, R.L., in *Electron Spectroscopy*, ed.: Shirley D.A. (North Holland, Amsterdam, 1972).
7. Plog, C., *et al*, *Surf. Sci.*, **67**(1977)565.
8. Tolk, N.H., *et al.*, *Elastic Ion-Surface Collisions* (Academic Press Inc., New York, 1977).
9. Fiermans, L., *et al.*, *Electron and Ion Spectroscopy of Solids* (Plenum Press, New York, 1978).
10. Hagstrum, H.D., *Phys. Rev. Lett.*, **43**(1979)1050.

11. Behrisch, G., *Sputtering by Particle Bombardment I* and *II* (Springer Verlag, Heidelberg, 1983).
12. Yamamura, Y. and Tawara, H., *Energy Dependence of Ion-Induced Sputtering Yields from Monoatomic Solids at Normal Incidence* (Mar. 1995, NIFS Research Reports: DATA-23).
13. Lindhard, J., *et al.*, *Phys. Rev.*, **124**(1961)128.
14. Sigmund, P., *Phys. Rev.*, **184**(1969)383.
15. Rol, P.K., *et al.*, *Physica*, **26**(1960)1009.
16. Czanderna, A.W., *Methods of Surface Analysis* (4th edition, North-Holland, Amsterdam, 1989).
17. Evans, R.D., *The Atomic Nucleus* (McGraw-Hill Inc., New York, 1955) Chapter 18-25.
18. Marmier, P. and Sheldon, E., *Physics of Nuclei and Particles* (Academic Press Inc., New York, 1969) Vol. 1, Chapter 4.
19. Fano, U., *Ann. Rev. Nucl. Sci.*, **13**(1963)1.
20. Northeliffe, L.C., *Ann. Rev. Nucl. Sci.*, **13**(1963)67.
21. Northeliffe, L.C. and Schilling, R.F., *Nuclear Data Table*, A **7**(1970) 233.
22. *SRIM 2003.* http://www.srim.org/.
23. Ziegler, J.F., Biersack, J.P. and Littmark, U., *The Stopping and Range of Ions in Matter* (Pergamon Press, New York, 1985).
24. Ziegler, J.F. and Manoyan, J.M., The Stopping of Ions in Compounds, *Nucl. Instr. Meth.*, B**35**(1988)215-228.
25. Anderson, H.H. and Ziegler, J.F., *Hydrogen Stopping Powers and Ranges in All Elements* (Pergamon Press Inc., New York, 1977).
26. Ziegler, J.F., *Helium Stopping Powers and Ranges in All Elemental Matter* (Pergamon Press Inc, New York, 1977).
27. Huang, W.D., *A Study of Mechanisms Involved in the Interactions between Low-energy Ions and Biological Molecules* (in Chinese, Ph.D. dissertation, Institute of Plasma Physics, Chinese Academy of Sciences, 1995).

FURTHER READING

1. Wang, Y.H., *et al.*, *Fundamentals of Ion Implantation and Analysis* (in Chinese, Aeronautical Industry Press, Beijing, 1992).
2. Zhang, G.H. and Zhong, S.L., *Ion Implantation Technology* (in Chinese, Engineering Industry Press, Beijing, 1982).
3. Nastasi, M., Mayer, J.W. and Hirvonen, J., *Ion-Solid Interactions: Fundamentals and Applications* (Cambridge University Press, Cambridge, 1996).
4. Gras-Marti, Alberto, *Interaction of Charged Particles With Solids and Surfaces* (Kluwer Academic Pub., New York, 1991).
5. Gibson, W.M., *Ion-Solid Interactions* (Institution of Electrical Engineers, London, 1980).

4

INTERACTIONS BETWEEN ENERGETIC IONS AND BIOLOGICAL ORGANISMS

The established theory of the interaction between energetic ions and metals or semiconductors is based on an accurate description of the bullet-target system. In this system, the arrangement of the target atoms is either ordered (crystalline) or disordered (noncrystalline/amorphous). For a disordered system, collisions of the energetic particle with successive atoms are, in a sense, uncorrelated and random, while for an ordered system there is a structural correlation between successive collisions.

For the objects discussed in this book, the situation is not so clear. A biological organism is a complex system composed of solid, fluid and gas. It is not as stable as a solid; it is instead a more-or-less continuously changing object. For example, living things grow and develop; a seed breathes slightly, even in its dormancy. The biological structure is separated spatially into many small rooms (cells), and a cell is further divided into many smaller compartments. Temporal and spatial changes in the biological structure greatly complicate studies of the interaction between energetic ions and the living biological target. However, from an atomic collision point of view, no matter when and where the target atoms are, once collisions occur the energy loss laws remain the same. In the case of low energy ion implantation, an ion transfers some of its energy to a target atom and is deflected through a large angle, and the energy transferred causes the target atom to be displaced; this is the nuclear collision energy loss. Again, the moving ion transfers some of its energy to the electrons of target atoms to cause ionization or excitation or electron capture; the energy loss in each collision in this case is relatively small and the ion undergoes only a small deflection; this is the electron collision energy loss. The complexity of biological organisms is due to its multiple layer, multiple phase, and multiple channel structure. After a collision between the incident ion and a target atom it is difficult to determine where and with what kind of target atom the next collision will occur. Now the energetic ions are indeed going into a black box. To establish a physical model for such a collision system, the available information seems to be too little.

Biological structure is closely related to function. Structural biology, as an important branch of life science, is witnessing increasing attention. Structural biology developed in several important stages along with developments in science and technology. Biological structures were initially described in terms of morphology and anatomy. The invention of the optical microscope made possible observation of biological microstructures at the cell level. The electron microscopes pushed the knowledge of biological structure deep into the sub-cell level. Applications of nuclear techniques revealed the 3-dimentional structure of biological molecules such as DNA and some proteins. We can predict with some confidence that the use of ion beams and other kinds of beams for selection of impact-induced optical emission and image reconstruction will allow detection and measurement of the fine structure of atomic and molecular cell composition. This will become one of the most active areas in the life sciences. Probably, at that future time, a physical interaction model for energetic ions with complex biological systems should be well understood.

The structure and function at every level inside the biological organism can be found in various topical textbooks. This chapter firstly provides a brief description of biological target structure that is related to ion implantation and the changes that can occur in biological structures during ion implantation. Then we discuss charge exchange and ion deposition effects on the physical and chemical characteristics of biological organisms. Finally we estimate the energy deposition features of ion implantation. Note that the primary physical process involved in the interaction between energetic ions and biological organism occurs within a very short time period, less than 10^{-13} seconds, and thus changes in the biological organism within this time – a special factor – can be neglected [1].

4.1. DESCRIPTION OF TARGET STRUCTURE AND COMPOSITION

Among the approximately 100 natural elements, only 30 elements are necessary for constructing life systems. It seems that in the biological evolutionary process, these elements were particularly suitable for maintaining life activities and were thus selected. Most of these elements that are essential to living organisms are relatively minor components of the Earth's crust. In order to extend life activities indefinitely, recycling of these elements is important. For example, it is estimated that the photosynthesis by living organisms solidifies carbon at a rate of up to about 5×10^9 tons per year. If not for recycling, the CO_2 existing in nature would be exhausted within 300 years, the oxygen concentration would be doubled in 2000 years, and the ocean water would be completely consumed in 2 million years. We know that the relative stability of the atmospheric gas composition must be due to complicated biological activities in which substance is returned to the crustal rock at the same rate at which it is removed from the crust. This equilibrium state relies on a delicate balance between synthesis and dissolution processes. Control of the two complementary processes is assumed by biological organisms with their specially complex structures.

The elements required by life systems do not exist uniformly inside the biological organisms. Instead they are structured into biological molecules or units based on certain rules, and many repeated units form biological macromolecules. For example, the basic unit of all proteins is amino acid; and the unit of nucleic acid is nucleotide. Terminology such as primary, secondary, tertiary and quaternary structures is commonly used to describe the proteins and nucleic acids. Each level of the structure is determined by the chemical bond types required. A feature of the primary structure is the formation of a linear series of units held together by covalent bonds. Secondary structures are formed by curly poly-chain series with the aid of non-covalent bonds (polar or hydrogen bonds). Tertiary structures are the result of folding of ordered polymers. Quaternary structures are formed by interactions among the folded polymer molecules and other aggregates. All structures higher than the primary structure are called the high-level structure sequence. These complicated high class structures make a homogeneous spatial distribution of the elements impossible. In fact, biological macromolecules themselves are heterogeneously distributed in biological organisms. For example, nucleic acid exists in a cell only inside the nucleus, mitochondrion and chlorophyll, and these subcellular structures occupy only a very small part of the total cell volume [2].

The largest component of a biological organism is water. The water contained in a living plant is normally greater than 70% of the total weight. Maintenance of sufficient water is necessary for the maintenance of life itself. Plants absorb water effectively from the soil in order to maintain that part that lives above the soil. Animals drink water to make up for water evaporation from the body. The water balance in biological organs is a kind of dynamic balance. Thus it has been proposed that "life is a hydraulic system". In radiation biology (including heavy ion biology), water is often used to simulate a real biological organism. This can provide a good approximation before the actual atomic and molecular arrangements are sorted out.

Both life on land and life in water continuously exchange substance and energy with the surrounding environment to maintain their normal living activities. Transport or delivery of external substance and energy to the interior of biological bodies, as well as waste excretion, must rely on intrinsic passages (channels). Biological channels are everywhere, such as pores on the surface, spacing between cells, double-layer phospholipid channels in membranes, micro-tube channels inside the cells, and so on. The venue of the respiration activities is in mitochondria, and photosynthesis takes place in chlorophyll. When green plants perform photosynthesis, they make nutrients. CO_2 diffuses through the cell wall via surface holes, then dissolves in the liquid phase in the capillaries, and finally diffuses to cytoplasm and then the location of chlorophyll carboxyl. The diffusion speed of the gas in the ventilating holes is ten thousand times faster than in the liquid phase. When the ventilating holes on the plant leaf surface are completely open, the total surface area of all the holes is about 1% – 2% of the total leaf area. This guarantees the necessary requirements for the plants to exchange gas with the external environment.

Living activities need nutrients in addition to sunlight, water and air. Animals acquire nutrients from food, and plants assimilate nutrients from soil through their roots. For plant growth, the nutrients normally include N, P, K, S, Ca, Mg and Fe as well as trace elements such as Mn, B, Zn, Cu, Mo and Cl. In some situations other minerals such as Na and Se are beneficial to plant growth but are not essential. Plants that have azotification nodules need small amounts of Co as nitrogen-fixing symbionts.

In summary, there are several points to be emphasized for the case when biological organisms are used as targets in atomic collision processes. First, maintenance of living activities requires a variety of necessary elements. Since the stopping power of collisions between energetic ions and the target atoms is dependent on the mass number and atomic number of the target atom, for a multi-element target, the stopping power is just an average along the incident particle trajectory. Second, the elements that compose the biological organisms are distributed heterogeneously, and specific elements may be concentrated in some places and rare elsewhere. Third, as biological holes or channels are present, they can be considered to be transparent to ions in vacuum; for low energy ion implantation, the dimensions of the holes or channels can be even greater than the ion range, and thus they should not be neglected.

4.2. STATE OF THE TARGET DURING ION IMPLANTATION

Target materials used in ion beam biotechnology have usually been, up to now, dry crop seeds, and sometimes cells. As ion implantation is carried out in vacuum, the changes that occur in the structural states for both types of the targets during ion implantation should be clarified.

4.2.1. Seeds

A seed is normally covered by a seed peel. Beneath the peel is the aleurone layer, and the plant embryo is yet deeper. Some seeds have albumen, and some not. A dry seed contains less than 12% water. For example, we often measured the water contents of skinless rice seeds to be about 12%. After ion implantation the water content generally decreases to 6% – 7%. This indicates that over a certain time period, about 50% of the water molecules in the rice seeds escapes into the surrounding vacuum. The loss of water from the seeds is realized through the complex gas holes or channels in the seeds.

The mass transfer due to diffusion per unit time is proportional to the gas density and the diffusion area. If the density has a gradient $d\rho/dZ$ in the Z direction, the diffused mass at $Z = Z_0$ for unit area dA is

$$dm = D(d\rho/dZ)_{z=z0}dA. \qquad (4.2.1)$$

Here dm is the mass of gas passing through the area dA per unit time, $(d\rho/dZ)_{z=z0}$ is the density gradient at the $Z = Z_0$ plane, and D is the diffusion coefficient. For high pressure, the gas self-diffusion coefficient can be expressed as

$$D = \tfrac{1}{3}\lambda v , \qquad (4.2.2)$$

where λ is the mean free path, and v is the mean velocity of the gas molecules. Thus, the diffusion equation can be written as

$$dm = \tfrac{1}{3}\lambda v \, (d\rho/dZ)_{z=z0}dA, \qquad (4.2.3)$$

where dm is in g/s, $(d\rho/dZ)_{z=z0}$ is in g/cm^2, and D is in cm^2/s. If the transferred quantity is expressed as the number of transferred molecules, then

$$dN = \tfrac{1}{3}\lambda v \, (dn/dZ)_{z=z0}dA. \qquad (4.2.4)$$

From the diffusion equation (4.2.1) it is clear that the greater the diffusion coefficient the greater the mass transfer rate; and the longer the diffusion distance the less the transferred mass. Let us suppose that a rice seed has a mass of 2.5×10^{-2} g, and contains 10% water with a mass of 2.5×10^{-3} g. Additionally, let us assume that the surface area of the rice seed is 0.5 cm^2, and that water is adhered to all of the surface. The saturated vapor pressure of water at 10°C is 1226 Pa and in vacuum the evaporation rate per unit area is 0.13 g/cm^2·s. Thus the evaporation time is

$$t = 2.5 \times 10^{-3} \text{ g}/(0.13 \text{ g/cm}^2 \cdot \text{s} \times 0.5 \text{ cm}^2) \approx 38.5 \times 10^{-3} \text{s}.$$

Thus in less than 40 milliseconds all the water will evaporate into the vacuum. Actually the diffusion of water in the seeds through the complex holes takes a much longer time. According to the diffusion equation (4.2.1), the quantity of water vapor transferred per unit area per unit time is

$$M = D(d\rho/dZ)_{z=0}. \tag{4.2.5}$$

The diffusion coefficient D of vapor into the air is known to be 0.241 cm^2/s and $(d\rho/dZ)$ can be estimated. The vapor pressure is 2493 Pa at 20°C, and the corresponding density is 1.98×10^{-5} g/cm^3. For a vacuum pressure of 0.013 Pa, the vapor density can be neglected. The diameter of a rice seed is about 0.2 cm. If all of the channels are assumed to originate at the center of the seed and terminate at the surface, the hole length is 0.1 cm. Thus

$$M = 0.24 \times (1.98 \times 10^{-5}/0.1) = 4.75 \times 10^{-5} \,(\text{g/cm}^2 \cdot \text{s}).$$

If we assume that the total cross sectional area of the channels is 1% of the seed surface area, the time needed to lose 1×10^{-3} g of water (corresponding to the loss of 40% of the total water content) due to diffusion is

$$t = 1.0 \times 10^{-3}/4.75 \times 10^{-3} \text{ g/cm}^2 \cdot \text{s} \times 0.005 \text{ cm}^2 \approx 4.2 \times 10^3 \text{ s}.$$

Thus 40% of the water in the seed is lost to vacuum through gas channels in about 70 minutes.

In fact, a rice seed contains about 12% water (measured in our experiments). After 2 hours of pumping the water content was measured to be about 8%, for a water loss of 1×10^{-3} g. The water-loss time was longer in the experiment than the above estimated value. This is because of the assumed values for channel cross sectional area and length. The actual vapor diffusion channels in dry crop seeds are far more complicated. We estimated above a vapor-diffusion channel length of 0.1 cm and a cross sectional area of 0.005 cm^2, for a total channel volume of 5×10^{-3} cm^3. A rice seed has mass of 2.5×10^{-2} g, density about 1.1 g/cm^3 and thus a volume of about 2.27×10^{-2} cm^3. Hence the seed internal volume can be compressed by more than 11% (Table 4-1). Thus we find that the above estimated volume of holes that connect to the surface is only about 20% of the total compressible volume.

The rice seed volume is compressed 11.25% at a pressure of 294 bar. The compressed volume includes some large voids such as gas holes, channels and free space. When the seed is subjected to ion implantation, the gas in the holes and channels is exhausted into vacuum and the bombarding ions do not collide with gas molecules as they would at atmospheric pressure; thus the ions suffer less energy loss per unit length of their path.

Table 4.1. Volume compression ratios of some materials for a pressure of 294 bar.

Material	Water	Metal	Rice
Volume compression (%)	0.31	0.003	11.25

4.2.2. Cells and Calluses

Cells and calluses contain a large quantity of water, and thus their state is completely different from that of seeds. For simplicity, the cell and callus can be considered to be water. At the liquid (solid) – vapor two-phase equilibrium, from the Clausius-Clapeyron equation, we have

$$\frac{dP}{dT} = \frac{\Delta H}{T(V_2 - V_1)},$$

(4.2.6)

where ΔH is the molar heat of vaporization, V_1 is the volume of one mole of liquid, and V_2 is the volume of one mole of vapor. If the vapor is considered as an ideal gas, then from the ideal gas equation we have

$$PV = MRT/\mu,$$

(4.2.7)

where R is the gas constant, P is the pressure, T is the temperature, M is the mass of gas, and μ is the molar weight. For one mole of vapor, since $V_2 \gg V_1$, Eqs. (4.2.6) and (4.2.7) yield

$$\frac{dP}{dT} = \frac{P\Delta H}{RT^2}.$$

(4.2.8)

Integration of this equation gives

$$\ln P = \int \frac{\Delta H}{RT^2} dt + C.$$

(4.2.9)

For small temperature change, since the heat of vaporization ΔH is a slow-changing function of temperature and ΔH can be taken to be constant, equation (4.2.7) can be solved as

$$\ln P = -\frac{\Delta H}{RT} + C,$$

(4.2.10)

or

$$\lg P = A - B/T,$$

(4.2.11)

where A and B are constants (to be determined by experiment), and T is the absolute temperature. This is the approximate expression usually used to relate vapor pressure to temperature. Figure 4.1 shows the $\lg P$ *vs.* $1/T$ linear relationship. When the vapor pressure is known at several temperatures, the vapor pressures at other temperatures can be determined. Conversely the temperature at the phase equilibrium can be determined from the vapor pressure.

When a cell or callus is suddenly exposed to vacuum, evaporation of water removes considerable heat and the temperature of the cell or callus rapidly decreases. When a phase balance is reached, Figure 4.1 indicates that the temperature at the interface of the two phases is below 0°C. Thus the water at the cell or callus surface will freeze to form an ice shell. For fixed vacuum pressure (i.e. the vapor pressure remains fixed), the temperature of the ice shell surface is constant. This is equivalent to cooling the cell or callus in a constant temperature cooling liquid. The water in the cell as a whole will also freeze after a certain time period that is determined by the thermal conductivity. Figure 4.2 shows a real time measurement of the relationship between the chamber pressure and the temperature of a cell population. In this experiment, 3×10^6 A_L (human-hamster hybrid) cells were collected in an *Eppendorf* tube by 1000 rpm centrifugation, and then all the supernatant was sucked out. The cell tube was put in the sample chamber. A T-type copper *vs.* copper-nickel thermocouple with the measurement point of a 0.4 mm in diameter was embedded into the center of the cell aggregate. After preliminary pumping, the sample was exposed to high vacuum. The sample temperature rapidly decreased to −28°C accompanied by a pressure drop from 10^3 to 10^2 Pa, indicating a dehydration dominating process. When the pressure continued to decrease, the temperature decrease became very slow, implying that either a complete dehydration or a frozen ice solid in the cells has been reached.

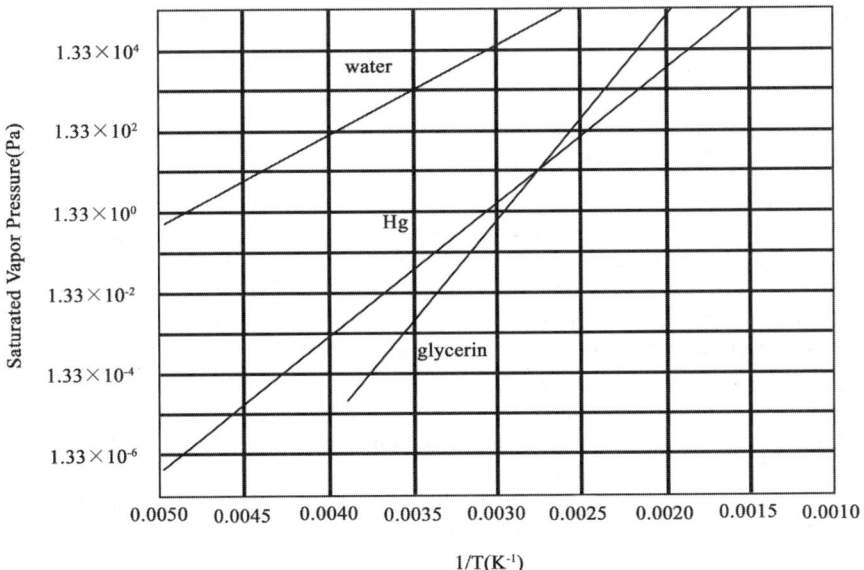

Figure 4.1. Relationships between the saturated vapor pressure and temperature for water, mercury and glycerin.

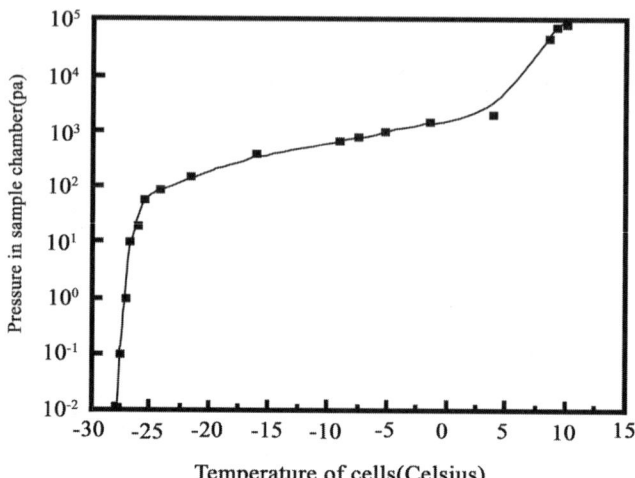

Figure 4.2. Relationship between the vacuum chamber pressure and the temperature of A_L cells.

In ion implantation experiments in living biological organisms, because the inner water is already frozen during the pumping process before ion implantation, the target can be thought of a crystalline. The crystal is not a perfect crystal; it is divided into many small units. Some units contain more water and thus can be frozen normally; some such as the gas diffusion channels contain less water and so may not be blocked by ice. Water molecules continually evaporate into the vacuum. Therefore during ion implantation living biological organisms suffer not only from freezing but also from a large loss of water. Thus the viability is ultimately affected.

4.3. CHARGE EXCHANGE EFFECTS

The surface of a biological sample, such as a cell, contains negative ion groups that can be observed to move to the anode during electrophoresis. When subjected to incident ionizing radiation, radiation damage changes the number of ionized groups on the surface. Dissociation and topography change the cell-cell interaction, the interaction between Ca^+ ions and the cell membrane, the ion permeability of the cell membrane, and other physiological changes occur. In the case of interaction between energetic ions and biological organisms, energy deposition plays the role of ionizing radiation in damaging the ionized groups on the biological organism surface. The important factor is the charge exchange in the electro-characteristic of the sample [3].

As the incident ion loses energy in the target material, every collision with target atoms has a certain probability for loss or capture of electrons from the target material. This phenomenon is called the charge exchange effect of the incident ion. If the ion has charge state i before the charge exchange collision and f after the collision, the reaction cross section for loss or capture of electrons is σ_{if}. The process of electron capture or loss in a collision is most important. The cross section is strongly dependent on the incident ion velocity and charge number, but almost independent of the charge number of the target material.

When the incident ion has a velocity much greater than the electron orbital velocity, a collision between the ion and the target material will rapidly strip electrons to leave an exposed nucleus. In this case, the probability for the ion to capture an electron is very small, and even when the ion captures an electron it easily loses it in next collision. After many collisions between the ion and the target material, the ion loses its energy. When its velocity is close to the orbital velocity of the electron that is to be captured, the capture probability increases and the electron loss probability decreases. The orbital velocities of the electrons in different shells of the ion decrease from the inner shells to the outer shells. When the ion velocity gradually decreases to velocities corresponding of the electrons in each shell, the ion continuously captures electrons in sequence. Finally, when the ion velocity decreases to less than the velocity of the electrons in the outermost shell that binds the electrons only loosely, the incident ion is neutralized. Now it loses its energy mainly through collisions with target nuclei (i.e. nuclear stopping then dominates), and very little energy is transferred to the target electrons. When the neutralized particle slows to thermal velocity, it either combines with the target atoms to form a compound, or becomes embedded in the target material as an interstitial. For a hydrogen ion (proton), the charge exchange reaction can be expressed as

$$H^+ + e^- \Leftrightarrow H^0. \tag{4.3.1}$$

The critical velocity of a proton for electron capture corresponds to an energy of 25 keV. When the ion energy is less than 200 keV, there is a charge exchange effect; when the energy is less than 100 keV, the probability of electron capture increases [4].

Let us take σ_{10} to be the cross section for electron capture by a H^+ ion, σ_{01} the cross section for electron loss by the fast hydrogen atom, π the target thickness, F_0 the neutralized fraction of the H^+ beam, and F_1 that part of the fast hydrogen atom beam that loses electrons to become hydrogen ions. Then we have

$$dF_0/d\pi = F_1\sigma_{10} - F_0\sigma_{01} \tag{4.3.2}$$

and

$$F_1 = 1 - F_0. \tag{4.3.3}$$

Combination of Eqs. (4.3.2) and (4.3.3) results in

$$dF_0/d\pi = \sigma_{10} - F_0(\sigma_{10} + \sigma_{01}), \tag{4.3.4}$$

and the solution is

$$F_0 = \sigma_{10}/(\sigma_{10} + \sigma_{01}) \ \{1 - \exp[-\pi(\sigma_{10} + \sigma_{01})]\}. \tag{4.3.5}$$

When $\pi \to \infty$, we have

$$F_{0\infty} = \sigma_{10}/(\sigma_{10} + \sigma_{01}) \tag{4.3.6}$$

and

$$F_{1\infty} = \sigma_{01}/(\sigma_{10} + \sigma_{01}).\tag{4.3.7}$$

For a 25 keV hydrogen ion the maximum neutralization efficiency is 80%. The target thickness is actually the area density of the target. When the target thickness is 1×10^{16} atoms/cm^2, the neutralization efficiency is already 95% of the maximum efficiency. This indicates that when a 25 keV H$^+$ ion is incident on just 1–2 single atomic layers of target material, it can reach a balance in neutralization efficiency.

When positive ions are implanted in biological organisms, due to the charge exchange effect, molecular groups in 1–2 single atomic layers on the top surface of the target lose electrons and become neutral or positive ion groups. If the ions are singly charged, the implantation dose is D, and the neutralization efficiency is F_o, the accumulated charge per unit area on the surface is

$$q = \frac{1}{6.25 \times 10^{18}} D \cdot F_o \text{ (Coulomb/cm}^2\text{)}.\tag{4.3.8}$$

For example, for a H$^+$ implantation dose of $D = 5 \times 10^{14}$ ions/cm^2 and a neutralization efficiency $F_o = 80\%$, then $q = 6.4 \times 10^{-5}$ Coulomb/cm^2.

Since biological organisms are not good electrical conductors, the accumulated surface charge is not immediately released. Instead, the charge is maintained for a time long enough for changes in the electrical characteristics of the sample surface to be examined by a capillary-electrophoresis method. For example, when positive ions are implanted in minute biological molecules, as expected the results show that as the implantation dose increases the electrical characteristics of the minute biological molecules are changed. Figure 4.3 shows the electrophoresis behavior in the capillaries of the tyrosine (Tyr) crystal after ion implantation [3]. The Tyr molecular crystal is electro-negative and moves to the anode during electrophoresis, but after implantation of positive ions, the speed of movement decreases. When the implanted ion dose reaches a critical value, the speed decreases to zero. If the ion dose is increased further, Tyr moves to the cathode.

As pointed out in the first section of Chapter 3, the interaction of energetic ions with a solid surface can cause surface secondary electron emission. This effect can accelerate changes in the electrical characteristics of biological organism surfaces. The accumulation of surface charge not only affects the electrical characteristics of biological organisms, but also may release biological molecular debris or groups from the surface because of local electrostatic expulsion. If the distance between two groups on the surface of the ion-irradiated biological organism is a_0, the average charge of each group is ne, and the permittivity is ε, the Coulomb repulsive force between them is then $n^2e^2/\varepsilon a_0^2$. Hence the local repulsive force per unit area is $n^2e^2/\varepsilon a_0^4$. If the electrostatic stress is greater than the mechanical strength,

$$n^2e^2/\varepsilon a_0^4 > Y/10,\tag{4.3.9}$$

where Y is the Young's modulus, electric groups will overcome the bonding force and be released.

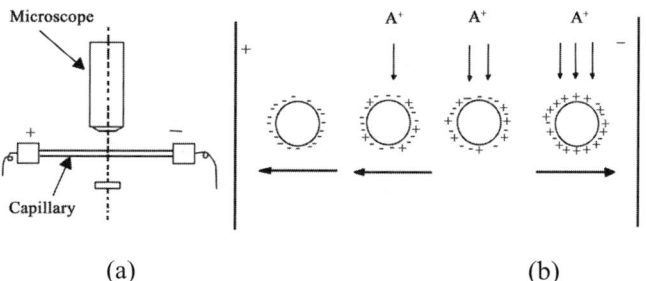

(a)   (b)
Figure 4.3.   Schematic of electrophoresis (a) and mode of change in electric characteristic of Tyr (b).

In experimental ion implantation into biological samples, the doses normally used are in the range $10^{15} - 10^{17}$ ions/cm^2. From Eq. (4.3.8) we can estimate that the q value is about $5 \times 10^{-5} - 5 \times 10^{-3}$ C/cm^2. If the distance a_0 between two electric groups is 5 Å, the charge carried by each cluster is $1.25 \times 10^{-19} - 1.25 \times 10^{-17}$ Coulomb, corresponding to $0.78 - 78$ electron charges.

According to Eq. (4.3.9), if n is taken as 1, that is each group carries one positive charge due to charge exchange, the electrostatic force $n^2 e^2 / \varepsilon a_0^2$ is about 2×10^5 N/cm^2. For example, for wood, Y $= 7 - 15 \times 10^5$ N/cm^2 and the bonding strength is $0.7 - 1.5 \times 10^5$ N/cm^2. This shows that if each group has only one charge exchange (n $= 1$), the relationship described by Eq. (4.3.9) is satisfied [5].

4.4. SURFACE ETCHING AND VOLUME DAMAGE

Etching is a complicated physicochemical process of the solid surface corrosion caused by physical and chemical factors. Energetic ions interact with the solid surface to cause emission of surface atoms or atomic groups so as to leave corrosion traces on the solid surface. This phenomenon is applied to the development of ion beam etching techniques as particularly used for surface precision processing of complicated components such as integrated circuits. In the past, ion beam etching has been limited to materials such as metals and semiconductors. Consideration about whether the interaction between energetic ions and the target surface results in the same processes for biological systems, and what differences there may be in the etching process for non-living materials and living biological organisms, will be the major topic discussed in this section.

Ion beam etching of biological organism surfaces was recognized shortly after the initial discovery of biological effects of ion implantation. Later, many scientists repeated the experiments and obtained similar results. Figure 4.4 shows an early SEM photograph of an ion beam etched biological surface [6]. The sample was the inner surface of a bean cotyledon. It was observed in the experiment that (1) all of the cell walls that faced the ion beam were thinned and a few cells were even cut in half, leaving an ordered traumatic surface; (2) corrosion of the cells was substantial, exhibiting deep holes and grooves; and (3) the sputtered substances collected included not only single atoms or molecules but also biological macromolecules and cell debris.

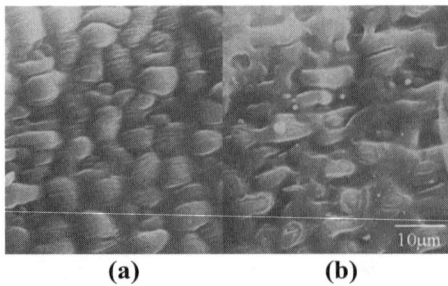

(a)   **(b)**

Figure 4.4. SEM photographs of inner surface of bean cotyledon. (a) Control, and (b) ion-beam etched surface.

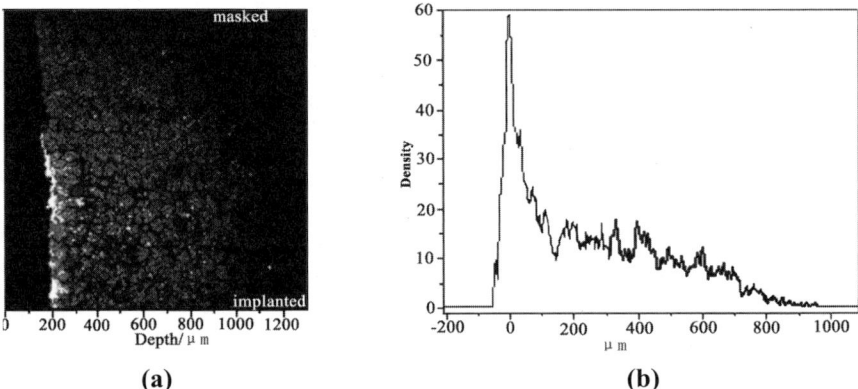

(a)   **(b)**

Figure 4.5. Results on V–ion depth distribution in peanut seed implanted with 200-keV vanadium ions. (a) TPLSM scanning fluorescence image, and (b) converted ion concentration-depth distribution curve.

Figure 4.5 shows a fluorescence image (a) from a 200-keV V-ion implanted peanut seed and the converted concentration-depth distribution curve (b). The mass density of the dry peanut seed was 1.01 g/cm^3, close to that of water. Before ion implantation, the peanut seed was cut into pieces in size of 4 mm × 8 mm × 12 mm. Only a half of the side of each piece facing to the ion beam was exposed to the beam while another half was masked by 100-μm thick aluminum foil. After ion implantation, the seed was cut along the ion incident direction. The freshly cut cross section surface was observed vertically under a Two Photon Laser Scanning Microscope (TPLSM) [7]. An incident laser at a wavelength of 850 nm scanned the surface in an area of 1.3 mm × 1.3 mm, which deliberately crossed the boundary between the implanted and unimplanted areas for analyzing differences in the fluorescence intensity between the two areas, and obtained the fluorescence image through a 460-600-nm optical band filter [8]. It can be seen from the figure that ion implantation causes an increase in the fluorescence intensity from the peanut seed and the peak is at about 50 μm. As the scanned depth increases, the fluorescence intensity gradually decreases and down to 800 μm the intensity is nearly the same as the background. This indicates that the volume damage to the peanut seed caused by 200-keV V-ion implantation is at least down to a depth of 800 μm from the top surface. The fluorescence from the masked area is relatively weak, and the farther away from the implanted area the weaker the fluorescence.

In studies of heavy ion bombarded solid surfaces, pits and voids can be observed using AFM. However, as shown in Figure 4.4, the size of the corroded pits on the bean is of the cellular scale, 10^4 times greater than the size of those observed on solid surfaces. We can estimate from Figure 4.4 that the thickness of the sputtered layer (at the cell wall) is about 0.5 μm when the dose is 10^{16} Ar^+/cm^2 at 20 keV. This same implantation dose results in a sputtered layer of only about ten to several tens of atomic layers in metals and semiconductors. Further, Ar^+ ion bombardment of metal surfaces can sputter atomic groups, but it is not possible to sputter large, massive molecules similar to cell debris as observed in this experiment. We conclude that high-dose, low-energy ion beam bombardment of biological organism surfaces can have a sputtering yield (or etching rate) 2–4 orders of magnitude greater than for metals.

A dry crop seed contains 10% or more free space. The free space consists of gas holes, channels and voids. Some of these features (such as gas holes) are connected to the surface and some are not. When grooves and holes are gradually formed on the biological surface due to ion etching, subsequent ions will penetrate to deeper layers beneath the surface and further repeat the same etching process as on the surface. With increasing dose, those free spaces that are originally not linked to the surface can become linked so that the incident ions can continuously etch the organism deep below the surface and damage several cell mono-layers near the surface. The example of ion beam induced gene transfer in rice seeds described in Chapter 12 is a good demonstration of this. The receptor is the mature embryo. In order to accomplish gene transfer, a prerequisite is to form channels on the mature embryo for the plasmid to pass through. Under normal conditions these channels do not exist. Only when the embryo is partly ion beam etched are pathways formed for transferring genes (depending on the ion etching direction). Let us roughly estimate the spatial scale of the pathway. For example, in the gene transfer experiment, the plasmid used was pBI222 in 2.7 kb, which had a ring diameter of 3,000 Å. No matter in what state (line, ring, or helix) the plasmid enters the rice embryo cell, the diameter of the pathway that allows the plasmid to pass should be 200–300 nm. The length of the pathway is at least 40 μm (through the seed skin), known from anatomy of the rice embryo. This indicates that ion etching forms pathways 200–300 nm in diameter and 40 μm in length on the mature embryo.

The ion bombardment induced abnormal etching rate of the biological organism surface and the production of pathways in the volume are dependent on the incident ion mass number, dose and dose rate, and more importantly on special properties of the biological target. As biological organisms are poor electrical conductors, the charge exchange process, besides atomic sputtering, causes release of groups or molecular debris and thus greatly increases the etching rate. Nearby the trajectory of the incident ion, suddenly explosive ionization of atoms causes simultaneous displacement of other atoms in the region and leaves a vacancy-rich space. Furthermore, the collision of the incident ion with the target can be considered as an adiabatic process. When the energy deposited in the collision volume is equal to the heat of solution of the volume, the instantaneous localized high temperature that is produced leads to sudden collapse of arrangement of the atoms in the collision volume and a concentrated vacancy region is formed (see Figure 3.9).

4.5. PRODUCTION OF FREE RADICALS

The term free radicals refers to those independently existing atoms, groups or ions, which contain one or more unpaired electrons that are able to have bonds with others. In radiobiology, free radicals play an important role because in many radiation processes the primary products are free radicals that can become an active factor in secondary reactions. These indirect interactions cause damage to biological molecules, particularly to genetic substances, and are one of the most important factors in genetic mutation.

Free radicals are produced in the collision volume when energetic ions are implanted into biological organisms, due to excitation and ionization of the target molecules. The ways in which free radicals are formed can be summarized as follows:

Dissociation of an excited molecule: $AB^* \rightarrow A^{\circ} + B^{\circ}$,

Dissociation of an excited ion: $(A^+)^* \rightarrow R^{\circ +} + S^{\circ}$,

Excited molecule – molecule reaction: $A^* + RH \rightarrow AH + R^{\circ}$,

Dissociation due to capture of a slow-electron: $AB + e \rightarrow A^{\circ} + B^-$,

Ion – molecule reaction: $RH^+ + RH \rightarrow RH_2 + R^{\circ}$,

Dissociation due to ion neutralization: $A^+ + A^- \rightarrow A^* + A$,

$$\rightarrow R^{\circ} + S^{\circ}.$$

It should be pointed out that these free radicals produced in primary reactions are usually "hot" free radicals, i.e., they have far more kinetic energy or excitation energy than neighboring atoms, molecules or original free radicals. Therefore they have higher activity and cause serious damage to organisms via secondary reactions.

The unpaired electrons of a free radical can pair up with the unpaired electrons of another free radical and thus a free radical – free radical combination reaction occurs. This reaction has a very low activation energy, almost zero, and thus the reaction can easily take place. Sometimes this kind of reaction produces again a new free radical, and this new free radical can continue to react with original reaction participants so as to form a sequence of reactions. The energy released from the free radical combination reaction can be quite considerable. If there is no loss or dispersion of this energy, it is possible to dissociate combined products again back to the previous free radicals.

Similar to the free radical combination reaction, two identical free radicals can produce two different molecules. For example, an ethyl free radical in the gas phase is disproportionated to produce ethane and ethylene: $2C_2H_5^{\circ} \rightarrow C_2H_6 + C_2H_4$. Due to the high activity of free radicals, they can easily take part in addition reaction, hydrogen-extraction and electron-transfer reactions with biological target molecules. Double free-radicals very easily react with target molecule free radicals to cause oxygen-fixing, resulting in target molecule damage that cannot easily be repaired.

Free radicals are very unstable. Except for a few special kinds, most free radicals have very short lifetime. For example, the radiation induced hydroxyl free radical has a half-life of only $10^{-10} - 10^{-9}$ s; the hydrated electron has a half-life of 2.3×10^{-4} s in neutral water and 7.8×10^{-4} s in basic water solutions. A few free radicals are stable. The stability of free radicals depends on their structure and environment. If a radical can delocalize unpaired electrons to reduce the spin density of the free radical center, the tendency of the free radical to reaction is very small and thus its stability increases. Spatial steric hindrance is also an important factor influencing free radical stability, because it prevents free radical disproportionation, dimerization and other reactions based on dimolecular reaction processes. For example, for the trityl free radical ph_3C^*, although its free radical center has a spin density greatly different from that of benzyl, it is much more stable because three benzyl rings play a role in the spatial steric hindrance of the unpaired electrons of the methyl central carbon atom. The ability of the free radical center to obtain a planar configuration is another important factor for free radical stability. A deviation from planar configuration implies a reduction in stability.

Biological organisms generate a large number of long-life free radicals when ion implanted. This type of free radical may have two sources: the first is a primary reaction product of the ion implantation process, and the second is free radical stabilization by spatial position-resistance due to deposition of implanted ions and displaced atoms. It should be noted that when the implantation dose increases, originally stable free radicals may become unstable, or non-free radicals due to progressive reactions. This is a dynamic reaction process. If intermediary reactions are not taken into account, free radicals are increased by an amount ΔR when the dose is increased by ΔD. The free radical increase ΔR should be proportional to ΔD as well as to the concentration of non-free radicals. The proportionality constant is the reaction constant of non-free radicals becoming free radicals. Since a collision volume contains free radicals, they may become non-free radicals under the action of ΔD and thus this part should be subtracted. Usually, an increase in free radical concentration in the reactive volume is equal to the number of non-free radicals becoming free radicals minus the number of free radicals becoming non-free radicals.

If the portion of the free radicals in the collision volume is P_1, that of long-life free radicals P_2, that of non-free radicals P_0, and the reaction coefficient for a radical to change its state from i to j is K_{ij} in units of cm^2, then

$$dP_1/dD = K_{01}P_0 + K_{21}P_2 - K_{10}P_1 - K_{12}P_1, \tag{4.5.1}$$

$$dP_2/dD = K_{02}P_0 + K_{12}P_1 - K_{20}P_2 - K_{21}P_2, \tag{4.5.2}$$

and

$$P_0 = 1 - (P_1 + P_2). \tag{4.5.3}$$

Let

$$a_1 \equiv K_{21} - K_{01}, \quad a_2 \equiv K_{01} + K_{10} + K_{12},$$

$$a_3 \equiv K_{12} - K_{02}, \quad a_4 \equiv K_{02} + K_{20} + K_{21}.$$

Eqs. (4.5.1) and (4.5.2) can then be rewritten to be

$$dP_1/dD = a_1P_2 - a_2P_1 + K_{01} \tag{4.5.4}$$

$$dP_2/dD = a_3P_1 - a_4P_2 + K_{02}. \tag{4.5.5}$$

The above equations can be solved for the long-life radicals:

$$P_2 = A \exp(-C_1D) + B \exp(-C_2D) + P_{2\infty}, \tag{4.5.6}$$

where

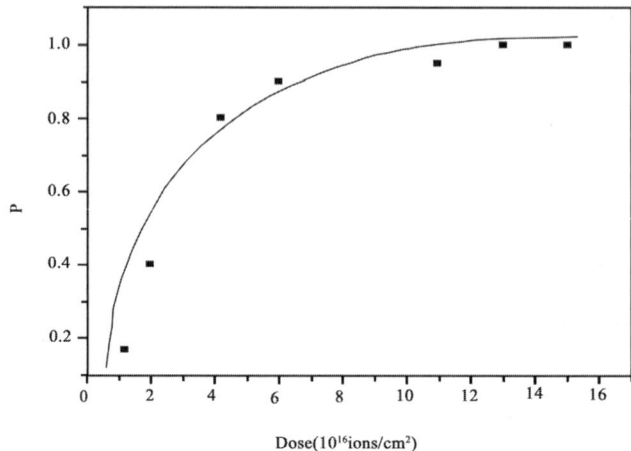

Dose(10^{16}ions/cm^2)

Figure 4.6.   Relationship between free radical content (peak-peak value) and ion dose for N-ion implanted rice dry seeds.

$$C_1 = \tfrac{1}{2}\{(a_2 + a_4) - [(a_2 - a_4)^2 + 4a_1a_3]^{1/2}\},$$

$$C_2 = \tfrac{1}{2}\{(a_2 + a_4) - [(a_2 - a_4)^2 - 4a_1a_3]^{1/2}\},$$

$$P_{2\infty} = (a_2K_{21} + a_3K_{12})/(a_2a_4 - a_1a_3)^{1/2}.$$

In Eq. (4.5.6), A and B are constants determined by

$$P_{2\infty} = \lim_{D \to \infty} P_2(D) \tag{4.5.7}$$

$$A + B + P_{2\infty} = 0 \qquad (D = 0, P_2 = 0). \tag{4.5.8}$$

It can be seen that $(A + B)$ has a negative value whose absolute value depends on the reaction coefficient. Up to now, the reaction coefficients have not been determined. If the content of the long-life free radicals is to be estimated, the values of A and B should be determined by experiment. Figure 4.6 shows the relationship between the free radical content (peak-to-peak) and the ion dose for N ion implanted rice seeds. The dots in the figure are the experimental data and the curve is the calculated result.

The free-radical storing effect induced by ion implantation in dry crop seeds is not obvious. After the seeds are stored in the dry condition for 18 months, the free radical content is found to be decreased by less than 10%. This indicates that under dry conditions the combination probability of long-life free radicals is very low (Figure 4.7 [9]). However, under wet conditions, the free radical concentration will have a fairly large change. Figure 4.8 shows the free radical concentration as a function of time during the germination period for ion-implanted rice seeds [10]. In the first few hours, the absorbed water in the seeds increases greatly. Water molecules combine with free electrons to form hydrated electrons so that the free radicals have increased their chances to combine. After 24 hours, the free radical concentration decreases to a minimum and then slowly increases up to 48 hours, when a balance is reached. For unimplanted seeds, the free radicals also change similarly during this period, probably related to some free radicals such as O° and OH° produced due to certain enzyme activity which is related to biological organism oxidation and the mitochondrion expiration chain. In biological organisms, the SOD (Super Oxide Dismutase) enzyme assumes cleaning of free radicals and exists widely inside both aerobic and aerotolerant biological tissue. Owing to the intrinsic adaptability of biological organisms, when the O_2° free radical concentration increases due to ion implantation, the SOD biological combination also substantially increases by induction of a higher substrate concentration. Hence the SOD concentration of the implanted seeds is higher than that of the unimplanted seeds. In the initial stage, the implanted seeds synthesize the SOD enzyme at a higher rate than unimplanted seeds. After 24 hours the rate of synthesis of SOD of both kinds of seeds becomes slow. Then in the unimplanted seeds an accumulation of the O_2° disproportionation product H_2O_2 results in a higher rate of SOD inactivation than the synthesizing rate, and thus the curve is slowly lowered. At 60 hours the curve is the lowest and then increases. But the synthesis rate of implanted seeds evidently does not decrease. A comparison of the free radical and SOD enzyme change curves between implanted rice seeds and the control clearly shows that free radical scavenging is an important factor influencing changes in free radical concentration.

In radiation studies the free radical concentration is used to measure the degree of radiation damage. In the case of ion implantation of dry crop seeds, the water content in the seed is very low and thus the free radical diffusion is much less. Thus the free radicals are relatively "fixed" at their initial sites. Because of this the location of ion-implantation-produced radiation damage can be examined qualitatively. In one experiment, the electron-spin resonance (ESR) signal was measured for ion implanted seeds (5×10^{16} N^+/cm^2, 30 keV), and the seeds were then peeled layer by layer from the ion implanted side. When the thickness of the peeled layers was about 200 µm, it was found that the free radical ESR signal intensity was basically the same as that of the control. This result indicates that the ion implantation induced radiation damage in the 30 keV 5×10^{16} N^+/cm^2 implanted seeds is present only in a thin layer of about 100–200 µm beneath the surface.

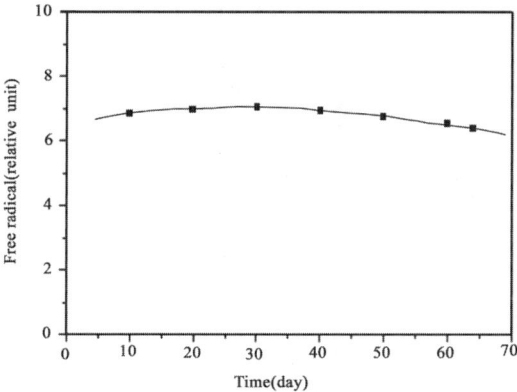

Figure 4.7. Storing effect of free radicals produced by ion implantation of dry rice seeds.

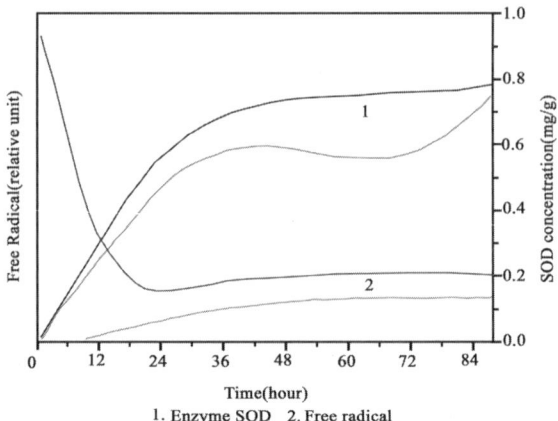

Figure 4.8. Changes in free radical and SOD as a function of time in the germination process of rice seeds. For each pair of the curves, the up one results from ion implantation and the low one is from the control.

4.6. MASS DEPOSITION REACTIONS

In radiobiology, generally only the radiation induced damage to the body is considered. Actually, cosmic ray particles from the universe are always deposited in biological organisms even in the absence of implantation. These particles are deposited and from a structural biology point of view they are a kind of damage, no matter in what form (interstitial, substitution or combination with nearby molecules).

It is an important feature of material modification by ion implantation that in principle, ion species can be introduced into the substrate material and the atomic composition formed is not restricted by the equilibrium metallurgical phase diagram. "Non-equilibrium" materials can be formed. For example, copper and tungsten are very difficult to dissolve in each other even in the liquid state, but W ion implantation into Cu can achieve a substitutional solid solution of 1% W in Cu. Simultaneous Ti- and C-ion implantation into steel can form TiC. This process is called ion implantation modification of the material surface. Nevertheless, it is a kind of damage in terms of the modified structure.

Based on the same principle, for ion implantation into organisms, when the ion energy is decreased down to less than 20 eV, ions are deposited. The implanted ions, displaced atoms, or interstitials may react with the nearby atoms and molecules. The reaction products may be stable or excited. The unstable products further participate in reactions and eventually a balance is reached.

4.6.1. Free Radical Reactions

As the deposit concentration increases, the deposits compete with the substrate for free radicals and competitive reactions occur:

$$R^\bullet + A \xrightarrow{\quad Ka \quad} P_a, \qquad\qquad (4.6.1)$$

$$R^{\bullet} + B \xrightarrow{\;Kb\;} P_b, \qquad (4.6.2)$$

where A is the displaced atom or the implanted ion, B is the substrate solute, P_a and P_b are the products, and K_a and K_b are the reaction coefficients. The relative reaction probability is

$$\eta = K_a[A]/(K_a[A] + K_b[B]). \qquad (4.6.3)$$

It can be seen that the competitive reaction is dependent on the deposited particle concentration. Here, the competitive reaction is actually the free-radical scavenging reaction. For example, in this book the implanted ion species is usually nitrogen. An important ion-molecule reaction of N^+ is: $N^+ + O_2 \rightarrow NO^+ + O$. NO^+ captures an electron and further reacts with oxygen: $2NO + O_2 \rightarrow 2N_2O$. Since N_2O very easily reacts with electrons, when the concentration is high enough, the electrons that originally recombine with positive ions may meet and react with N_2O molecules to produce N_2:

$$e^- + N_2O \rightarrow N_2 + O^-.$$

Therefore N_2O is often used as an effective scavenger for hydrated electrons.

4.6.2. Acid-Alkali Balance

For a living environment of cells, in addition to water as well as organisms and inorganic materials dispersed in the water, the acid-alkali balance should also be considered. In most cases an almost neutral acid-alkali level is maintained between inside and outside of the life body. However, under ion implantation the original acid-alkali balance is most probably disturbed and can cause serious damage to the organism.

The acid-alkali level of a solution is normally described by the pH value. The solution pH value is equal to the negative logarithm of the hydrogen activity in the solution. In diluted solutions, for most conditions encountered in biochemistry, the activity approximately equals the hydrogen concentration. Thus, the pH can be expressed by the negative logarithm of the hydrogen concentration:

$$pH = -\log [H^+].$$

If the initial concentration in the cell is $[H^+]$ and after implantation of hydrogen ions the concentration has a change of $[\Delta H^+]$, then

$$pH_1 = -\log [H^+ + \Delta H^+],$$

and the pH increment is

$$\begin{aligned} \Delta pH &= pH_1 - pH \\ &= \log [H^+] - \log [H^+ + \Delta H^+] \\ &= \log \frac{1}{1 + \dfrac{[\Delta H^+]}{[H^+]}} < 0. \end{aligned} \qquad (4.6.4)$$

This indicates that hydrogen ion implantation causes a decrease in the pH value in the cell. It should be pointed out that if hydrogen implantation increases the solution $[H^+]$ concentration, only a small portion possibly comes from the implanted ions and most comes from the dissociation of the free radicals OH° and HO_2° during ion implantation or from the products of the charge exchange between the neutral hydrogen and the solution particles. As long as the acid-alkali balance in the cell is disturbed, the biochemical process will be accelerated to rebuild an acid-alkali balance for a normal pH value of the organism. For example, the ion implantation induced increase in the $[H^+]$ concentration of cytoplasm can deliver protons and electrons to excited oxygen to form water through the oxidation respiratory chain of the mitochondrion, so that the pH value is recovered to a nearly neutral level.

4.6.3. Mass Deposition Reactions

A living organism can be defined as a system which can decrease its own entropy through the environment. The "reactor" of this system is the cells. The cells organize the elements necessary for life into a highly ordered structure as well as controlling the inflow of required reagents. When elements are deposited in a small compartment in the cell the reaction rate will change according to the mass interaction law of chemical reactions.

The reaction equation can be written

$$mA + nB \Leftrightarrow pC + qD.$$

The positive reaction rate is $V^+ = k^+ [A]^m \cdot [B]^n$, and the negative rate is $V^- = k^- [C]^p \cdot [D]^q$. When the concentration of substance A increases by ΔA due to ion implantation, the balance shifts to the right hand side and the positive reaction rate increases.

Suppose that the cell cytoplasm solution is an inert electrolyte with a neutral pH value. The deposited element A and the reactor ion B can form an activated complex with a reaction rate proportional to the activated complex concentration. The reaction between A and B can be expressed as

$$A + B \xrightarrow{K} (AB)^* \xrightarrow{K'} P. \tag{4.6.5}$$

The reaction rate V is

$$V = \frac{d[P]}{dt} = K'[(AB)^*]. \tag{4.6.6}$$

Because

$$K = \frac{a(AB)^*}{a_A a_B},$$

where a is the activation, and for the case of no dissociation of the solvent, the activation equals the product of the concentration and the activation coefficient, then

$$K = \frac{[(AB)^*]f_{(AB)^*}}{[A]f_A \cdot [B]f_B},$$

where f is the activation coefficient. Therefore,

$$[(AB)^*] = K[A][B] \cdot \frac{f_A f_B}{f_{(AB)^*}} \, ,$$

and

$$V = K' [(AB)^*]$$

$$= K'K[A][B] \cdot \frac{f_A f_B}{f_{(AB)^*}}$$

$$= K_0 [A][B] \frac{f_A f_B}{f_{(AB)^*}} = K[A][B],$$

namely,

$$K = K_0 \frac{f_A f_B}{f_{(AB)^*}} \, . \tag{4.6.7}$$

Substitution of the Debye-Huckel limitation formula for the activation coefficient,

$$\log f_i = - CZ_i^2 \sqrt{I}$$

into Eq. (4.6.7) gives

$$\log K/K_0 = - C[Z_A^2 + Z_B^2 - (Z_A + Z_B)^2] \sqrt{I}$$

$$= 2CZ_A Z_B \sqrt{I} \, , \tag{4.6.8}$$

where Z_A and Z_B are the charges of ions A and B, respectively, I is the ion intensity in the solution, and coefficient C equals 0.509 for the water solution at 298 K. This shows that the reaction rate for similarly-charged ions increases with increasing ion intensity, while the reaction rate for the oppositely-charged ions decreases with increasing ion intensity. If there is at least one type of uncharged ions in the reactors, the reaction rate is independent of the ion intensity. In the case of A being the deposited ion, no matter whether A is a necessary element in the cell or not, the reaction between ions A and B may be either accelerated or decelerated (depending on the signs of the ion charges of A and B).

One of the essential differences between low energy ion implantation effects and radiation damage resides is mass deposition. This book will often use the idea of mass deposition in the following chapters, and particularly in Chapter 5, studying possible processes and products of the reaction between deposited ions and minute biological molecules.

4.7. ENERGY LOSS FEATURES

In Chapter 3 we found that the total energy loss per unit path of a single incident ion in the target material is given by

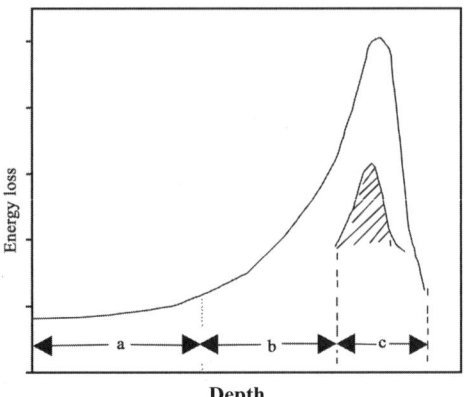

Figure 4.9. Depth distribution of energy loss of implanted ions.

$$-\frac{dE_1}{dx} = N\left[S_n\left(E_1\right) + S_e\left(E_1\right)\right], \qquad (4.7.1)$$

where E_1 is the incident ion energy, N is the number density of the target material, and $S_n(E_1)$ and $S_e(E_1)$ are the nuclear stopping and electronic stopping powers, respectively. From the relationship between S_n and S_e and the ion energy shown in Figure 3.5, one can qualitatively describe the spatial distribution of the energy loss and ion deposition of the ions from their first penetrating the target to deposition (Figure 4.9). In the early stage when the incident energy is high, the ion loses its energy mainly through electronic stopping, i.e., energy loss of the incident ion is via ionization and excitation of target atoms. If the energy loss due to nuclear stopping is added, in this stage (path section *a*) the total energy loss of the ion is approximately constant. As the ion energy gradually decreases, the energy loss due to nuclear stopping plays a progressively dominant role, and with decreasing ion energy there is a corresponding increase in energy loss. This occurs in the path section *b*. As the incident ion energy decreases further, nuclear stopping power rapidly increases and the energy loss reaches a maximum (the Bragg peak in section *c*). Now, due to decreasing energy, the incident ions eventually come to rest with a Gaussian distribution. In the case of low energy (the nuclear stopping process), the energy deposition does not have a pronounced Bragg peak but the ions are still distributed with a Gaussian distribution (the shaded area in the figure).

The energy loss distribution of the incident ion along its trajectory has only a macroscopic average meaning. In fact, the energy deposition fluctuates around a certain mean value in time and space. This fluctuation depends on two geometrical and physical factors. The physics factor is the probability of the interaction between radiation and material. It is assumed that in a differential volume the deposited energy equals the absorbed energy. However if the incident ion causes displacement of the target atoms, the displaced atoms can escape from this volume without consuming all of its energy, resulting in a deposited energy smaller than the absorbed energy. On the other hand, it is possible that some recoil atoms outside the unit volume can enter the volume so as to increase the absorbed energy. Because the production of recoil atoms is a statistical process, their contribution to the energy absorption is also statistical.

Up to now, the basis of our discussion about incident ion energy loss is the assumption of a continuous, homogeneous distribution of the atomic volume density, which is much greater than the ion range. This is actually not the real situation. In nature no material is ideal. Inside materials there are always impurities, defects and channels. For metals and semiconductors, these defects are of the atomic scale and are describable. Thus the ion implantation theory built on the basis of atomic collisions and solid state physics basically agrees with experiments.

For biological organisms, a homogeneous and continuous distribution of atoms in space does not exist. The actual distribution is unknown. Perhaps in the future when structural biology is very well developed, the structure of biological organisms may be precisely described at the atomic scale. Possibly then a theory might be developed for the interaction between energetic ions and complicated biological systems with the aid of knowledge of atomic collisions and structural biology. Now, the incident ions surely indeed enter a black box. The energy loss and distribution of the ions in the box depend on our knowledge of the inner structure of the black box and how this type of the structure is described.

In the first section of this chapter we pointed out that the complex spatial configuration of biological organisms is multi-layered, multi-phase, and multi-channeled, as well as having some free space and volume. The biological channels and voids are not on the atomic scale; in vacuum, as the gas inside is pumped out, they are "transparent" to the incident ions.

In the second and third sections of this chapter, the erosion of biological organisms by ion sputtering and electron sputtering is described. This actually provides subsequent incident ions with passages in which the ions have no energy loss. Based on the above two points, for dry crop seeds the target atom volume density along the ion implantation direction after implantation to a certain dose can be expressed as

$$N(x) = \begin{cases} 0 \text{ at vacancies } (-l, \alpha l), \\ \\ N \text{ at non-vacancies } (\alpha l, l). \end{cases} \qquad \text{(for } \alpha, \text{ see below)} \quad (4.7.2)$$

In the interval $(-l, l)$, expansion of the above expression as a Fourier series gives

$$N(x) = a_0 / 2 + \sum_{n=1}^{\infty} (a_n \cos n\pi x / l + b_n \sin n\pi x / l) \qquad (4.7.3)$$

where

$$a_0 = N(1 - \alpha)$$

$$a_n = -\frac{N}{n\pi} \sin n\pi\alpha,$$

$$b_n = \frac{N}{n\pi} [\cos n\pi\alpha - (-1)^n].$$

With these expressions, Eq. (4.7.2) becomes

$$N(x) = \frac{(1-\alpha)}{2} N + \sum_{n=1}^{\infty} \frac{N}{n\pi} \left\{ [(\cos n\alpha\pi) - (-1)^n] \sin n\pi x / l - \sin n\alpha\pi \bullet \cos n\pi x / l \right\} \quad (4.7.4)$$

In the above expression, $(-l, l)$ is the distribution period of the atomic density of the biological organism in the direction of the ion trajectory. It should be pointed out that this biological material density distribution is not a periodic function. As for the initial implantation dose, the distribution of the vacancies and biological channels formed in the direction of the ion trajectory is statistical. If the ions do not lose their energy when they are travelling in the vacancies (provided that the geometrical dimension of the vacancies is much greater than the ion radius so that the interaction between the moving ions and the vacancy walls can be neglected), then only vacancies along the ion trajectory are considered, no matter where in the trajectory they are.

The parameter α is dimensionless, indicating the fraction of vacancies along the trajectory, and its value is determined by Eq. (4.7.2) between $-l$ and l. The value of α is closely related to the biological organism (such as the dry crop seed) structure, implanted ion dose, sputtering coefficient, and so on. In the process of ion implantation, as the implanted ion dose increases the dimension of the vacancies in the biological organism may increase, and new combined vacancies may also be produced, as has been experimentally observed. The ion beam acts like a milling cutter, eventually possibly linking vacancies in the region near the surface of the organism. This provides subsequent ions with a section of unimpeded passage, which, for example as in the gene transfer process, can allow plasmid DNA that is greater than the implanted ion radius to pass through smoothly.

Based on the above analysis, the physical meaning of α can be given. If N is the atomic volume density (not including vacancies) of the biological organism and A is the ion implantation area, the number of total target atoms in a distance $2l$ along the implantation direction is $N \cdot A \cdot 2l$. If these atoms are moved to area A, the target area density is then

$$N_A = 2l \cdot N. \quad (4.7.5)$$

In the above volume, the vacancy volume intrinsically in the organism is specified as η. If the atoms that can be contained in these vacancies are moved to area A, the number of vacant atoms per unit area is then

$$N_C = 2\eta l \cdot N. \quad (4.7.6)$$

Now the effects of ion and electron sputtering and chemical sputtering are taken into account. The number of the target atoms emitted per unit area is

$$N_S = rD, \quad (4.7.7)$$

where D is the implantation dose, and r is a coefficient related to the particle emission. Hence the total number of lost target atoms per unit area in terms of distance in the ion incident direction is

$$l' = \frac{N_s + N_c}{N} = \frac{rD}{N} + 2\eta l. \tag{4.7.8}$$

From Eq. (4.7.2), l' is known as the section in $(-l, l)$ with zero density, namely,

$$l' = l + \alpha l, \quad \alpha = \frac{l'}{l} - 1.$$

Use of l and l' in Eq. (4.7.5) and (4.7.8) in the above expression results in

$$\alpha = \frac{\dfrac{rD}{N} + 2\eta l}{\dfrac{N_A}{2N}} - 1 = \frac{2rD}{N_A} + \frac{2\eta \bullet 2Nl}{N_A} - 1.$$

Since

$$-\frac{dE_1}{dx} = \frac{(1-\alpha)}{2} + \sum_{n=1}^{\infty} \frac{N}{n\pi} \left\{ \left[\cos n\alpha\pi - (-1)^n \right] \sin n\pi x/l - \sin n\alpha\pi \bullet \cos n\pi x/l \right\}$$

and

$$2Nl = N_A,$$

thus

$$\alpha = \frac{2rD}{N_A} + 2\eta - 1. \tag{4.7.9}$$

Since $\eta \ll 1$, when $D = 0$, α is very close to -1; when rD is close to N_A, then $\alpha \to 1$. Eq. (4.7.9) gives a physical meaning to α and relates it to the ion implantation dose and the free space fraction in the biological organism.

Substitution of Eq. (4.7.4) into Eq. (4.7.1) results in

$$-\frac{dE_1}{dx} = \left\{ \frac{(1-\alpha)}{2} + \sum_{n=1}^{\infty} \frac{N}{n\pi} \left\{ \left[\cos n\alpha\pi - (-1)^n \right] \sin n\pi x/l - \sin n\alpha\pi \bullet \cos n\pi x/l \right\} \right\}$$

$$\times [S_n(E_1) + S_e(E_1)]. \tag{4.7.10}$$

It can be seen that fluctuation of the incident ion energy loss along the direction of the ion trajectory is a feature of beaded energy deposition.

For low energy ion implantation, electronic stopping can be neglected and the total stopping power can be replaced by S_n (see Eq. (3.2.8)). Integration of Eq. (4.7.10) gives

$$\int_0^x N(x)dx = -\int_E^0 dE_1 / S_n = E_1 / S_n^0.$$

The left hand side of the above expression can be evaluated by integrating Eq. (4.7.4):

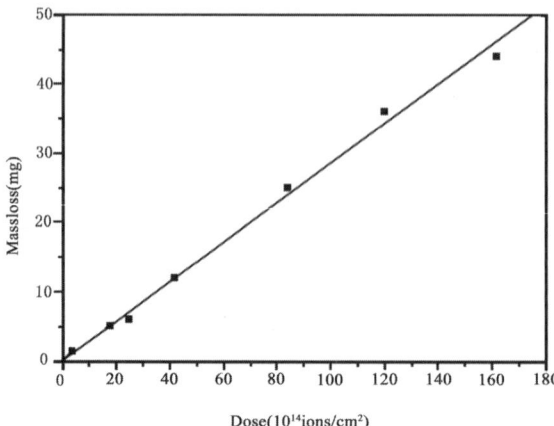

Figure 4.10. Mass loss of lactamine as a function of ion dose.

$$\frac{(1-\alpha)}{2}x + \sum_{n=1}^{\infty}\frac{l}{n^2\pi^2}\left\{\left[\cos n\alpha\pi - (-1)^n\right](1-\cos n\pi x/l) - \sin n\pi x \bullet \sin n\pi x/l\right\}$$

$$= E_1/NS_n^0 \qquad (4.7.11)$$

Neglecting the higher-order terms in Eq. (4.7.11) gives

$$X = \frac{1-\alpha}{2}E_1/NS_n^0 .$$

Using the α value in the above expression results in

$$X = \frac{N_A}{(1-\eta)N_A - rD}\bullet\frac{E_1}{NS_n^0} . \qquad (4.7.12)$$

Eq. (4.7.12) indicates the possible penetration depth in the target material at the moment after implantation of the initial ion dose. For metals and semiconductors, as $\eta \approx 0$, the unit-area energy loss (rD) can be neglected compared to N_A. Thus Eq. (4.7.12) simplifies to Eq. (3.2.12C):

$$X = E_1/NS_n^0 . \qquad (3.2.12C)$$

The formula above is an expression for the range of low energy ions implanted in metals or semiconductors (see Chapter 3). For the case of an ion implanted organism, η in Eq. (4.7.12) is considerable (e.g. for crop seeds, $\eta > 10\%$), and more importantly, the sputtering yield (here described by the emission coefficient r) is much greater than for metals and semiconductors. Therefore, using Eq. (3.2.12C) to estimate the range of low energy ions in crop seeds loses its practical meaning.

The sputtering yield of ions etching an organism can be measured experimentally. For example, lactamine ($C_3H_7O_2N$) is a very good bio-equivalent material which can be used for such an experiment. This material was formed into a film with thickness 1.1 mg/cm^2 in an oven at 40°C. The film was then implanted with 30 keV N$^+$ ions and the mass loss of the film then measured as a function of implantation dose. The results are shown in Figure 4.10 [11]. Details of the mass loss are not taken into account. The mass loss is used to represent the sputtering yield (it can be imagined that the sputtering yield is greater than the mass loss, because some of the sputtered heavier pieces may fall back onto the film and its substrate and then be measured). One can calculate from Figure 4.10 that the emission coefficient r is 290 lactamine-molecules/ion or 3.8×10^3 atoms/ion. This is 3–4 orders of magnitude greater than the sputtering yields for the elements of C, H, N and O calculated using classical collision cascade theory, and also about 3 orders in magnitude greater than those for metals (see Figures 3.2 and 3.3).

The classical theory looks at the target atom as an isolated target when considering ion sputtering. This is correct for metals and semiconductors. For organisms, atoms and radicals bound together by covalent bonds, the implanted ion energy deposition causes organic molecules and groups to be lost from the surface and even whole molecules to be broken up. These atoms, groups and molecular fragments form rich radiolysis products through chemical processes. A large number of minute volatile molecules will escape from the sample and be pumped out from the vacuum. This process can be called "chemical sputtering". This joins together with ion sputtering and electron sputtering to cause the sputtering yield to increase significantly.

Analysis of the transmission energy spectrum of α particles (^{241}Am emitted particles at 5.484 MeV) from ion-implanted organisms yields information on the organism structure and the thinning effect of low energy ion sputtering on the organism. Figure 4.11 shows the α-particle transmission energy spectra from a piece of tomato skin of thickness of 50 μm before and after implantation with 30 keV nitrogen ions to 1×10^{17} N^+/cm^2. Before implantation, the energy spectrum of the experimental system has a very broad distribution. This indicates a heterogeneous distribution of the spatial structure of the tomato skin, where areas with very small mass-thickness exist. This broad spectrum is equivalent to a piling up of the energy spectra from a Mylar film (a bio-equivalent material with chemical formula $C_{10}H_8O_4$) with an α-particle transmission thickness of 11–20 μm. In other words, tomato skin does not look as thick as it appears, and there must be pores and voids in it for the α particles to pass through without losing energy. After ion implantation, the energy spectrum obviously shifts to the high energy side. The spectral distribution is equivalent to the sum of the energy spectra from a Mylar thin film with thickness 9–17 μm. This indicates that ion beam etching thins the tomato skin, equivalent to a 2–3 μm Mylar film thickness, or an apparent thickness of the tomato skin by 7.5–9.1 μm. For ion beam etching of metal samples, using the same parameters we can only thin the metal surface by 10 to several tens of single atomic layers. The difference between the two cases is three orders of magnitude.

It can be seen from Figure 4.11[12] that there appears to be a completely free α-particle peak in the transmission energy spectrum after ion implantation. The freely passing α particles are approximately 4×10^{-5} of the total transmitted particles. This means that if the embryo of a crop seed is covered by a 50-μm (apparent thickness) seed skin and implanted with a dose of 1×10^{17} ions/cm^2, and even at the last second of implantation the seed skin is penetrated, there are 4×10^9 ions/cm^2 in this second passing through the seed skin to interact directly with the embryo part.

The above results were obtained using equivalent biomaterials and isolated samples. Application of nuclear analysis techniques can directly measure the implanted ion distribution in the dry crop seeds.

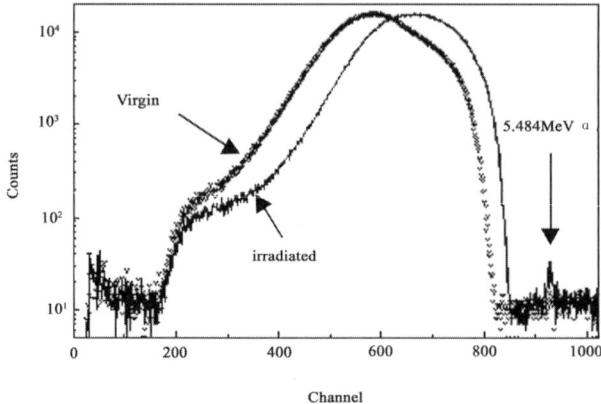

Figure 4.11. α-particle transmission energy spectrum of tomato fruit skin after N-ion implantation to a dose of 1×10^{17} ions/cm^2.

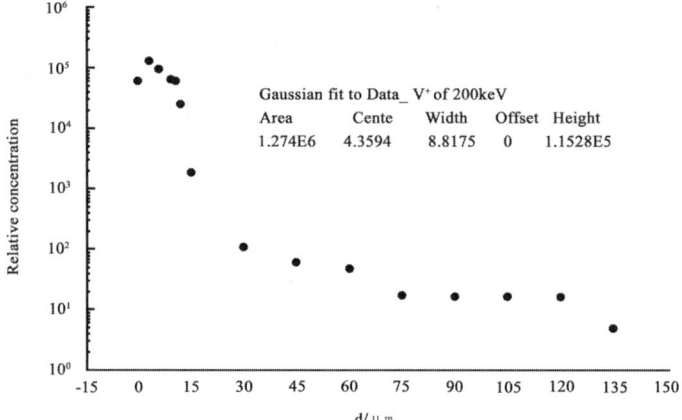

Figure 4.12. The relative concentration-depth distribution of 200-keV vanadium ions implanted in peanut seed measured by using PIXE technique.

    Figure 4.12 shows the vanadium ion concentration-depth distribution in peanut seed implanted with 200-keV V ions [13]. In this experiment, a dry peanut seed was fixed in a specially made modulus and put onto a slicer to cut off a part of the seed. The seed was placed in the target chamber with the cut side facing the ion beam and then implanted. After ion implantation, the sample was packed with graphite, put onto the slicer again at the original position and sliced into 15-μm pieces. Each piece was analyzed using proton induced X-ray emission (PIXE) technique for V-ion concentration. To investigate in detail the vanadium concentration distribution in the first piece (0 – 15 μm), after PIXE analysis the first piece was analyzed again using a scanning electron microprobe. The preliminarily obtained V ion concentration as a function of depth was normalized. It is seen from the figure that the V peak in the first 0 – 15 μm is very high and about 98% of vanadium is deposited in this region. In the second 16 – 30-μm slice, the V peak rapidly decreases and continues to decrease in subsequent slices. In the region of 121 – 135 μm, the V concentration is close to the PIXE detecting limit. From this result we learn that the range of 200-keV V ions implanted in the peanut seed is of about three orders of magnitude greater that that in water which has almost the same mass thickness as that of a peanut seed.

4.8. DOSE AND FLUENCE

In previous chapters we have used the parameter ion implantation dose many times and related it to physical and chemical changes of materials (including living materials). It should be noted that the dose used in this book is different from the concept of dose used in radiobiology. For convenience in description, in this section the ion implantation dose is called fluence, but in later applications, due to the traditional appellation, the noun "dose" is still used.

4.8.1. Radiation Dose

The radiation energy absorbed per unit mass is defined as the radiation dose (absorption dose) D

$$D = \Delta E / \Delta m. \tag{4.8.1}$$

This definition (absorbed energy and absorbed dose) is actually a macroscopic parameter, a mean value for the radiated object mass. At a point A, the dose D_A is a limit

$$D_A = \lim_{\Delta V \to \infty} \frac{\Delta E}{\rho \Delta V} \tag{4.8.2}$$

where ΔV is the volume element surrounding point A, and ρ is the density.

Dose is accumulated with radiation time. The accumulation rate is expressed by the dose rate:

$$P = dD/dt. \tag{4.8.3}$$

Generally, the dose rate is dependent on time t. The dose in the time interval from t_1 to t_2 equals to

$$D = \int_{t_1}^{t_2} P(t) dt \tag{4.8.4}$$

4.8.2. Linear Energy Transfer

The energy transferred from the charged particle to the material per unit path is called the linear energy transfer (LET), namely,

$$L_\Delta = (dE/dl)_\Delta, \tag{4.8.5}$$

where Δ is a small energy loss, and dE is the energy loss due to electronic collisions smaller than Δ.

The LET should be distinguished from the stopping power defined in Chapter 3. The stopping power specifies the energy loss of a charged particle per unit path in the material. The particle's lost energy may not be equal to the material's absorbed energy. For example, parts of the energy loss of a fast charged particle may be carried away from the volume due to Bremsstrahlung. If the energy loss due to Bremsstrahlung can be neglected, LET equals the stopping power.

LET is a mean macroscopic parameter. The unit is J/m, but traditionally keV/μm (1 keV/μm $=$ 1.602 $\times$ 10^{-10} J/m). Using different methods to calculate LET produces different results. One is the trajectory method, which partitions the trajectory equally, calculates the deposited energy in every length, and then finds the mean value. Another is the dose-average method, which partitions the trajectory according to equal energy, and then divides the total energy deposited on the trajectory by the trajectory length. Generally speaking, the LET for heavy ion radiation is dependent on the heavy ion velocity as well as proportional to the square of the charge number. For the target material, the higher the density and the atomic number are, the greater the LET is.

4.8.3. Particle Fluence

This is defined as the number of incident particles passing through unit-cross-sectional differential spheres. The unit is m^{-2}.

4.8.4. Particle Flux Density

This is defined as the number of incident particles passing through unit-cross-section differential spheres per unit time. The unit is $m^{-2}s^{-1}$. The particle flux density equals the fluence per unit time.

4.8.5. Radiation Energy Flux Density

This is defined as the energy carried by the incident particles that pass through unit-cross-section differential spheres per unit time. The radiation energy flux density is normally called radiation intensity.

For a mono-energetic particle beam with energy E, the relationship between fluence F, flux density $\Phi(t)$ and intensity $I(t)$ from time t_1 to time t_2 is

$$F = \int_{t_1}^{t_2} \Phi(t) \, dt = \int_{t_1}^{t_2} \frac{I(t)}{E} \, dt. \qquad (4.8.6)$$

In this book, taking the biological effect of mass deposition into account, we introduce a new specific parameter: deposited mass. The deposited mass D_0 is defined as ion mass M_i absorbed per unit volume:

$$D_0 = M_i/\Delta V. \qquad (4.8.7)$$

The deposited mass D_0 should include two parts – implanted ions and displaced atoms. If it is assumed that the number of displaced atoms is related to the implantation dose in the volume element ΔV, the number of displaced atoms is

$$D_p = pD_\Delta,$$

so the deposited mass is

$$D_0 = D_\Delta + p\,D_\Delta = (1+p)\,D_\Delta, \tag{4.8.8}$$

where p is the proportionality coefficient, $p > 0$ indicating entrance of displaced atoms into the volume element, and $p < 0$ indicating exit of displaced atoms from the volume element.

The concept of deposited ion mass is convenient for description of the mass deposition effect in biology. Due to the participation of the deposited ions, in reactors such as cells, changes in concentrations of some substances result in an entropy change as well as affecting the chemical reaction speeds and processes. All of the changes will influence the eventual biochemical and biological results of the reactors.

The radiation effects normally observed are not as expected, because they are always accompanied by harmful consequences, such as cell death, unbalancing the normal functions in the cells and their organs. In these cases, "injury" or "damage" is used instead of radiation effects. Of course, there exist some radiation biology effects that may be beneficial. The task of radiation biology is to take advantage of the benefits of radiation effects and avoid the harmful consequences. It should be noted that "beneficial" and "harmful" are relative. For example, X-rays themselves can function to induce cancer, but under some conditions they can be applied to cure cancer.

For any changes due to radiation effects, the primary reason is that the radiated object absorbs radiation energy. Therefore it is very natural to link the investigated radiation effects with the absorbed energy quantitatively. In radiation biology, the quantitative biological effect is expressed by η. It can be related to dose in a dose-effect function:

$$\eta = f(D), \tag{4.8.9}$$

where $f(D)$ is a function of dose.

In low energy ion beam biology, induction of biological effects is not only due to absorption of energy from the energetic ions by biological organisms, but also to ion deposition. Therefore, the dose-effect relation can be expressed as

$$\eta = f_1(D) + f_2(D_0) \tag{4.8.10}$$

or

$$\eta = f_1(D) + f_2[(1+p)D_0]. \tag{4.8.11}$$

It goes without saying that the biological effect-dose relation for low energy ion beams can be expressed as the sum of the absorbed energy and the deposited mass. The "effect" defined here is a probability concept, such as the cell death rate, or mutation rate, and so on. The energy-deposition induced cell "damage" and the mass-deposition induced "damage" are independent of each other. According to the basic probability relationships, the total "damage rate" is the sum of both. It should be noted that the sum in the above expressions is an algebraic sum. Because in low energy ion beam biology there must exist this probability, in which energy deposition causes some atoms of some elements to be removed from biological molecules and ions (if just those of that element) deposit in the location where the elemental atoms are removed, chemically combining with the molecule. In this case, ion deposition is a restoration to energy damage.

REFERENCES

1. Lu, G.Y., *Biophysics* (in Chinese, Wuhan University Press, Wuhan, 2000).
2. Yu, Z.L., *et al. Anhui Agri. Sci.*, **1**(1989)17.
3. Shao, C.L., Xu, A., Yu, Z.L., Charge Exchange Effect of Ion Implantation to Biomolecules, *Nuclear Techniques*, **20**(1997)70-73.
4. Zhang, H.S., *Ion Sources and Great Power Neutral Beam Sources* (in Chinese, Science Press, Bejing, 1987).
5. Xia, S.X., *Radiobiology* (in Chinese, Military Medical Science Press, Bejing, 1998).
6. Yu, Z.L., Deng, J.G. and He, J.J., Mutant Breeding by Ion Implantation, *Nucl. Instr. Meth.*, B**59/60**(1991)705-708.
7. Denk, W., Piston, D.W. and Webb, W.W., Two-Photon Molecular Excitation in Laser-Scanning Microscopy, in *Handbook of Biological Microscopy*, edited by Pawley J.B. (Plenum Press, New York, 1995).
8. Xie, J.W., Zhou, H.Y., Ding, X.J., Liu, Z.G., Song, H., Lu, T., Zhu, G.H., *J. of Physics*, **52**(10)(2003)2530-2534.
9. Deng, J.G., *et al.*, Free Radicals Study on Seeds Irradiated with Ion Beam. *J. Anhui Agri. Univ.* (in Chinese with an English abstract), **18**(4)(1991)263-268.
10. Wu, Y.J., Wu, M.D., *et al.*, Study on Free Radicals and Enzymes in the Germination Process of Ion-beam Irradiated Rice Seeds, *Proceedings of the 2nd National Conference on Biological Effects of Ion Implantation* (in Chinese), Sept. 8-11, 1993, Hefei, China.
11. Han, J.W. and Yu, Z.L., Dose Response of Alanine on keV-Ions Irradiation. *Acta Biophysica Sinica*, **14**(2)(1998)341-345.
12. Han, J.W. and Yu, Z.L., Study on Etching and Penetrating of Low Energy Ions into Tomato Peel, *Acta Biophysica Sinica*, **14**(4)(1998)757-761.
13. Zhu, G.H., Zhou, H.Y., Wang, X.F., *et al.*, Measurement of Low Energy Ion Implantation Profiling in Seeds by PIXE and SEM with Slicing-up Technique, *Nuclear Techniques*, **24**(6)(2001)456-460.

FURTHER READING

1. Lu, G.Y., *Biophysics* (in Chinese, Wuhan University Press, Wuhan, 2000).
2. Gao, B.H. and Cui, S.Y., *Vacuum Physics* (in Chinese, 1st ed., Scientific Publishing, Beijing, 1983).
3. Xia, S.X., *Radiobiology* (in Chinese, Military Medical Science Press, Beijing, 1998).
4. Halliwell, B. and Gutteridge, J.M.C., *Free Radicals in Biology and Medicine* (Oxford University Press, New York, 1985).
5. Qiu, G.Y. and Feng, S.Y., *Radiobiophysics* (in Chinese, Wuhan University Press, Wuhan, 1990).
6. Hall, E.J., *Radiobiology for the Rradiologist* (4th ed., Lippincott, Philadelphia, 1994).
7. Zhu, R.B., *et al.*, *Radiobiology* (in Chinese, Science Press, Beijing, 1990).
8. Yu, Z.L., Low Energy Ion Biology, *Science* (Chinese), **4**(1993)36-39.

5

REACTION PROCESSES OF ION IMPLANTED BIOLOGICAL SMALL MOLECULES

Studies of the primary processes involved in interactions between implanted ions and biological organisms encounter both the multiplicity of factors related to the implanted ions as well as the complexity of the organisms themselves. Ion implantation, as distinct from ionizing electromagnetic radiation, is characterized by not only the energy exchange process but also the mass deposition and charge transfer processes [1,2]. These three factors act simultaneously on the biological organism and induce damage. The effects may be described in terms of formation, reaction processes, and final products. In this chapter we consider the individual biological small molecules, or monomeric units of biological macromolecules, that compose the organism and look at the chemical processes for which ion implantation induced damage is physicochemically repaired. First, however, we discuss damage effects to biological molecules in solution caused by ionizing radiation.

5.1. RADIATION DAMAGE TO BIOLOGICAL SMALL MOLECULES IN SOLUTION

Aqueous solutions subjected to ionizing radiation form free radicals such as OH• and H•. The former is the major cause of damage to biological molecules. For the free bases, pyrimidine is generally more sensitive to radiation than purine, and OH• radicals cause damage in different ways. In water solution, the radiation-dissociated products of thymine and cytosine mainly come from the addition reaction of OH• to the C=C double bonds at positions 5 and 6 (the position numbers are defined as below).

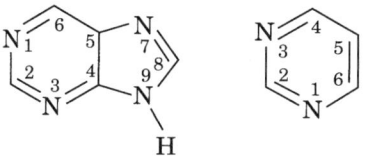

Purine   **Pyrimidine**

The radiation-dissociated products of adenine are mainly due to OH$^\bullet$ radicals added to carbon atoms at position 8 and subsequently causing the double bonds at positions 7 and 8 of the imidazole ring to break.

To nucleotides, ionizing radiation first causes the glycoside bond to break, then releases bases and finally damages the glycone part of the molecule. Damage to the base part of nucleotide is similar to that in free bases. The ribose part is damaged due to the dehydrogenation of OH$^\bullet$ radicals and the subsequent release of a phosphate radical. For example, the dephosphorylation of 5'-GMP is mainly a consequence of OH$^\bullet$ dehydrogen from the C_4' position.

Ionizing radiation can also cause DNA strands to break. The direct reason may be damage to deoxypentose or a break of the phosphodiester bond, and the indirect reason may be damage to or dropping out of bases. But the radical factor is mainly a consequence of OH$^\bullet$ radicals dehydrogenating deoxypentose.

OH$^\bullet$ can dehydrogenate at C_1', C_2', C_3', C_4' and C_5' in deoxypentose causing damage to the deoxypentose. OH$^\bullet$ dehydrogen at C_3' and C_5' will directly bring about the elimination of phosphoric acid. The dephosphorylation rate is higher at C_3' than at C_5'. After C_1', C_2', and C_4' are attacked by OH$^\bullet$, they can form alkali labile sites (ALS). These sites can cause breaking of the DNA strand by alkali treatment.

In addition, a DNA strand under the action of OH$^\bullet$ can form apurine or apyrimidine sites (APS) due to the de-alkali-base or damaged bases cut by enzymes. These sites will also result in breaking of DNA strands under action of specific enzymes.

Most DNA strand breaks are single-strand breaks (SSB). Only about 1/10–1/20 of the SSBs can form double strand breaks (DSB). The oxygen reaction effect (ORE) can increase the yield of OH$^\bullet$ and thus increase DNA strand breaking.

Protein is composed of various amino acids in the form of a peptide strand and will produce a variety of damage under ionizing radiation. For example, the reaction process of breaking of the peptide strand due to the ionizing-radiation-produced OH$^\bullet$ radicals is

$$
\begin{array}{c}
\underset{\displaystyle \underset{\text{H}}{|}}{\overset{\displaystyle \overset{\text{O}}{\|}}{-\text{C}}}-\underset{\underset{\text{H}}{|}}{\text{N}}-\underset{\underset{\text{H}}{|}}{\text{C}}- \ + \ \dot{\text{O}}\text{H} \longrightarrow \ -\overset{\overset{\text{O}}{\|}}{\text{C}}-\underset{\underset{\text{H}}{|}}{\text{N}}-\underset{\underset{\text{H}}{|}}{\dot{\text{C}}}- \ + \ \text{H}_2\text{O}
\end{array}, \qquad (5.1.1a)
$$

$$
-\overset{\overset{\text{O}}{\|}}{\text{C}}-\underset{\underset{\text{H}}{|}}{\text{N}}-\underset{\underset{\text{H}}{|}}{\dot{\text{C}}}- \ + \text{O}_2 \longrightarrow \ -\overset{\overset{\text{O}}{\|}}{\text{C}}-\text{N} \ + \ -\overset{\overset{\text{O}}{\|}}{\text{C}}- \ + \ \text{H}_2\text{O} \qquad (5.1.1b)
$$

Another important reaction of protein under the action of OH$^\bullet$ is sulfhydryl oxidation:

$$\text{RSH} + \text{OH}^\bullet \rightarrow \text{RS}^\bullet + \text{H}_2\text{O}, \qquad (5.1.2)$$
$$\text{RSH} + \text{OOH}^\bullet \rightarrow \text{RS}^\bullet + \text{H}_2\text{O}, \qquad (5.1.3)$$
$$2\text{RSH} + \text{H}_2\text{O} \rightarrow 2\text{RS}^\bullet + 2\text{H}_2\text{O}, \qquad (5.1.4)$$
$$\text{RS}^\bullet + \text{RS}^\bullet \rightarrow \text{RSSR}. \qquad (5.1.5)$$

The oxidation product, RSSR, of the sulfhydryl reaction can be decreased by the action of H radicals:

$$RSSR + H \rightarrow RSH + RS$$
$$\begin{array}{c} | + 2H \\ \xrightarrow{\hspace{3cm}} RH + H_2O. \end{array} \qquad (5.1.6)$$

Furthermore, ionizing radiation can cause damage to peptide strands such as dissociation and ionization. This kind of damage will lead to further changes in the primary structure of the protein and thus changes in function, such as enzyme inactivation. It should be noted that after ionizing radiation, changes in protein structure are closely related to the concentration of some amino acids as well as the radiation environment (such as the oxygen content).

5.2. DAMAGE TO ION-IMPLANTED BIOLOGICAL SMALL MOLECULES

Ion implantation in bio-samples is normally performed under dry vacuum conditions. Although ion implantation may produce molecular radicals, it is not possible, when water solutions participate, for the OH° radicals formed from the radiation dissociation of water to bring about indirect damage. For solid-state biological small molecules, ion implantation plays a direct role in causing damage. The results of this action includes (1) breaking of bonds or strands by ionization, excitation or displacement of bio-molecules, (2) formation of new molecules or groups from the recombination of implanted ions, displaced atoms or substrate elemental atoms, and (3) production of free radicals.

5.2.1. Structural Damage

A nucleotide molecule, a basic unit of a nucleoside, has a phosphate radical as well as a base, and is an ampholite, which is unstable in acid solution but very stable in alkaline and neutral solution. One of the important damage mechanisms to a nucleotide molecule under ion implantation is dephosphorylation. Various nucleotides have different sensitivities to ion implantation induced damage, namely, 5'–dTMP > 5'–CMP > 5'–GMP > 5'–AMP. Generally, pyrimidine nucleotides are more sensitive to ion implantation damage than purine nucleotides [3-6]. For nucleosides, the sensitivities to ion beam irradiation are found in an order of dTR>GR>AR>CR>UR (Figure 5.1) [7].

Under ion implantation, in nucleotide the pentose or base is subjected to damage and produces some unstable phosphate esters, which dissociate under strong alkali treatment releasing inorganic phosphorus. For example, after ion implantation of 5'-dTMP and followed by treatment with 0.1N NaOH, about 60% of the phosphorus is released as inorganic phosphorus, and the amount is related to the duration of alkali treatment. The longer the treatment time, the greater the amount of inorganic phosphorus released (Figure 5.2). Unstable phosphate esters can also dissociate and release inorganic phosphorus with heat treatment. Figure 5.3 shows the release of inorganic phosphorus from three ion-implanted nucleotides before and after heat treatment in a 90℃ water bath for 1.75 hours. After heat treatment, the phosphorus release increases by about 60%, equivalent to treatment with 0.1N NaOH (Figure 5.2).

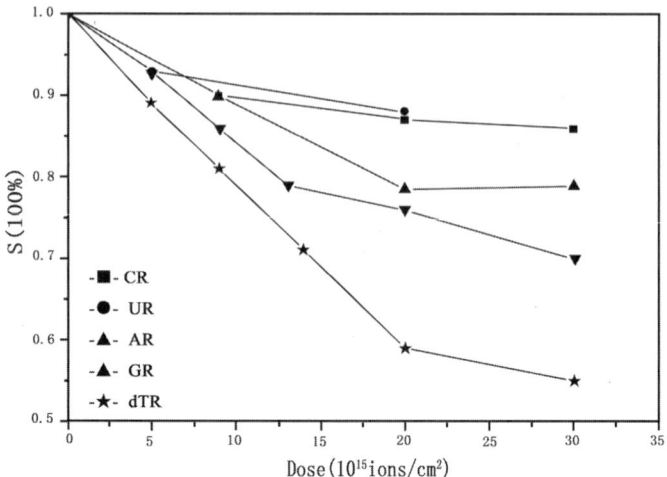

Figure 5.1. Survival rates, S, of five nucleosides after ion implantation as a function of ion dose.

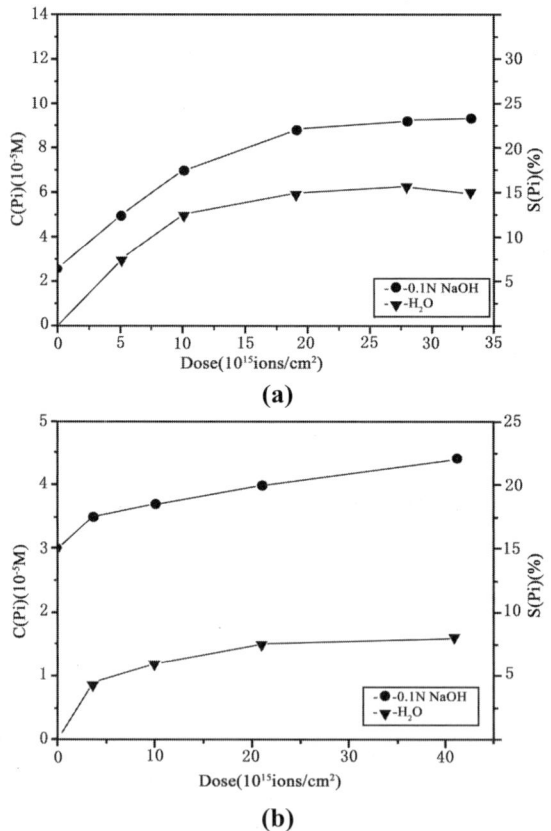

Figure 5.2. Effect of alkali treatment on dephosphorus of ion-implanted 5'-dTMP. (a) As-treated, and (b) alkali treated for 40 min.

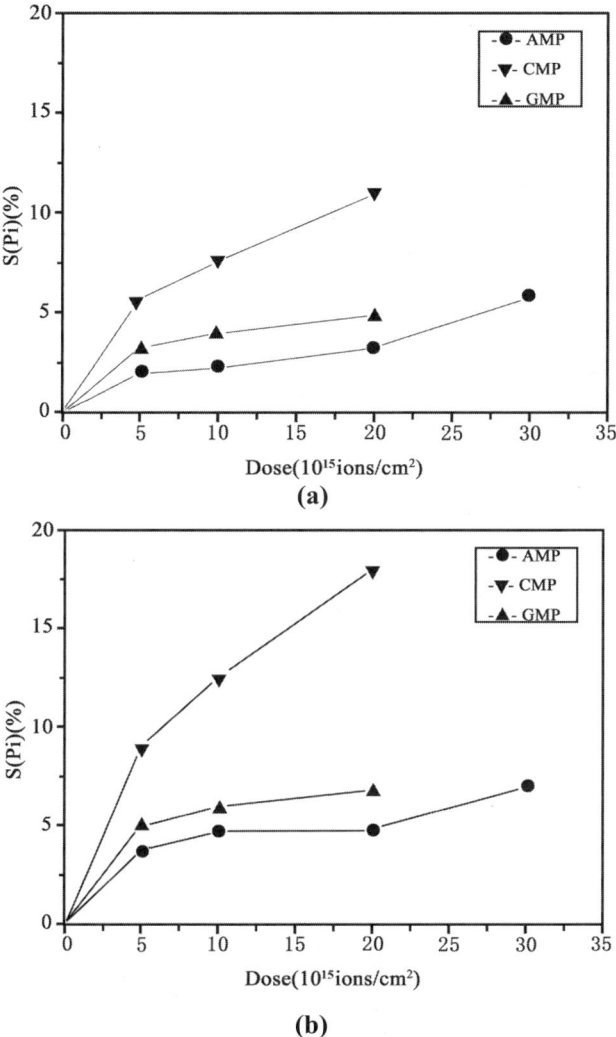

Figure 5.3. Effect of heat treatment on dephosporization of ion-implanted AMP, CMP and GMP. (a) Before heat treatment, and (b) after heat treatment.

For nucleotide 5'-dTMP under ion implantation, breaking of the glycosidic bond releases bases and the quantity of base released increases with increasing dose (Figure 5.4). But alkali treatment decreases the base concentration, probably because the bases in the nucleotide molecules that contain the damage spots are broken by treatment with strong alkali.

It can been seen by comparing Figure 5.2 with Figure 5.4 that the concentration of free bases is higher than that of inorganic phosphorus after the ion-implanted 5'-dTMP solid dissolves in water. This implies that nucleotide releases bases first and then inorganic phosphorus, as a consequence of ion implantation,

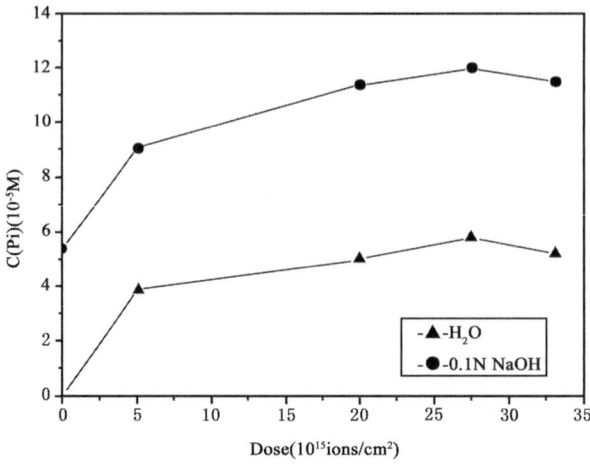

Figure 5.4. Ion implantation induced base release for 5'-dTMP.

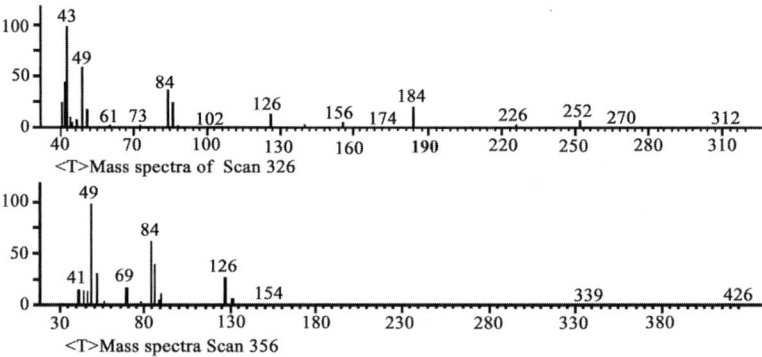

Figure 5.5. Mass spectra of derivatives of two products of ion-implanted glycine.

Amino acid under ion implantation is also dissociated into a series of products. For example, two new products are obtained from ion implanted glycine samples by combined chromato-mass-spectrometric analysis (Figure 5.5). These two products are $HOOC-CH_2CH(NH_2)-COOH$ and $HOOC-CH(NH_2)-CH(NH_2)-COOH$. Their formation proceeds according to the following reactions [8]:

$$^+NH_3 - CH_2 - COO^- + e \ \rightarrow\ NH_3 + \ ^\cdot CH_2COO^-$$

$$^\cdot CH_2COO^- + NH_2 - CH_2COOH \xrightarrow{\ H+\ } HOOC - CH_2 - CH(NH_2) - \ COOH$$

$$^+NH_3 - CH_2 - COO^- + H^+ \ \rightarrow\ H_2^+ + CH(N_3^+) - COO^-$$

$$2\ ^\cdot CH(NH_3^+) - COO^- \rightarrow HOOC - CH(NH_2) - CH(NH_2) - COOH$$

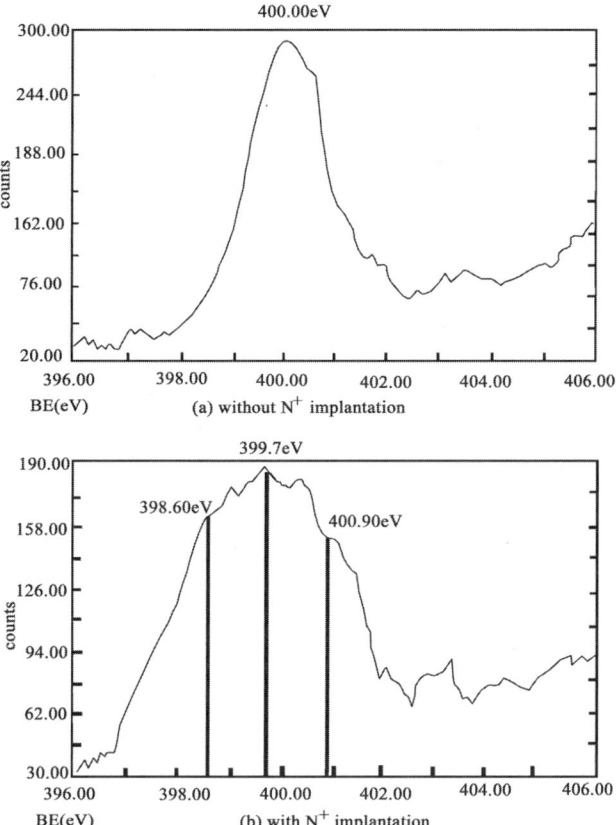

Figure 5.6. XPS spectra of N_{1s} in thymine before (a) and after (b) N-ion implantation.

5.2.2. Compositional Changes

The electron binding energy in the 1s orbit of an atom reflects the surrounding chemical environment of the atom. A tiny change of the binding energy, which can be analyzed by X-ray photo-electron energy spectroscopy (XPS), indicates changes in elements that bind with this element. Figure 5.6 shows the N_{1s}-electron binding energy spectra of thymine before and after ion implantation [9]. As the chemical environments at HN$\langle \genfrac{}{}{0pt}{}{C}{C}$ of the 1 and 3 positions are basically similar, N_{1s} has a single peak and the binding energy is 400eV before the sample is ion implanted. After ion implantation, N_{1s} has three peaks at 398.60 eV, 399.70 eV and 400.90 eV, and the N_{1s} peak area increases by about 30%. This shows that the implanted N^+ ions not only change the surrounding chemical environment of $-N-$, but also increase the relative concentration of nitrogen. Since cascade collisions between the implanted ions and the target atoms displace a large number of atoms, rearrangements and combinations of the displaced atoms, implanted ions and substrate atoms can possibly form groups such as NH_2, $-N_2H_3-$, $-(CH_3)NH-$, $-HCN-$, and $-NO-$. According to the order of electronegativity,

O>N>C>H, and chemical calculations, on the lower binding energy side there should mainly be amino groups, while on the higher binding energy side mainly groups of combined nitrogen and oxygen or carbon. As N_2H_4 is fairly close to –HN–, it is possible to form a combined peak and N_2H_4 on the lower binding energy side. Thus the main peak shifts to lower binding energy by 0.3 eV and the new peak is at 399 eV.

Figure 5.7 shows the XPS spectra of C_{1s} of thymine before and after ion implantation. Both have three peaks, corresponding to the groups C–CH₃, HN–C–C, $O= C\langle{}^{C}_{N}$ or $O= C\langle{}^{N}_{N}$, respectively. The peaks before ion implantation are at 284.50 eV, 286.18 eV and 288.18 eV, respectively, but after ion implantation they occur at 284.48 eV, 286.05 eV, and 288.37 eV, respectively, shifting to the lower or higher binding energy side by -0.02 eV, -0.13 eV and +0.19 eV, respectively. After ion implantation, the relative intensities of the peaks obviously decrease, particularly the last peak which suffers a comparatively greater decrease, indicating that the O=C⟨ structure suffers from heavy damage. The spectrum of O_{1s} also has similar changes.

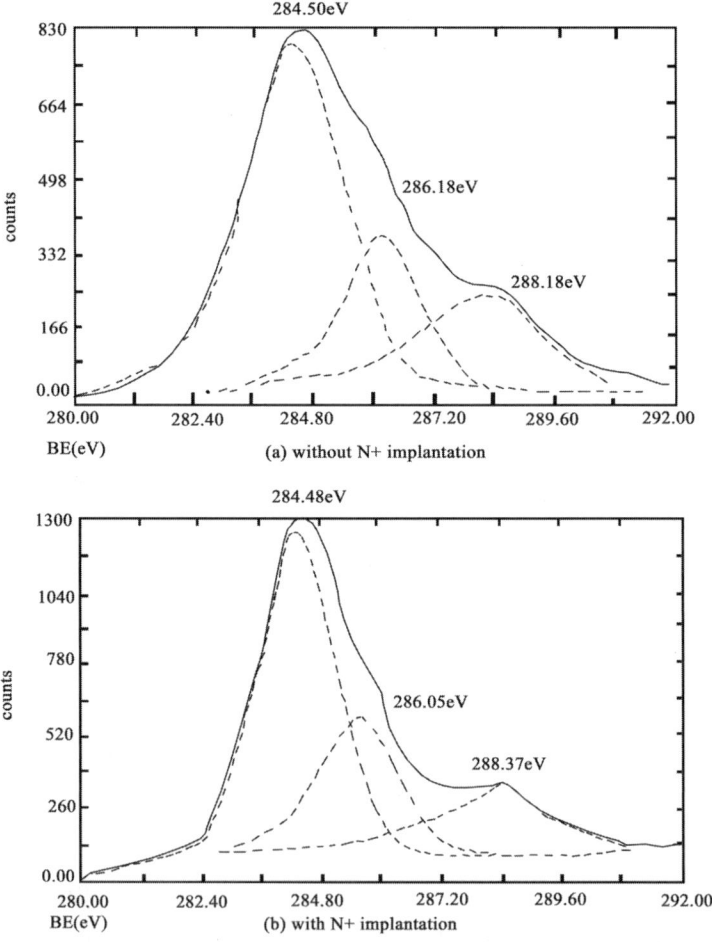

Figure 5.7. XPS spectra of C_{1s} in thymine before (a) and (after (b) N-ion implantation.

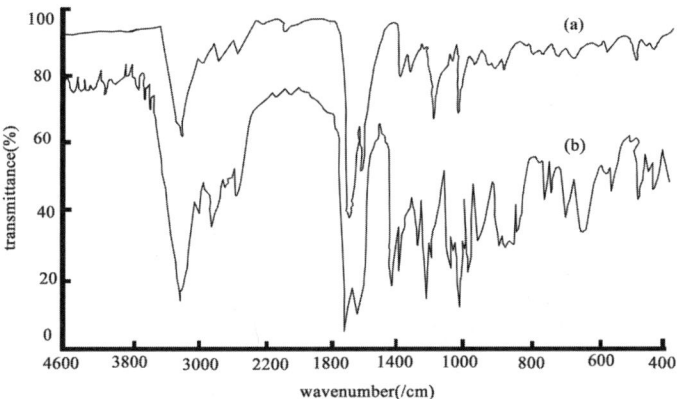

Figure 5.8. FT-IR spectra of thymine before (a) and after (b) N-ion implantation.

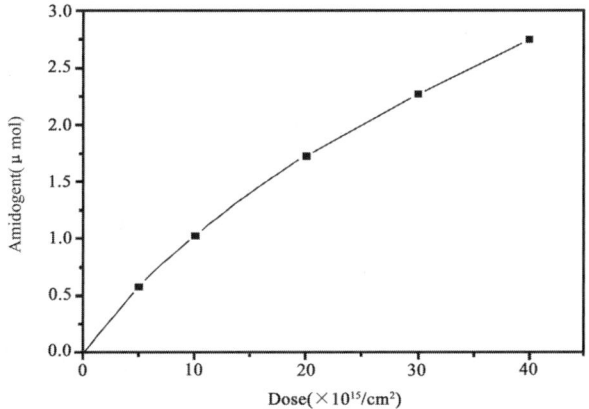

Figure 5.9. Quantity of NH_2 produced from N-ion implanted α-NAA as a function of ion dose.

Fourier-transform infra-red (FT-IR) spectra also show the structure of thymine being damaged due to ion implantation [10]. Figure 5.8 shows the FT-IR spectra of thymine before and after N ion implantation. After implantation the C=O peak at 1,696 cm^{-1} is split into two peaks and a new peak appears at 1,712 cm^{-1}. This demonstrates that ion implantation results in changes in the C=O group. The new peak occurring at 2,360 cm^{-1} after ion implantation implies that the deposition of implanted N^+ ions produces new C=N groups.

Experiments with N-ion implantation of α-naphthalene acetic acid provide an example showing the deposition reaction of implanted ions in bio-molecules [11]. α-naphthalene acetic acid (α-NAA) is a consistent aromatic compound that has a structure similar to that of amino acids and nucleotide, which are the basic units of biological organisms. Its molecule is composed of three elements: C, H and O. Studies have found that 30 keV N ion implantation into solid α-naphthalene acetic acid can form amino groups that do not exist in the original molecules. From measurements using the method of reduction of indene triketo, it is known that the content of amino groups in α-naphthalene acetic acid after N ion implantation increases with increasing ion dose (Figure 5.9) and finally reaches a saturation value.

5.2.3. Production of Free Radicals

Another important kind of damage in ion implanted base and amino crystals is production of free radical molecules [12,13]. The free radicals produced by ion implantation at room temperature from five biological bases that compose DNA and RNA (ribonucleic acid) are all single peaks [12]. Their intensities increase gradually to saturation values with increasing implantation dose (see Chapter 4). In the solid state, ion-implantation produced free radicals are localized and the diffusion distance is very short. However, in biological organisms, due to the participation of water in life activities, the diffusion of free radicals and the attack of new radicals produced from the reaction with water on neighboring molecules are important, and damage the organisms.

5.3. ION DEPOSITION REACTION PROCESSES

From ultraviolet measurement of 30-keV N-ion implanted solid tyrosine (Tyr), it was found that before and after ion implantation the maximum wavelengths (λ_{max}) of ultraviolet absorption of the samples were unchanged, both at 274 ± 1 nm [14,15]. But for different doses, the specific absorption intensities changed (Figure 5.10). It should be expected that due to ion implantation damage, the number of Tyr molecules decreases logarithmically with increasing dose, and thus the ultraviolet absorption of Tyr decreases with increasing dose. But the fact is that the absorption value is initially small, then grows, and finally decreases again as the implantation dose is increased. This phenomenon implies that N-ion-implanted Tyr samples might produce a new substance which has a λ_{max} also at about 274 nm but with an extinction coefficient greater than that of Tyr.

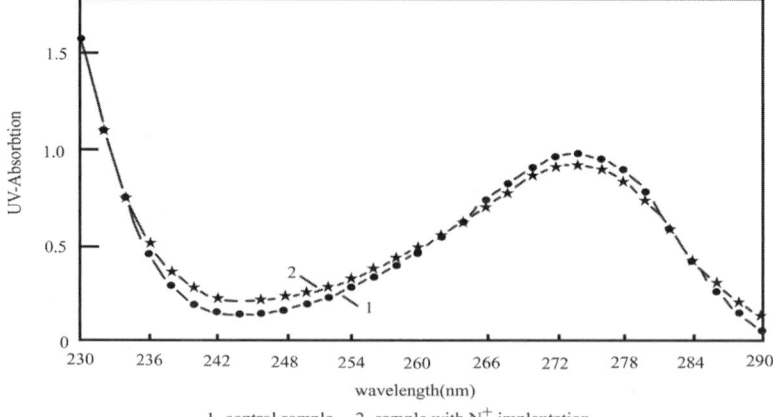

1. control sample   2. sample with N$^+$ implantation

Figure 5.10. UV spectra of Tyr before and after N-ion implantation.

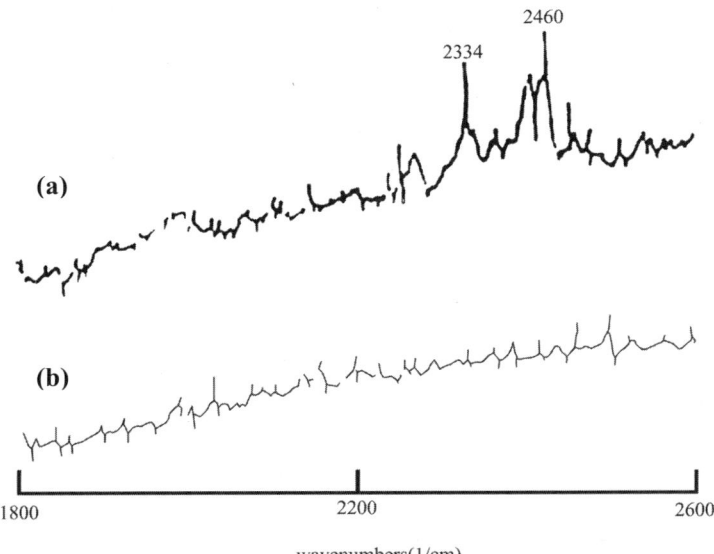

Figure 5.11. Raman spectrum of the N-ion-implanted Tyr sample (a) compared with that of the control (b).

Some new weak absorption peaks in the KBr-tablet laser Raman spectrum of the ion-implanted Tyr sample appear (Figure 5.11). These peaks can be attributed to the nitrogen heterocyclic V_{NH} vibration, but the Tyr molecule does not contain this structure. Hence they must be caused by the reaction between the implanted N ions and the benzene ring in Tyr [14]. Then, which atom or group in the benzene ring can most easily react with the implanted ion? Answers to questions such as this can be determined by quantum biology.

Quantum biology uses quantum mechanics principles to study the electron structure, energy transfer and chemical reaction of biological molecules through mathematical computation [16]. The principal equation of the electron wave function of a bio-molecule described by quantum biology is the Schrödinger equation:

$$H\Psi = E\Psi, \tag{5.3.1}$$

where H is the Hamiltonian and Ψ is the orbital wave function of the molecule. A linear composition of atomic orbitals (LCAO) can be used to express the molecular orbital wave function as Ψ_i:

$$\Psi_i = \Sigma C_{iu} \Phi_u. \tag{5.3.2}$$

For the aromatic amino acids, electron structure can be calculated using the Huckle molecular orbital (HMO) method. This method applies the Born-Oppenheimer adiabatic approximation, orbital approximation (neglecting the interaction between electrons) and LCAO approximation. The method additionally separately considers the π electron and the σ electron and simultaneously assumes that the σ electron is localized whereas the π electrons move in the framework formed by the nucleus and the σ bonds. Thus, the Schrödinger equation to a very good approximation becomes related only to the π electrons:

$$H_\pi \Psi_\pi = E_\pi \Psi_\pi. \tag{5.3.3}$$

H_π can be approximately expressed as

$$H_\pi = \sum_{i=1}^{n} H_i^{\text{eff}}, \tag{5.3.4}$$

where $H_i^{\text{eff}} = -\frac{1}{2}\nabla_i^2 - \sum_{u=1}^{n} \frac{Z_u{}'}{r_{iu}}$ is the single electron Hamiltonian, where $Z_u{}'$ is the effective nuclear charge number which takes into account the mean screening effect from all the σ electrons and π electrons.

If we define $\Psi = \prod_{i=l}^{n} \Psi_i$, and $E = \sum_{i=1}^{n} \varepsilon_i$, equation (5.3.3) then becomes

$$H_i^{\text{eff}} \Psi_i = \varepsilon_i \Psi_i. \tag{5.3.5}$$

According to linear variational calculus, combining equation (5.3.2) we obtain the Hartree-Fock-Roothaan equation:

$$\sum_v (H_{uv}^{\text{eff}} - \varepsilon_i S_{uv})C_{iv} = 0, \quad u = 0, 1, 2, \dots n . \tag{5.3.6}$$

From this equation, the secular equation is obtained

$$D | H_{uv}^{\text{eff}} - S_{uv}\varepsilon_i | = 0. \tag{5.3.7}$$

The following approximations are used for the conjugate alkanes:

(1) The Coulomb integral of every carbon atom in the molecule is the same as α;
(2) The exchange integral for every pair of bonded atoms is the same as β, but for unbonded atoms is 0;
(3) Overlap integrals among the atomic orbits are all 0.
That is
$H_{uv}^{\text{eff}} = \alpha$
$H_{uv}^{\text{eff}} = \beta$ (bonded u and v)
$H_{uv}^{\text{eff}} = 0$ (unbonded u and v)
$S_{uv} = \delta_{uv}$ (when $u = v$, $\delta = 1$; when $u \neq v$, $\delta = 0$).
From these relationships the coefficients C_{iu} can be obtained

$$(\alpha - \varepsilon_i) C_{iu} + \sum_v \beta_{uv} C_{iv} = 0, \quad u = 1,2,\dots,n. \tag{5.3.8}$$

In the above equations, $\beta_{uv} = \beta$ (u and v are bonded), or $\beta_{uv} = 0$ (u and v are not bonded).

For the tyrosine molecule, its conjugate structure is

This conjugate structure contains a heteroatom O. The Coulomb integral α_x of the impurity atom and the exchange integral β_{cx} of the C–heteroatom bond must be considered separately. Generally, the following form is used

$$\alpha_x = \alpha + \delta\beta, \quad \beta_{cx} = \eta\beta. \tag{5.3.9}$$

For the structure $=\overset{|}{C}-\ddot{O}-$, it is assumed that $\delta = 2$, $\eta = 0.9$. If we define $\alpha - E = \beta X$, a set of linear equations among the coefficients C_{iu} can be obtained as

$$
\begin{aligned}
C_1 X + C_2 + C_6 &= 0, \\
C_1 + C_2 X + C_3 &= 0, \\
C_2 + C_3 X + C_4 &= 0, \\
C_3 + C_4 X + C_5 + 0.9 C_7 &= 0, \\
C_4 + C_5 X + C_6 &= 0, \\
C_5 + C_6 X + C_1 &= 0, \\
0.9 C_4 + (X + 2) C_7 &= 0.
\end{aligned} \tag{5.3.10}
$$

The X values can be derived from the above equations' secular determinant that is set equal to 0. Use of these X values in the coefficient equations results in all values of C_{iu}:

C_{iu}	X_u						
	-2.353	-1.799	-1.0	-0.792	1.0	1.088	2.037
C_{1u}	0.108	0.481	0.0	0.523	0.0	0.578	0.383
C_{2u}	0.137	0.432	0.5	0.208	0.5	-0.315	-0.390
C_{3u}	0.239	0.297	-0.5	-0.361	0.5	-0.236	0.412
C_{4u}	0.468	0.102	0.0	-0.493	0.0	0.573	-0.448
C_{5u}	0.239	0.297	0.5	-0.361	-0.5	-0.236	0.412
C_{6u}	0.137	0.432	-0.5	0.208	-0.5	-0.315	-0.390
C_{7u}	0.789	-0.456	0.0	0.367	0.0	0.160	0.100

From these values, all molecular orbital wave functions ψ_i can be derived. Among these functions, $\psi_1 \sim \psi_4$ are the occupied orbitals, and $\psi_5 \sim \psi_7$ are the empty orbitals. Therefore, some quantum biological indices of the conjugate structure of the Tyr molecule can be found.

(a) Electron density

The electron density at r in an atom is

$$q_r = \Sigma N_i C_{ir}^2, \tag{5.3.11}$$

where N_i can be 0, 1, 2. Normally, for the ground state and the occupied orbitals, $N_i = 2$, so

$$q_r = 2 \sum_i^{occ} C_{ir}^2. \qquad (5.3.12)$$

The greater the charge density is, the easier for the electrophilic reaction to occur; whereas smaller q_r improves the chances of attack by a nucleophilic reagent.
(b) Bond order
The total bond order between two atoms, r and s, is defined as

$$P_{rs} = 2 \sum_i^{occ} C_{ir} C_{is}. \qquad (5.3.13)$$

The bond order value is considered to be correlated to the bonding ability of the bond. The greater the bond order is, the greater the bond strength and the shorter the bond length.
(c) Free valence
The free valence is the difference between the maximum bonding intensity N_{max} and the real bonding intensity N_r of an atom:

$$F_r = N_{max} - N_r = N_{max} - \sum_s P_{rs}. \qquad (5.3.14)$$

Here, for the π bond order, $N_{max} = \sqrt{3}$ (for C atoms), or $N_{max} = 1$ (for O atoms). Generally, the greater F_r value of the atom, the greater the residual bonding ability, and thus a higher the reaction activity expected.
(d) Electrophilic superdelocalizability
The electrophilic superdelocalizability is expressed as

$$S_r^E = 2 \sum_i^{occ} \frac{(C_{ri})^2}{\lambda_i}, \qquad (5.3.15)$$

where λ_i is the wavelength corresponding to energy $\varepsilon_i = \alpha + \lambda_i \beta$. The electrophilic superdelocalizability is not only a measurement of the reaction difficulty inside the molecule but also an index for comparing the reactivity of molecules.

After calculation, the quantum biological indices of the Tyr molecule can be expressed graphically as follows [16]:

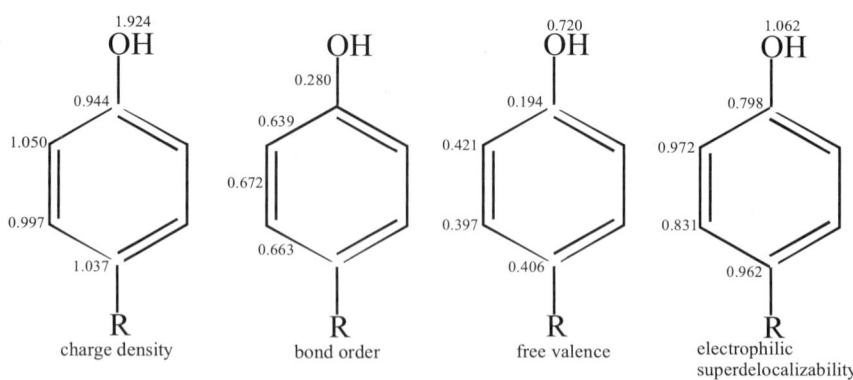

charge density   bond order   free valence   electrophilic superdelocalizability

It can clearly be seen from the above that in the benzene ring of the Tyr molecule, the charge density on the carbon atoms depends on the position of the carbon atom – the C_5 or C_3 atom has the highest charge density, the greatest free valence and electrophilic superdelocalizability, indicating that the implanted N^+ reacts easily with these atoms. However, C_3-C_4 or C_4-C_5 has the smallest bond order, indicating that the implanted N^+, when slowed down, can have a displacement reaction with the $=C_5$— or $=C_3$— radical to form new nitrogenous heterocycles. After the benzene ring is azacyclized, the ultra-violet absorption of the sample will increase greatly, but the specific wavelength changes little. This agrees with the results shown in Figure 5.11.

It is found by measurements using Fourier-transform infra-red spectroscopy (FT-IR) that after N-ion implantation of Tyr samples, a new medium-strong peak appears at 1.728 cm^{-1} [14] (Figure 5.12). This peak belongs to the C=O group. Furthermore, measurement of the ion implanted sample using mass spectroscopy found changes in the mass spectrum. In the new mass spectrum, there is not only a peak at $m/e = 181$ which reflects the Tyr molecular mass, but also some new peaks such as $m/e = 167$, 182, etc. [14] (Figure 5.13), which belong to new substances. According to the "nitrogen rule", the new substance molecule should contain an even number of nitrogen atoms. Since tyrosine originally contains one nitrogen atom, at least one nitrogen atom in the new substance must come from the implanted ions. It can be inferred that N ion implantation of Tyr molecules can produce a new carbonyl-containing azacyclic compound that has a molecular mass of 182. A proposed mechanism is as follows [14]:

$$\text{HO}-\overset{\displaystyle \nearrow\text{CH}-\text{CH}\searrow}{\underset{\displaystyle \searrow\text{CH}=\text{CH}\nearrow}{\text{C}}}\text{C}-\text{R} \xrightarrow{-\text{H}\cdot} \dot{\text{O}}-\overset{\displaystyle \nearrow\text{CH}-\text{CH}\searrow}{\underset{\displaystyle \searrow\text{CH}=\text{CH}\nearrow}{\text{C}}}\text{C}-\text{R}, \quad (5.3.16)$$

$$\longrightarrow \text{O}=\dot{\text{C}}\overset{\displaystyle \nearrow\text{CH}-\text{CH}\searrow}{\underset{\displaystyle \searrow\text{CH}=\text{CH}\nearrow}{}}\text{C}-\text{R}, \quad (5.3.17)$$

$$\xrightarrow[-\text{CH}]{+\text{N}^+} \text{O}=\text{C}\overset{\displaystyle \nearrow\dot{\text{N}}-\text{CH}\searrow}{\underset{\displaystyle \searrow\text{CH}=\text{CH}\nearrow}{}}\text{C}-\text{R}, \quad (5.3.18)$$

$$\xrightarrow{+\text{H}\cdot} \text{O}=\text{C}\overset{\displaystyle \nearrow\text{NH}-\text{CH}\searrow}{\underset{\displaystyle \searrow\text{CH}=\text{CH}\nearrow}{}}\text{C}-\text{R}. \quad (5.3.19)$$

The intermediate products of the reactions from (5.3.17) to (5.3.18), namely, the product with molecular mass 167 from the product of reaction (5.3.17) after the —CH— group is removed, and the product from reaction (5.3.19) with molecular mass of 182, are both shown in the mass spectrum. It should be noted that the actual reaction process may be much more complicated than the above description.

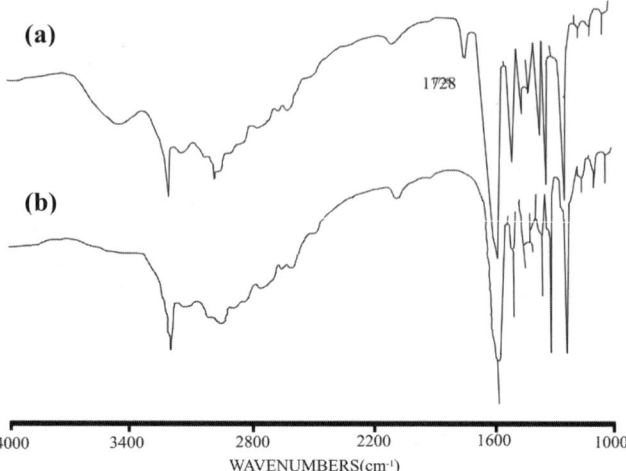

Figure 5.12. IR spectrum of the N-ion implanted Tyr sample (a), compared with that of the unimplanted control (b).

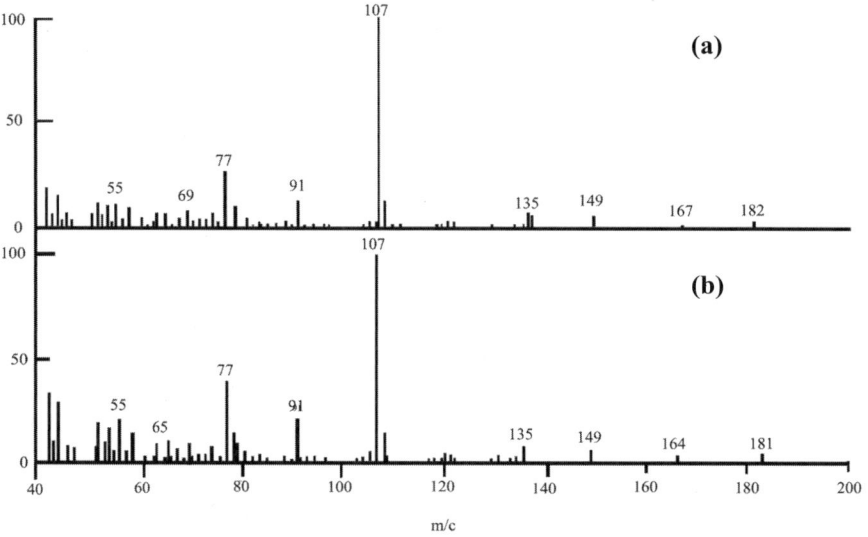

Figure 5.13. Mass spectrum of the N-ion implanted Tyr sample (a), compared with that of the unimplanted control (b).

5.4. PHYSICOCHEMICAL REPAIR

Damage to high level living structures can be restored to normal by repairing the DNA (see Chapter 6). The time needed for this restoration is called the biological relaxation time. Organisms may be disturbed by physical and chemical factors. The repair takes place before biological repairing occurs can be called physicochemical repair or primary repair. How to calculate physicochemical repairing (now without the biological repairing process) effects of the ion implanted biological small molecules is the topic to be discussed in this section.

It is assumed that when a volume element absorbs 100 implanted ions, the number of chemical changes, H, occurring due to reactions between implanted ions or secondary products (such as displaced atoms and free radicals) and particles in the reaction system is

$$H = 100 \times n/N, \tag{5.4.1}$$

where n is the number of chemical changes occurring in the reaction system due to the action of implanted ions in a certain time period. For example, some groups form or molecules dissociate, and so on. In Eq. (5.4.1), N is the total number of implanted ions absorbed by the reaction system. Thus the H value obtained is a mean over this time period. At every moment we have

$$h(t) = 100 \times \mathrm{d}n(t)/\mathrm{d}N(t) . \tag{5.4.2}$$

The mean H is

$$\overline{H} = \frac{1}{t_0} \int_0^{t_0} h(t)\mathrm{d}t. \tag{5.4.3}$$

The initial H can be expressed as

$$H = 100 \times \lim_{D \to 0} (\mathrm{d}n/\mathrm{d}D), \tag{5.4.4}$$

where D is the number of implanted ions absorbed by the volume element. In experiments, the repairing effect can be estimated if the H value is known.

Let us consider deaminization induced by ion implantation of threonine as an example (Figure 5.14). It is natural that N ion implantation into threonine leads to aminolysis. The reverse process also possibly exists. In this process, the deposited N^+ ions, displaced atoms, and atoms in the system are re-arranged and combine to form new amino-groups.

Deaminization damage is considered to result from the first collision of the ion. Thus the amino survival rate should follow a logarithm decay law. If the deaminization probability of a threonine molecule is a, the total number of threonine molecules is N_0, and the number of the molecules within the ion effect range is $c \times N_0$, then the number of molecules outside the effect range is $(1 - c)N_0$. Thus, the number of surviving amino molecules in the sample, N_1, is the sum of the number of molecules outside the range and the number of molecules which are not effectively hit by the ions inside the effect range, namely,

$$N_1 = (1 - c)N_0 + cN_0 \exp(- aD). \tag{5.4.5}$$

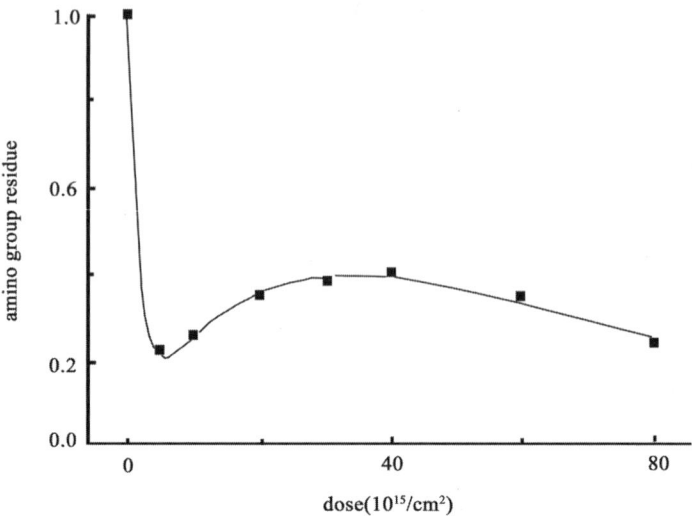

Figure 5.14. Survival of amino group of N-ion implanted threonine as a function of ion dose.

If the newly formed amino-groups are products from the first effective collision (the first collision that can form the amino group), the number of formed amino molecules having amino groups, N_2, is given by the Poisson equation as

$$N_2 = cbDN_0 \exp(- bD), \qquad (5.4.6)$$

where b is the probability of forming new amino-groups due to an ion interacting with each threonine molecule. Therefore the number of molecules actually surviving, N, is

$$N = N_1 + N_2 = (1 - c)N_0 + cN_0 \exp(-aD) + cbDN_0 \exp(-bD). \qquad (5.4.7)$$

Then the survival rate of the amino-groups, S, is

$$S = N/N_0 = (1 - c) + c \exp(-aD) + cbD \exp(-bD). \qquad (5.4.8)$$

Based on the experimental curve in Figure 5.14 [17], the coefficients can be derived using a least squares fit: $a = 0.537 \times 10^{-15}$, $b = 0.0298 \times 10^{-15}$, and $c = 0.96$.

Since the number of threonine molecules in the sample is $5.364 \times 10^{17}/cm^2$ and the number of amino formed is given by eq. (5.4.6), the initial H value is

$$H = 100 \times \lim_{D \to 0}\{d\,[N_0\,cbD\,\exp(-bD)]/dD\}$$

$$= 100\,N_0\,cb$$
$$= 100 \times 5.364 \times 10^{17} \times 0.96 \times 0.0298 \times 10^{-15}$$
$$= 1534.5.$$

This shows that in the primary stage, the implantation of a nitrogen ion into solid threonine forms more than 15 amino-groups. This effect comes from recombination in the ion implantation process, and not only from deposition of implanted ions. An implanted N ion can form one amino-group at most. It is estimated from Figure 5.9 that N ion implantation into α-naphthalene acid forms amino-groups with a maximum H of 0.88. After subtraction of the portion of the implanted-N-ions formed amino-groups, the amino-groups formed from the reaction and recombination of free radicals, displaced atoms, and substrate atoms occupy the overwhelming majority of the newly formed amino-groups. It was pointed out in the previous chapter that implantation of a heavy ion at several tens of keV can result in several hundreds of atoms being displaced, and it is quite possible to form more than ten new groups by recombination.

Now let us differentiate Eq. (5.4.8) and put $dS/dD = 0$:

$$dS/dD = cb \exp(-bD)(1 - bD) - ac \exp(-aD) = 0. \qquad (5.4.9)$$

When $aD \gg 1$, $D_0 = 1/b$. This indicates that the dose at the zero derivative of the curve in Figure 5.14 determines the parameter b:

$$b = 1/D_0.$$

In the experiment shown in Figure 5.14, the dose at $dS/dD = 0$ is $D_0 = 40 \times 10^{15}$ N^+/cm^2, and thus

$$b = 1/(40 \times 10^{15}) = 2.5 \times 10^{-17}. \qquad (5.4.10)$$

The result calculated here is similar to the b value obtained from a least squares fitting (2.98×10^{-17}). According to definition, b is the probability of a certain effect (such as the loss of an amino group or the repair of an amino group) from one implanted ion to each molecule in the reaction system. Therefore b can also be designated as the physicochemical repairing coefficient.

REFERENCES

1. Yu, Z.L., Qiu, L.J., and Huo, Y.P., Progress in Studies of Biological Effect and Crop Breeding Induced by Ion Implantation, *J. Anhui. Agric. College* (in Chinese with an English abstract), **18**(1991)251-257.
2. Yu, Z.L., Shao, C.L., Dose Effect of the Tyrosine Sample Implanted by a Low Energy N^+ Ion Beam. *Radiat. Phys. Chem.*, **43**(4)(1994)349-351.
3. Shao, C.L., Yu, Z.L., Research into Releasing Inorganic Phosphate and Base from 5'-dTMP Irradiated by a Low Energy Ion Beam, *Radiat. Phys. Chem.*, **44**(6)(1994)651-654.
4. Shao, C.L., Wang, X.Q., and Yu, Z.L., Phosphate Release from N^+ Ions Irradiated 5'-CMP Nucleotide and Its Kinetics, *Radiat. Phys. Chem.*, **50**(6)(1997)561-565.
5. Shao, C.L., Yu, Z.L., Dose Effects of N^+ Ion Beam Irradiation-induced Damage to 5'-AMP and Its Components, *Radiat. Phys. Chem.*, **49**(3)(1997)337-345.

6. Shao, C.L., Yu, Z.L., Studies on Mass Deposition Effect and Energy Effect of Molecules Implanted by N^+ Ion Beam, *China Nuclear Science & Technology Report*, 1994, CNIC-00851, ASIPP-0042.

7. Shao, C.L., *Quantitative Study of Low-energy Ion Beam Biology*, Ph.D. dissertation, Institute of Plasma Physics, Chinese Academy of Sciences, June 1995.

8. Huang, W.D., Wang, X.Q., Yu, Z.L., and Zhang, Y.H., Research into keV N^+ Irradiated Glycine, *Nucl. Instr. Meth.*, B**140**(1998)373-379.

9. Yu, Z.L., Studying of Ion Implantation Effect on the Biology in China, *China Nuclear Science & Technology Report*, 1993, CNIC-00746, ASIPP-0036.

10. Yu, Z.L., Deng, J.G., He, J.J., and Huo, Y.P., Mutation Breeding by Ion Implantation, *Nucl. Instr. Meth.*, B**59/60**(1991)705-708.

11. Huang, W.D., Yu, Z.L., Jiang, X.Y., and Yang, X.H., Study on Mass and Energy Deposit Effects of Ion Implantation of NAA, *Radiat. Phys. Chem.* **48**(3)(1996)319-323.

12. Deng, J.G., Yu, Z.L., Qiu, L.J., ESR Study on Ion-beam Irradiated DNA and Its Bases, *Nucl. Techni.*, **15**(10)(1992)600-605.

13. Huang, W.D., Han, J.W., Wang, X.Q., Yu, Z.L., and Zhang, Y.H., KeV-Ion Irradiation of Solid Glycine: an EPR Study, *Nucl. Instr. Meth.*, B**40**(1998)137-142.

14. Shao, C.L. and Yu, Z.L., Mass Deposition in Tyrosine Irradiation by a N^+ Ion Beam, *Radiat. Phys. Chem.*, **50**(6)(1997)595-599.

15. Yu, Z.L. and Shao, C.L., Dose-effect of the Tyrosine Sample Implanted by a Low Energy N^+ Ion Beam, *Radiat. Phys. Chem.*, **43**(4)(1994)349-351.

16. Liu, C.Q., *et al.*, *Introduction to Quantum Biology* (Science Press, Beijing, 1989).

17. Huang, W.D. and Yu, Z.L., The Preliminary Study on Low Energy Ion Irradiation of Threonine, *Acta Biophysica Sinica,* **13**(2)(1997)250-254.

6

DAMAGE AND REPAIR OF ION-IMPLANTED DNA

Completeness of nucleic DNA molecules carrying genetic information is necessary for normal cell proliferation. Breakage of DNA double strands disrupts the completeness of the cell and, thus, is a most important kind of damage to biological organisms. In radiobiology, radiation energy transfer can lead to DNA double-strand breakage, and various biological end-sites are directly related to the DNA damage. Thus DNA is considered a radiation-sensitive location, or key target molecule in ion implantation. In ion implantation an important damaging mode of low energy ions to the nucleotide is dephosphorylation and release of inorganic phosphorus. Another damaging mode is momentum-exchange induced atomic displacement and recombination that can cause more complicated DNA damage. If the damage occurs inside the organism body, mutations can be expected.

The mutation process is important to genetics. In theory it is closely related to understanding of genetic substances as well as the evolution of life forms; in practice, mutation studies form the theoretical basis of modern breeding techniques. Earlier chapters of this book have described the primary processes involved in interactions between low energy ions and biological organisms and molecules, and some important concepts have been introduced such as ion-beam-induced bio-molecular displacement, release and rearrangement. These concepts have important significance to the interpretation of primary mutation mechanisms, and in practice underlie ion-beam-induced rearrangement mutation and gene transfer. From this chapter onwards, the emphasis of our discussion will shift to biological processes, including genetic mutations and chromosomal aberrations.

Genetic mutation is a site mutation. The main task of this chapter is to consider plasmid DNA as a research object with simple biological structure and clear DNA sequences and *E. coli* as a specific DNA host for studying the effects of implanted ions on plasmid DNA activity, single- and double-strand rupture and subsequently mutated gene sequences.

6.1. PLASMID DNA NORMALLY USED IN ION IMPLANTATION

A DNA molecule is composed of two poly-deoxyribonucleotide strands (Figure 6.1). The backbone of each strand is made from a connection between phosphodiester and two nucleoside furan β-D-deoxyriboses via 3' and 5' bonds. Both strands are dextrogyrate. They wind in opposite direction around the same axis, forming a dextrogyrate double helical structure. The helix rise per base pair is 3.4 Å. Every ten nucleotides form a helical turn and the height of each turn is 34 Å. The bases are inside the helix with their planes perpendicular to the helical axis. Phosphates are outside. The average diameter of the helix is 20 Å. The two strands are linked by hydrogen bonds between base pairs [1].

The spatial arrangement of base pairs is remarkably specific. Each adenine residue is paired with a thymine residue (A=T) by two hydrogen bonds, and each guanine residue is paired with a cytosine residue (G≡C) by three hydrogen bonds. Since the bases from two strands are on the same plane, they must be adapted to the phospho-deoxyribose backbone. Thus a base pair must be composed of a purine and a pyrimidine. At the positions opposite to purine must be pyrimidine, and at positions opposite to pyrimidine must be purine. However, note that adenine (A) and cytosine (C) cannot form a base pair, neither can guanine (G) and thymine (T) form a hydrogen-bonded pair. This means that on the position opposite to A must be T and on the position opposite to G must be C.

The total number of the purine bases in a DNA molecule is the same as the number of pyrimidine bases. This is determined by the strict pairing rule between two single strands in the DNA molecule. But on each strand the base types in front and behind are not controlled by any rule. Hence in a DNA molecule containing a large number of bases, four bases are arbitrary and their arrangements are unlimited, and thus the genetic information that can be carried is extremely large. However, the versatile structure is maintained in the replication process only if the DNA molecule suffers no damage

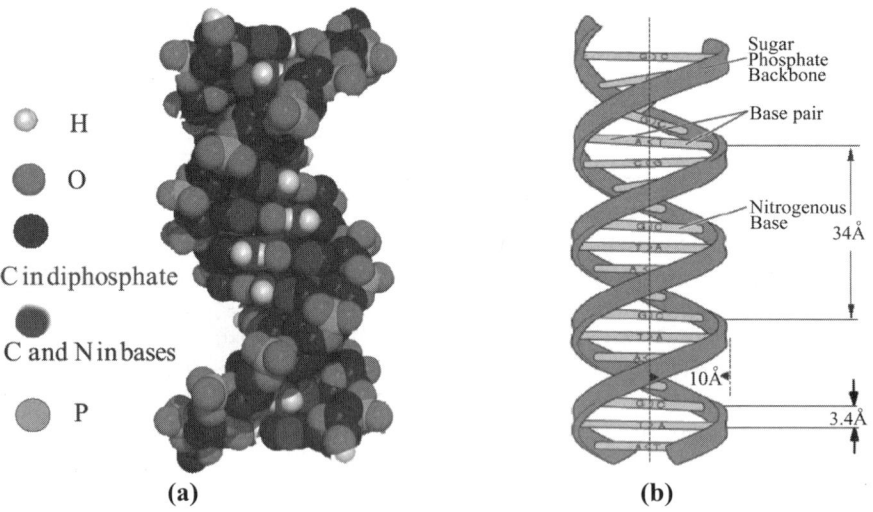

Figure 6.1. The model of DNA double helix (a) and the scheme of the macromolecule (b).

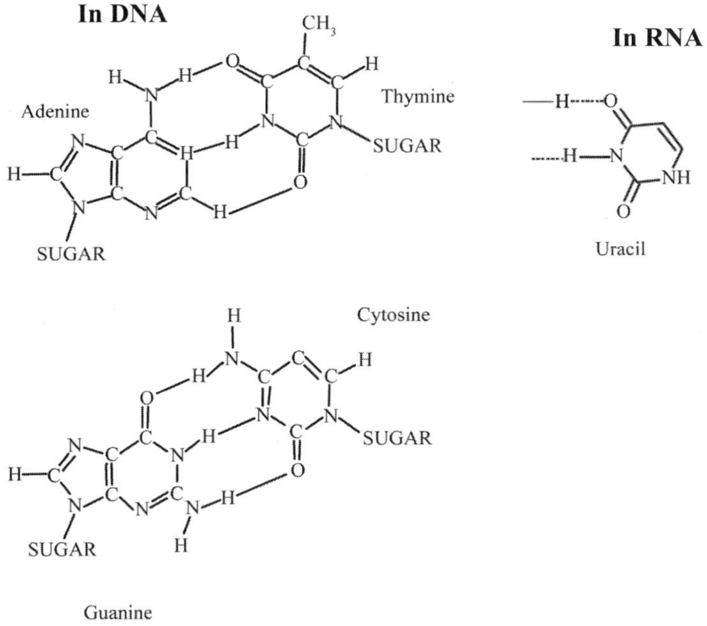

Figure 6.2.   Base pairing in DNA and RNA.

The structure of RNA is different from DNA, first due to uracil (U) replacing T and second due to ribose replacing deoxyribose. Additionally there is another important difference, in that most RNA is present in the form of a single strand, but can be folded to form several double strand domains. In these domains, all complementary bases can form hydrogen bonds. The base pairs in DNA and RNA molecules are shown in Figure 6.2.

Plasmid DNA is a kind of circular DNA molecule that can autonomously replicate in the host cells and is often used as a gene bearer. This nude DNA can suffer from radiation damage under ion implantation. The damaged DNA can then be reintroduced into cells in order to study the effects produced so as to understand the molecular basis of ion beam mutation at the gene level. Several types of commonly-used plasmid DNA are described below.

6.1.1. pBR$_{322}$

pBR$_{322}$ is a double-strand plasmid DNA clone carrier with a length of 4,362 bp (base pairs) and can be used for simple and quick preparation of cloning-recombined DNA segments. It has two antibiotic resistant genes (tetracycline and ampicillin), a replication starting site, and many restriction enzyme cleavage sites used for cloning or subcloning restriction enzyme fragments. When *E. coli* (*Escherichia coli*) that lacks antibiotic resistant genes is successfully transferred with the plasmid, the bacteria derive antibiotic resistance from the plasmid clone carrier. If ion implantation causes aberrations of a resistance gene of this plasmid to result in the bacteria losing this resistance, another antibiotic can be used for screening. Figure 6.3 is the restriction map of plasmid pBR$_{322}$.

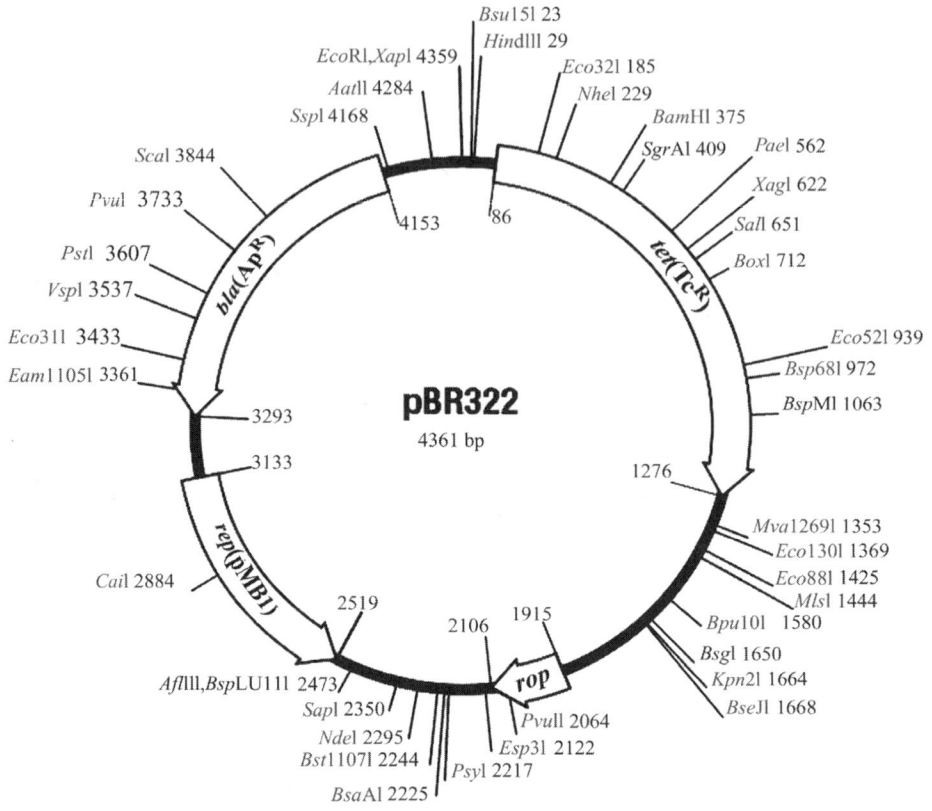

Figure 6.3. Restriction endonucleases sites in plasmid pBR₃₂₂.

6.1.2. M₁₃

M₁₃ is a kind of nemeous male-specific coliphage. Its genome is a 7.2-kb single-stranded circular DNA molecule. The clone carrier of the M₁₃mp series is made from modified wild-type M₁₃ to clone various DNA segments into the phage carrier. At the amino end of the *lacZ* gene, a multiple clone site (called a restriction enzyme cleavage site cluster) is introduced. *lacZ* gene is a structural gene of β-galactosidase in the metabolic process of the coding galactose. When the M₁₃mp phage that does not have inserted segments or induced aberrations infects β-galactosidase defective bacteria gal⁻, such as JM103 and JM109, the host can be complemented to becomes *gal*⁺, which can metabolize a galactose-like chemical, X-gal (5–Br–4Cl–3–indole–β-D–galactoside), to release a blue color. If cloning stops the *lacZ* gene or if the *lacZ* gene has aberrations, after infecting the host, the plasmid cannot metabolize X-gal to release the blue color. Thus white plaques represent carrying integrated segments or phage with aberrated *lacZ* genes, while blue color represents wild-type non-recombinants. Figure 6.4 shows a wild-type M₁₃ phage map.

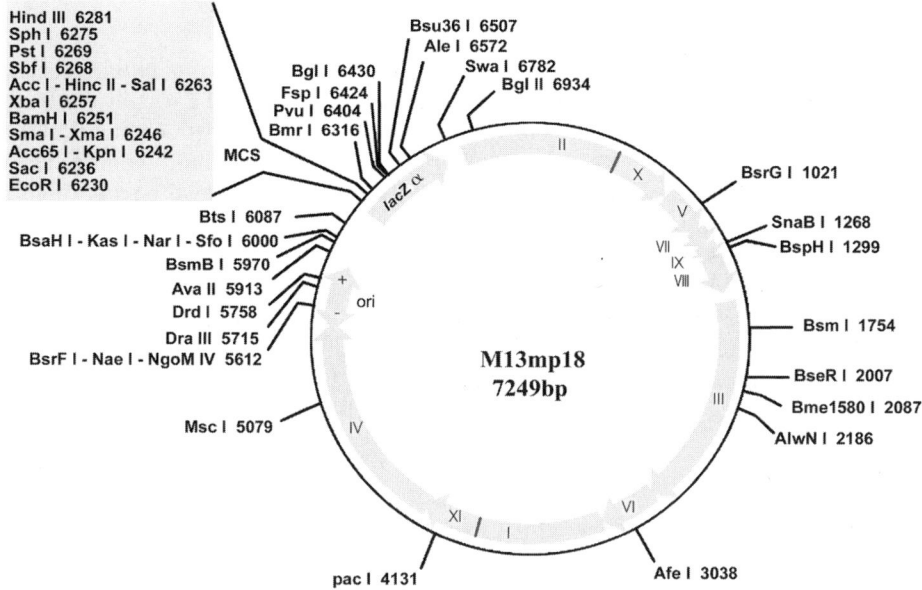

Figure 6.4. Map of $M_{13}mp18$ bacteriophage.

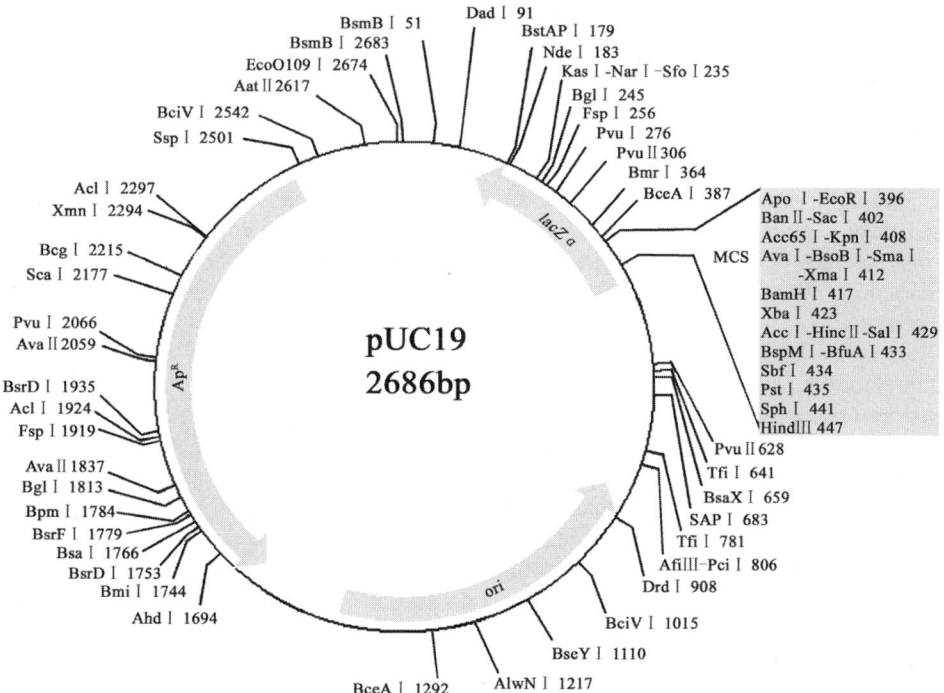

Figure 6.5. Map of pUC_{19}, a representative pUC plasmid.

6.1.3. pUC

The pUC-series plasmid with molecular weight 2.7 kb has the ampicillin resistance gene derived from pBR$_{322}$, the replication starting site of pBR$_{322}$ and a part of the *E. coli lacZ* gene. In the *lac* gene domain there is a restriction enzyme (RE)-recognized single-nick-site multi-joint sequence, which is completely the same as the multi-cloning sites of M$_{13}$. There are many copies of the pUC plasmid. The plasmid can also be amplified by chloramphenicol. When DNA segments are cloned into this pUC region, the *lac* gene is inactivated. When these plasmids are transferred into proper *lac-E. coli* such as JM103 or JM109 and grown in IPTG (isopropyl β-D-thiogalactoside) and X-gal contained culture media, white colonies can be produced. But when pUC that does not contain the inserted segments is transferred into the same host, colonies will be blue. As for M$_{13}$, the pUC carriers are also in pairs, and the sequence relative to the restriction enzyme cleavage sites in the cloning region of the *lacZ* promoter is also opposite. Figure 6.5 shows a map of this plasmid.

6.2. BREAKS OF DNA SINGLE STRANDS AND DOUBLE STRANDS

Helical strands are the basic structure of all regular linear polymers in nature. This spatial configuration provides every individual with the same spatial orientation in its molecules, namely every individual forms the same secondary bond. Otherwise, if some secondary bonds in the structure were stronger than others, instability would occur. At first glance, a DNA molecule is a fairly irregular polymeric strand and seemingly thus cannot form regular helical strands. But most DNA molecules contain two complementarily structural polynucleotide strands. The helical structure is stabilized by secondary internal and external bonds. Two strands are bonded by hydrogen bonds produced from complementary purine and pyrimidine base pairs. This arrangement stacks their flat planes on each other. A strand alone cannot form a regular backbone. Because pyrimidine is smaller than purine, the angle of helical spatial orientation would change with base sequence. This is obviously not a proper arrangement. As every base pair (–purine–pyrimidine) is the same in size, the double helical DNA formed from the complementary base pairs can have a regular structure and thus is also the most stable structure.

Double helical DNA molecules are quite stable at physiological temperatures. To disintegrate a DNA molecule, the hydrogen bonds between the double helical strands must be broken. But this is not to say that the DNA double helical structure is a totally fixed strong structure. In some cases, such as being impacted by energetic ions, DNA molecules and atoms can be displaced and rearranged to cause breaking of the double helical DNA single strand and double strands, despiralization and cross-linking.

The molecular diameter of nude DNA is 20 Å and the stacking distance between base pairs is 3.4 Å. For an ion implantation dose of 10^{15} ions/cm^2, a rough estimate reveals that the number of ions implanted into a 20 Å × 3.4 Å projected area is 6.8. The deposited energy of a 30 keV ion passing through a 20 Å diameter DNA molecule is sufficiently high to cause all of the atoms with which the ion collides along its path in the impacted DNA molecule to be completely displaced. If sputtering is considered and we take a sputtering yield as 1 atom/ion, then 6.8 atoms are sputtered from the projected plane of the DNA in the implantation direction. If the displaced or sputtered atoms, such as P, are on a phosphodiester strand backbone, they will cause single strand breaches. If the same event happens on another strand nearby, a double strand break occurs. If polymerization takes place at two or more than two DNA breaches or damage locations, DNA will be cross-linked to each other. The probability of ion implantation causing breakage of DNA molecule single strand and double strands and cross-linking increases with increasing dose.

6.2.1. Evidence Observed under the Nucleic-Acid Electron Microscope

Ion-implantation-induced breakage of DNA single strands or double strands can be directly observed using an electron microscope particularly suitable for observing nucleic acid chains. In an experiment, pUC_{18} plasmid was extracted using a chemical cleavage method, purified using cesium chloride density-gradient centrifugation, and observed under an electron microscope. It was found that most of the plasmids exhibit supercoiling (SC). When the pUC_{18} plasmid was bombarded with a low dose ($0.5–1.5 \times 10^{15}$ N ions/cm^2) ion beam, most of it was transformed into loosened open-circle (OC) structures (Figure 6.6b). When the ion dose was increased to over 2.5×10^{15} N ions/cm^2, linear molecules could be observed under the electron microscope (Figure 6.6c), as could also transformed DNA (cross-linked). Plasmid pUC is in a double-strand closed state due to a twisting angle between two neighboring nucleotides. Thus the tension caused by the negative supercoil twists the entire molecule to a supercoiling structure. If one of the strands is broken and the hydrogen bonds between two strands are also broken, the tension is released through rotation of the strand, and the DNA configuration becomes an open circle. With increasing ion dose, the open-circle DNA molecule is broken again. If the breaking sites are multiple, linear plasmid or DNA segments can be produced. The linear plasmid, fragmented and damaged DNA may be cross-linked. Hence the ion implantation effects on the pUC plasmid configuration intrinsically reflect the laws of induced DNA atomic displacement, rearrangement and polymerization.

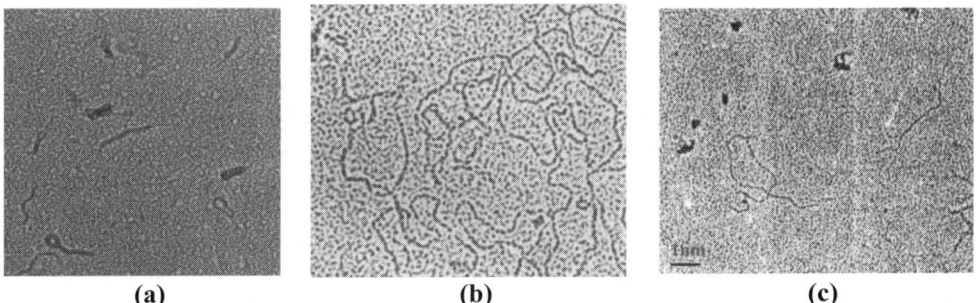

(a)   (b)   (c)

Figure 6.6. TEM photographs of ion implanted plasmid pUC and control.

6.2.2. Electropherogram

Ion implantation breaks the single strand and double strands of isolated DNA. This has been demonstrated by microscopic observation experiment of ion implanted pUC plasmid. It is known experimentally that for a very low dose (e.g., 5×10^{14} N ions/cm^2), ion implantation can break all supercoiled single strands and convert them into open-circle DNA. Normally, even after boiling for a short-time, this kind of small plasmid only has a part of the supercoil transformed in this way. DNA can suffer various cross-linkages due to ion implantation, besides strand breaking, to form products insoluble in TE [Tris-EDTA (mixture solution of trismethylaminomethane and ethylene-diamine-tetra-acetic acid)] buffer. This product stays in the spot holes in electrophoresis. The phenomenon is more noticeable when the plasmid product is mixed with chromosomal DNA. As the alkali method used to separate plasmid firstly transforms the chromosomal DNA of the host bacteria, the chromosomal DNA mixed in the plasmid sample is actually partly annealed DNA (namely, a part of DNA having the same or similar base series which forms double strands again). This part of the double-strand DNA is very sensitive to ion implantation. For low dose ion implantation, electrophoresis bands can completely be converted to tailing bands. There are two reasons for this. The first is that once strands are broken in the domain of some double-strand DNA which is associated with a large number of single strands, the molecular weight will immediately change and thus exhibit weakening or disappearance of the electrophoresis band lightness. The second is that some double-strand DNA suffers cross-linking reaction quite easily and remains in the spot holes during electrophoresis.

The pBR$_{322}$ plasmid DNA in JM109 *E. coli* was implanted with nitrogen ions at 30 keV to a dose of 3.5×10^{15} N ions/cm^2, followed by re-extraction of plasmid. In electrophoresis it was found that two bands, one from open-circle DNA and the other super helical DNA, coexist. But the band intensities were always weaker than that of the control, indicating that implanted ions broke some of the single strands of double-strand DNA. Concerning why ion implantation did not break all of the super helices of pBR$_{322}$ DNA in the bacteria, there may be two reasons. The first one may be that the bases exist at the DNA centers while the double-strand DNA main stem is outside of them. Viewed as an atomic displacement process, for naked DNA the above-mentioned dose is equivalent to removing at least 4 to 30 atoms from the nucleotide main stem so that DNA single strands or double strands can be completely broken. For DNA *in vivo*, this effect takes place with very low probability. Second is that once DNA *in vivo* is damaged by ion implantation, the repairing system is activated to play its role in partly eliminating ion implantation effects of breaking the single and double strands of plasmid DNA.

6.3. REPAIR OF DNA DAMAGE

It used to be thought that whether man-made or natural, DNA structural changes would all result in death or aberrations of organisms. In fact, DNA structural changes may not exhibit aberrations but may be the prelude to a complicated series of induced processes. Only when repair is ineffective or mistakes occur in the repair will aberrations or individual death follow [2]. An aberration process has the following stages: mutagen contacting with DNA $\rightarrow$ premutation $\rightarrow$ mutation $\rightarrow$ occurrence of genotypes. Radiation damage refers to causing changes in the DNA structure and is thus the premutation instead of mutant genotypes. After formation of premutation, DNA recovers to its original normal structure due to the functioning of the organism repair system [3]. Only a few DNA molecules are transformed to the mutant state to bring about the occurrence of genotypes. Therefore, before discussion of how ion implantation induces the process from DNA premutation to mutation, first the DNA damage repair systems must be understood.

6.3.1. Photoreactivation

After ultraviolet irradiation, exposure to visible light can bring about recovery from ultraviolet-induced damage to the organism; this phenomenon is called photoreactivation. The photoreactivation effect was first discovered in microbes. Microbes, after ultraviolet irradiation, were exposed to visible light again; their survival rate increased dramatically and the mutation rate decreased. Progressively more evidence has demonstrated that photoreactivation is a general phenomenon, observed in bacteriophages and higher organisms. Research has demonstrated that an enzyme that can be activated by visible light plays a critical role. This enzyme can combine with the ultraviolet-irradiated DNA in darkness; if the combined product is exposed to visible light, then the enzyme and DNA are dissociated again. In the dissociated DNA molecules, the original DNA damage, such as that caused by the thymine dimmer, can no longer be found (Figure 6.7a).

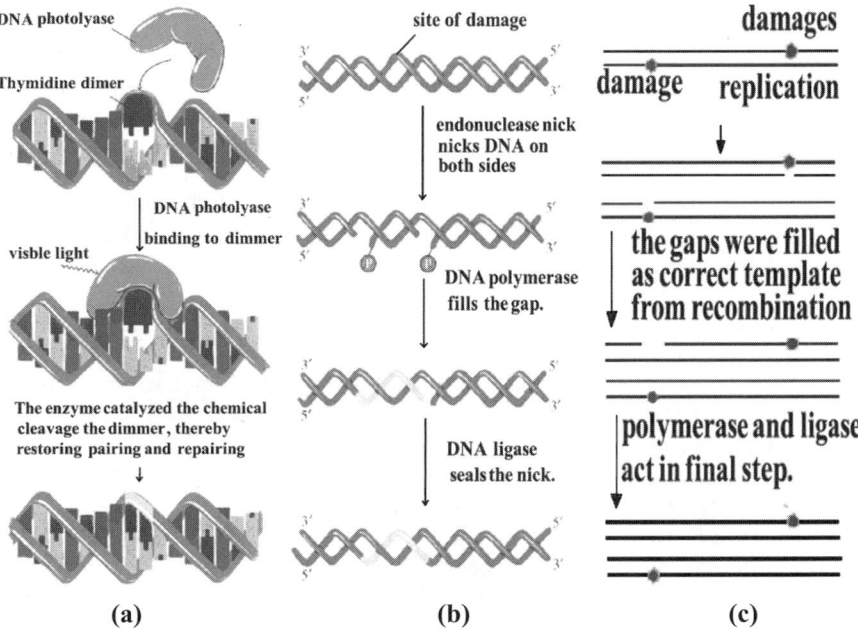

Figure 6.7. Three DNA repair processes. (a) Photoreactivation, (b) excision repair, and (c) recombination repair.

6.3.2. Excision Repair

Excision repair is DNA damage repair under coordination of three enzymes (Figure 6.7b). Initially, at the 5' side of the thymine dimer, nucleate endonuclease breaks the single strand and excises the thymine dimer. Then, under the action of DNA polymerase I, the endonuclease continues to excise a part of nucleotide and carries out repair synthesis. Finally, under the action of a joining enzyme, a complete double-strand structure is formed.

Microbes such as *E. coli* have at least three genes related to dimer excision in excision repair. The excision repairing enzyme system in the wild-type *E. coli* cell can repair not only its own ultraviolet damage but also DNA ultraviolet damage to extraneous bacteriophages (such as T1 and T3, etc.). For example, T1 and T3 are treated by ultraviolet and subsequently infected with *E. coli*, and the results show that a part of the radiation-lethal bacteriophages can be reactivated.

6.3.3. Recombination Repair

Recombination repair relies on genetic recombination of normal and damaged DNA molecules to repair damaged DNA molecules. This process is described as follows.

6.3.3.1. Replication

DNA that contains the thymine dimer or other damage can still be normally replicated. The replication starts from one end, ends at the damage location, and surmounts the damage location to continue replication again. In this way, breaches remain at the DNA daughter-strand and the corresponding damage locations of the mother-strand and thus a discontinuous daughter-strand is formed.

6.3.3.2. Recombination

The complete mother-strand and discontinuous daughter-strand recombine. Due to chromosome exchange, the breach locations are no longer facing the thymine dimers but a normal single strand.

6.3.3.3. Resynthesis

After recombination, due to the action of the DNA polymerase and the joining enzyme, the breach parts are repaired.

In the recombination process, the dimer on DNA is not removed. In the second replication, the damage remaining on the mother-strand cannot make the replication be carried out normally, and the breaches formed are remedied in the same recombination way. After many such replications, although dimers always exist, damaged DNA strands are gradually diluted and their fraction becomes less and less until they are not obstacles to normal physiological processes, and so damage is repaired.

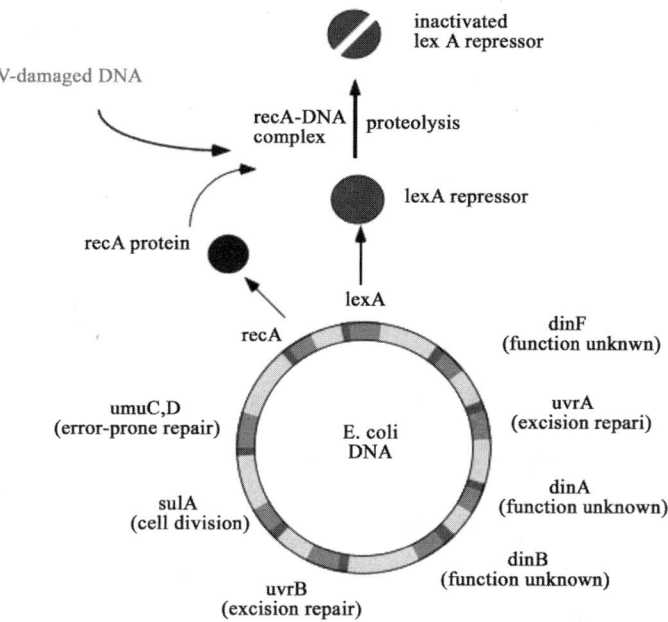

Figure 6.8. Regulation of the SOS repair system in *Escherichia coli*.

6.3.4. SOS ("Save Our Ship") Repair System

There is another repair mechanism existing in *E. coli,* suppressed in the normal physiological state. But once DNA suffers damage, a regulatory signal can be produced to lift the suppression from many genes and thus increase the survival rate of the bacteria. For example, ultraviolet-irradiated bacteriophages are infected with *E. coli*. A part of the bacteriophages may be killed and the surviving part may suffer mutation. If the ultraviolet-irradiated bacteriophages are infected with lightly ultraviolet-irradiated *E. coli*, the number of both surviving and mutated bacteriophages is seen to increase. This demonstrates that in *E. coli* there is an induced system which can repair damage in bacteriophage DNA.

The process of the SOS repair system in *E. coli* is shown in Figure 6.8. In the normal case, *lexA* gene, *recA* gene and other inducible genes are all suppressed by the *lexA* aporepressor. When DNA is damaged, *recA* protein is activated to produce an active protein enzyme and dehydrate the *lexA* aporepressor and also simultaneously lift the suppression of other related genes. These genes release their coded proteins, participate in repair activities, and DNA molecules are successfully replicated.

This repair process occurs due to the induction of damage signals and so is also called induced DNA damage repair. Wrong repairs can easily occur in this process.

6.3.5. Repair of Strand-Breaking and Cross-Linking

Radiation causes DNA single-strand breaking, which can be repaired and rejoined. Rejoining of the single-strand break is catalyzed by DNA polymerase and ligase and supplied with energy by adenosine triphosphate (ATP). Double-strand breaks cause more serious damage to DNA, originally thought to be non-repairable, but it has now been demonstrated to be repairable. Cross-linking can be repaired via excision. The endonuclease first excises one of two strands of DNA at the cross-linked location to leave spaces on this strand to be recombination-repaired, and then excises the cross-linking at the opposite strand in the same way.

6.4. GENE MUTATION

Following ion implantation, the transfection activity of plasmid DNA decreases with increasing implantation dose. For the pUC_{18} plasmid, when the implanted N ion dose was up to 5×10^{14} ions/cm^2, the invasive activity was still maintained at 50%. For this dose, nucleic acid electron microscopy showed that plasmid DNA mainly had single-strand breaks. When the ion dose was higher than 2.5×10^{15} ions/cm^2, the electron microscope revealed linear DNA molecules and double-strand breaks at one or more locations in the plasmid DNA. Now the transfection activity was less than 10%. Thus we see that ion implantation induced structural damage to plasmid DNA is the major reason for decreasing transfection activity.

Damaged plasmid DNA will be repaired in the host cell. Wrong or ineffective repair may produce mutations. This process can be inferred from phenotype variations of the host. For example, ion-implanted pUC_{18} plasmid produces mutations of *lacZ* gene on some plasmids that have the transforming ability to form white colonies from the transformed host [on the isopropylthiogalactoside (IPTG) and X-gal contained plate]. A comparison between ion beam and ^{60}Co-γ-ray treatments shows that the mutation rate of the *lacZ* gene on the pUC_{18} plasmid due to ion implantation is higher than from γ-rays (Table 6-1). According to the concept of mutation efficiency (mutation efficiency = mutation rate / biological damage), the ion-beam mutation efficiency is obviously greater than for γ-rays under most experimental conditions when the inactivation rate of the plasmid is used to indicate the damage degree.

Table 6.1. Comparison of the mutation effects of ion implantation and ^{60}Co-γ-radiation of pUC_{18} on *lacZ* gene.

Ion implantation (N⁺, 30 keV)				^{60}Co-γ-rays			
Dose (10^{14} ions/ cm^2)	Survival rate (%)	Muta- tion rate (%)	Muta- tion efficien- cy (10^{-3})	Dose (GY)	Survival rate (%)	Muta- tion rate (%)	Muta- tion efficien- cy (10^{-3})
5	53	0.1	2.2	10	39	0.06	0.93
10	23	0.62	7.9	20	26	0.17	2.2
15	15	0.24	2.8	30	14	0.25	2.9
25	9	0.29	3.2	40	8	0.26	2.8

There are various types of ion-implantation-induced *lacZ*⁻ gene mutations. Color reactions of the host bacteria examined on IPTG and X-gal contained plate can be basically classified into two types. One type is the no-indigo reaction. Another type is a reaction exhibiting light blue but delayed for 36 hours to one week. Both reactions can be inherited stably. This result indicates that ion implantation not only changes the structural genes but also causes variation of the regulating sequence. The ion-implantation-induced *lacZ*-gene-mutation transformed bacteria are treated for reverse mutation. It has been found that 4% of the bacteria can be reversed by ion beam compared to 1%–2% reversed by nitrous acid or ultraviolet irradiation, and particularly those mutated bacteria reversed by ultraviolet or nitrous acid can all be reversed by an ion beam. This shows that as a mutagen, ion beams are more effective than a single mutation factor such as energy deposition in irradiation mutation and mass deposition in chemical mutation. Ion beam incorporates multiple factors such as energy deposition, mass deposition, momentum exchange, and charge exchange, and is, in a sense, equivalent to a combined mutagen. It has been well demonstrated in radiobiology that the efficiency of combined mutation is greater than that for a single mutation factor.

The behavior of ion implanted M_{13} plasmid DNA has the same tendency as that of the implanted pUC_{18} plasmid DNA. But, at the same dose, the inactivation rate is greater than that of pUC_{18}, and the *lacZ*-gene mutation rate is obviously lower than that of pUC_{18}. For example, when 30-keV N ions are implanted at a dose of 1×10^{15} N ions/cm^2, the M_{13} survival rate is 20%, lower than pUC_{18} by 23% (Table 6-1), and the mutation rate is 0.18%, or less than 1/5 that of pUC_{18}. Since M_{13} is a kind of single-strand circular DNA molecule and under ion implantation the single strand is easily broken and hard to be repaired by rejoining, the survival rate is lower at the same dose. Furthermore, the molecular weight of M_{13} plasmid is 2.7 times that of pUC_{18} and thus *lacZ* gene occupies a smaller portion in the M_{13} molecular volume (target volume) than in pUC_{18}. Therefore, when ions are implanted in the target, the *lacZ*-gene segments are more easily shielded by other segments (such as folded DNA) and thus interact less directly with ions.

6.5. MOLECULAR MECHANISMS OF ION-IMPLANTATION-INDUCED MUTATION OF PLASMID DNA

Experiments have indicated that in most cases, the *lacZ*-gene mutation rate and mutation efficiency of ion-implantation-induced plasmid DNA are greater than that due to ^{60}Co-γ-ray radiation. This is a statistical result obtained from the phenotypic mutations of the host bacteria invaded and infected by the plasmid. But the molecular basis of genetic mutation is change in the DNA structure. It is not enough to describe the characteristics of ion-implantation-induced gene mutation only from phenotypic mutations. The DNA sequences must be analyzed at the molecular level to clarify where and how the changes take place, so that clear information about the molecular mechanisms of genetic mutations induced by ion implantation can be derived.

In one experiment [4], plasmid M_{13}mp18 was ion implanted, transferred into JM103 suspension (50 mmol/l CaCl$_2$) containing host bacteria *E. coli*, water-bathed for 30 min, and heat-shocked at 37°C for 5 min. After transfection was completed, 3 milliliters of pre-molten YT agar culture medium (which also contained 0.8-mg 5-bromo4-chloro-3-indoly-β-D-galactopyranoside, 0.95-mg isoproplyl-β-D-thiagalactopyranoside, and 0.2-

ml JH103 bacterial liquid in OD660 nm = 1) were added, immediately spread on a M9 plate and cultured at 37°C for 24 hours. Ten bacteriophage plaques were selected. After further repeated verification, M_{13}mp18DNA was picked up for sequence analysis (Figure 6.9). The sequence measurement was performed by a DNA auto-sequence analyzer. The reading started from the end 6282 of the polymer connectors on M_{13}mp18DNA, and 250 bases on the *lacZ* segment were read each time. Ten mutation samples at the mutation sites in the range of 250 bases are given in Table 6-2.

Table 6.2.   Mutation ranges and base mutation sites of various mutants.

Sample No.	Mutation range	Codons and corresponding amino acid mutations and occurring sites	
I_{5-1}	Mutations occur beyond the detected range		
I_{5-2}	Occur in the range 2-site base mutation	AGC 6377 AGT Ser35 → Ser	GAC 6316 GGC Aps15 → Gly
I_{7-1}	Occur in the range 3-site base mutation	AGC 6377 AGA Ser35 → Arg GAC 6316 GGC Aps15 → Gly	AGC 6316 AGT Ser73 → Ser
I_{7-2}	Mutations occur beyond the range		
I_{7-3}	Occur in the range 1-site base mutation	GTC 6526 GCC Val85 → Gly	
I_{7-4}	Mutation occur beyond the range		
I_{9-1}	Occur in the range 2-site base mutation	GCC 6454 GGC Ala61 → Gly	GTC 6519 GCC Thr83 → Ala
I_{9-2}	Occur in the range 6-site base mutation	GCG 6475 GAG Ala68 → Glu GAA 6483 AAA Glu71 → Lys CGC 6449 CGG Arg59 → Arg	CCT 6508 CAT Pro79 → His CAT 6360 AAT His30 → Asn GCC 6454 GGC Ala51 → Gly
I_{9-3}	Occur in the range 5-site base mutation	GCA 6355 GTA Ala28 → Val GAC 6317 GAT Asp15 → Asp T deleted (6298)	TGG 6380 TGT Typ36 → Cys TTG 6286 TCG Leu5 → Ser
I_{9-4}	Occur in the range 1-site base mutation	GCC 6454 GGC Ala61 → Gly	

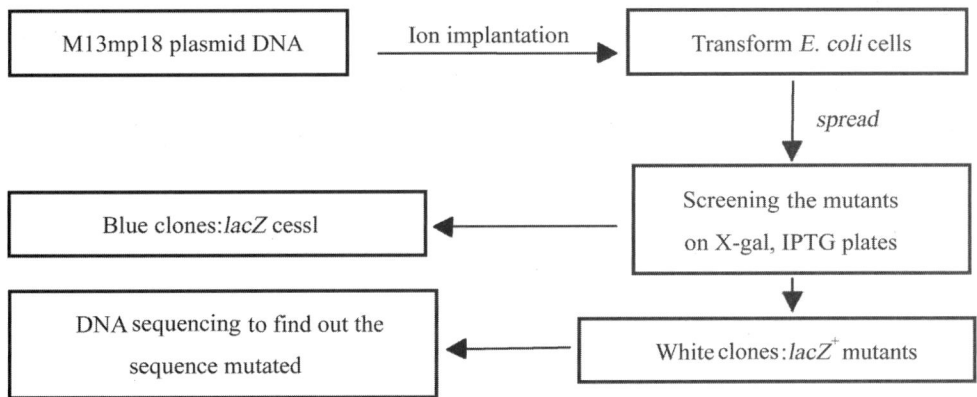

Figure 6.9. Schematic diagram of *lacZ* mutation induced by ion implantation of M13mp18 DNA.

It can be seen from Table 6-2 that among 10 mutation samples, 7 samples have base mutations in the detected base range (from 6282 to 6539), and another 3 samples have no base mutations detected in the measurement range but white bacteriophage plaques indicate occurrence of mutations in the phenotype. This is not strange. Theoretically speaking, for whether *lacZ*-gene segment itself or its corresponding regulating sequences (promotor/operator), any base mutation occurring related to the *lacZ* expression on M_{13}DNA, affecting the regulation of gene expressions or the compositions of the zymoprotein amino acids determined by the structural genes, can be exhibited as $lacZ^-$ mutations. Therefore it can be inferred that the base mutation sites of the three samples are beyond the detected range.

Pronounced characteristics of ion-beam-induced $lacZ^-$ gene mutations are high density and multi-site base mutations. Among 7 samples, 5 have more than two sites where mutation was detected, some samples have more than 6 mutation sites, and on average there occurs one mutation per 40 bases. This phenomenon is very infrequently seen in experiments using conventional ionizing radiation and non-ionizing radiation for direct mutation of DNA. As noted above, ion beam mutation is equivalent to mass-energy-charge combined mutations. In interaction with biological target molecules such as DNA, breakage of single strand and double strands, dropping out of bases, and release of phosphorus and bases can be produced. Implanted active ions and displaced atoms or groups can also be imbedded in DNA molecules to replace lost atoms or groups, causing a variety of base damage or modification to the DNA molecule. This kind of damage or modification of DNA will inevitably have wrong pairs of bases at the damaged locations in the repair process after being introduced into the host. So in the new replication, the base replacement may be maintained. The more kinds of DNA damage or modifications there are, the more mistakes occur in repairing, and the more multi-site and high-density base mutations are caused. In the measurement range, 7 samples have 20 base mutations, averaging 3 mutation sites per sample. It can be expected that for the whole M_{13} plasmid DNA (7.2 kb), each plasmid may have an average of 86 mutation sites following ion implantation.

```
              C                           -T
      A  GC  TTG  GCA  CTG  GCC  GTC  GTT  TTA  CAA  CGT  CGT
      6282                            6300
              G
              GT
      GAC  TGG  GAA  AAC  CCT  GGC  GTT  ACC  CAA  CTT  AAT  CGC
          6320                        6340
                                        A
             -T         A               T         T
      CTT  GCA  GCA  CAT  CCC  CCT  TTC  GCC  AGC  TGG  CGT  AAT
               6360                              6360

         T
      AGC  GAA  GAG  GCC  CGC  ACC  GAT  CGC  CCT  TCC  CAA  CAG
                    6400                                  6420
                                                    G
                                                    G
                                              G     G
      TTG  CGC  AGC  CTG  AAT  GGC  GAA  TGG  CGC  TTT  GCC  TGG
                    6440

                              A                    A
      TTT  CCG  GCA  CCA  GAA  GCG  GTG  CCG  GAA  AGC  TGG  CTG
      6460                        6480

                         A                    G         C
      GAG  TGC  GAT  CTT  CCT  GAG  GCC  GAT  ACT  GTC  GTC
      6500                        6520
```

GTC CCC TCA AAC

Figure 6.10. Mutation sequence map of *lacZ* base. The capital letters above the bases indicate the base mutations. "-" shows deletion.

When the mutated bases are arrayed on the measured *lacZ⁻* sequence (Figure 6.10), it is not difficult to see that ion implanted M_{13} plasmid DNA has a few mutation hot spots. For example, at the 6454 cytosine there occurs C→G transversion in 3 tested samples; and at the 6316 adenine A→G transition occurs in 2 samples. From a comparison of the surrounding sequences of these two hot spots, it is found that two bases at left and right are both TG and CT (Table 6.3). Previously, ^{60}Co-γ-rays were used to radiate M_{13} plasmid DNA and it was found that the bases nearby TGCT more easily became mutation hot spots. This also reflects differences between ion implantation mutation and radiation mutation.

The dominant kind of ion-beam-induced *lacZ⁻* mutation is substitution (95%) whereas base deletion is only 5%. Insertion or replication of bases is not detected. (Table 6.4). In base substitution, transition and transversion take almost half for each, respectively. Four bases that compose DNA can all be mutated by ion implantation, but cytosine is most sensitive, taking about 60% of the total mutations. Cytosine can be replaced by any of the other three bases (Table 6.3). This depends on the kind of damage to cytosine affected by ion implantation. It is generally thought that deaminization can cause cytosine to produce uracil compounds and pair wrongly with adenine residues so that the C→T transition can occur in the next duplication. If cytosine has a double-strand break between the carbon-positions 5 and 6 under the interaction from the OH° free radicals to form 5,6–dihydroxy–5,6–dihydrouracil, the pairing with adenine can also be wrong and thus further produce C→T transition. If the radiated DNA is of the double-strand structure, guanine that is opposite to cytosine may be converted into 8–hydroxyguanine by ion implantation. Thus, the cis-8–hydroxyguanine can be mispaired with guanine and so finally C→G transversion can occur. Additionally, guanine can be damaged and destroyed due to physical and chemical factors to form unstable N-glycosyl-bonds, which can be identified and excised by specific DNA glycosylases in the cell to form non-purine sites. In SOS repair, positions opposite to the non-purine sites are wrongly introduced with adenine finally to form C→A transversion. In the past, ^{60}Co-γ-rays were used to radiate single-strand M_{13}mp10 DNA, and it was found that C→T transition accounted about 50%, while for radiation of double-strand M_{13}mp10 DNA, the C→G transversion was dominant (70%). The result of ion implantation is that C→A, C→T and C→G occur at almost the same rates. This reflects the versatility of ion beam induced base mutation. Additionally, ion implantation in plasmid DNA also produces high-rate A→G transitions, which are fairly seldom seen with other kinds of ionizing radiation. It is normally thought that this A→G transition is due to wrong pairing of cytosine and hypoxanthine (H) that is formed from the adenine deaminization. Since ion implantation in DNA can more effectively induce C→A transversion and A→G transition, which are both related to group dropping (non-purine site or deaminization), than traditional ionizing (or non-ionizing) radiation, group dropping is an important reason for ion implantation induced DNA base mutation. This should not be difficult to understand from theory and has been verified by experiment as well (see Chapter 5). As discussed in Chapter 4 of this book, in the interaction between energetic ions and organisms, target atom displacement, rearrangement, group deletion (dropping) or mispairing via cascade collisions or Coulomb explosion inevitably follow.

Table 6.3. Mutated base positions and surrounding sequences.

Template base	Mutation type	Surrounding sequence			Position	Number of occurrence
C	C-A	AGCA	C	ATCC	6360	1
		CCAG	C	TGGC	6377	1
		GAAG	C	GGTG	6475	1
		CTTC	C	TGAG	6508	1
	C-G	GGCG	C	TTTG	6449	1 / 3
		TTTG	C	CTGG	6454	1
		GTGA	C	TGGG	6317	1
	C-T	CTTG	C	AGCA	6355	1
		CCAG	C	TGGC	6377	1
		ATAG	C	GAAG	6389	1 / 8
T	T-C	AGCT	T	GGCA	6286	1
		GTCG	T	CGTC	6526	1
	-T	GCCG	T	CGTT	6298	1 / 3
A	A-G	CGTG	A	CTGG	6316	2
		CGAT	A	CTGT	6519	1 / 3
G	G-A	GCCG	G	AAAG	6483	1 / 1
	G-T	GGTG	G	CGTA	6380	2

Table 6.4. Types of ion-implantation-induced *lacZ*⁻ base mutations.

Mutation type	Number of occurrence	Frequency (%)
Base substitution	19	95
Transition	10	50
A → G	3	15
G → A	1	5
C → T	4	20
T → G	2	10
Transversion	9	45
C → A	4	20
C → G	4	20
G → T	1	5
Base deletion		
-T	1	5

Table 6.5. Distribution of specific types of base substitutions in 100 mutants induced by 10 keV low energy nitrogen ion beam.

Type of base substitution	Position	Target sequence (5'→3')	Amino acid change	Occurrence	Total
					89
	1592	GTATCT<u>C</u>CGCACT	Ser531 Phe	12	
	1576	ATTACG<u>C</u>ACAAAC	His526 Tyr	8	
CG→TA	1565	CGCTGT<u>C</u>TGAGAT	Ser522 Phe	5	
or	1692*	AAACCC<u>C</u>TGAAGG	Pro564 Leu	2	
GC→AT	3921***	TAAAAA<u>C</u>ATCGTG	No change	1	
Transitions	1546	TTTATG<u>G</u>ACCAGA	Asp516 Asn	10	
	1600	GCACTC<u>G</u>GCCCAG	Gly534 Ser	7	
	1586	AACGTC<u>G</u>TATCTC	Arg529 His	1	
AT→GC	1552	GACCAG<u>A</u>ACAACC	Asn518 Asp	9	
or	1547	TTATGG<u>A</u>CCAGAA	Asp516 Gly	7	
TA→CG	1598	CCGCAC<u>T</u>CGGCCC	Leu533 Pro	2	
Transitions	1598	CCGCAC<u>T</u>CGGCCC	Leu533 Pro	3	
	1547	TTATGG<u>A</u>CCAGAA	Asp516 Val	8	
AT←→TA	1577	TTACGC<u>A</u>CAAACG	His526 Leu	5	
Transversions	1714	GGTCTG<u>A</u>TCAACT	Ile572 Phe	1	
	1598	CCGCAC<u>T</u>CGGCCC	Leu533 His	1	
GC←→CG Transversions	1551**	GGACCA<u>G</u>AACAAC	Gln517 His	6	
One T insertion	1983_1984	GTGGTA<u>+T</u>CCGTC	frameshift	1	

Note: The mutation sites were underlined. *: new nucleotide site leading to Rif^r was identified in this study but its synonymous mutation had been reported. **: new Rif^r-determining site. ***: not Rif^r-determining site.

6.6 MOLECULAR MECHANISMS OF CHROMOSOMAL DNA MUTATION INDUCED BY ION IMPLANTATION OF *E. COLI*

Research has demonstrated that the adaptive mutations of *E. coli* on the chromosome and extrachromosomal elements are different [5-8]. And also interaction modes of physical and chemical mutagens with DNA inside cell and outside cell are different. Therefore, it is necessary to investigate chromosomal DNA mutation induced by ion implantation in living *E. coli* cells and provide basis for mutation mechanism analysis.

A methodology for determining the specificity of base substitutions in chromosomal DNA has been already established based on *rpoB* gene in *E. coli* [9]. Rifampicin is an antibiotic that inhibits the function of RNA polymerase in eubacteria. Mutations affecting the beta subunit of RNA polymerase β subunit, which is encoded by the *rpoB* gene, can confer resistance to rifampicin [10,11]. Increased mutagenesis to rifampicin resistance reveals that base substitutions in *rpoB* confer *E. coli* cells this capacity [10,12]. Even though the full length of this gene is 4029 bp, the Rif^r responsible region was only 177 bp. Previous studies [11,13,14] identified 47 single-base substitutions at 29 sites and distributed among 21 coding positions. Most of them (46/47) lie in the region A defined in this research, whereas the mutations of the base substitution type leading to Rif^r in *rpoB* covered all kinds of transitions and transversions. Here, we designed two pairs of primers for specific regions of the *rpoB* gene. The corresponding regions were PCR-amplified and sequenced for analyzing the base substitutions. By using this experimental system we analyzed the base substitution mutations induced by either low-energy nitrogen-ion-beam implantation (Table 6.5) or ^{60}Co-γ-ray radiation (Table 6.6).

Table 6.6. Distribution of specific types of base substitutions of 70 mutants induced by 10-Gy ^{60}Co-γrays.

Type of base substitution	Position	Target sequence (5'→3')	Amino acid change	Occurrence	Total
					51
GC→AT or TA→CG Transitions	1546	TTTATGGACCAGA	Asp516 Asn	8	
	1600	GCACTCGGCCCAG	Gly534 Ser	6	
	1586	AACGTCGTATCTC	Arg529 His	3	
	1576	ATTACGCACAAAC	His526 Tyr	7	
	1565	CGCTGTCTGAGAT	Ser522 Phe	5	
	1692*	AAACCCCTGAAGG	Pro564 Leu	3	
AT→GC or CG→TA Transitions	1547	TTATGGACCAGAA	Asp516 Gly	5	
	1552	GACCAGAACAACC	Asn518 Asp	3	
	1598	CCGCACTCGGCCC	Leu533 Pro	2	
	1538	TGTCTCAGTTTAT	Gln513 Arg	1	
AT←→TA Transversions	1577	TTACGCACAAACG	His526 Leu	3	
	1598	CCGCACTCGGCCC	Leu533 His	3	
	1714	GGTCTGATCAACT	Ile572 Phe	1	
AT→CG or GC→TA transversions	1714	GGTCTGATCAACT	Ile572 Leu	1	

Note: The mutation sites were underlined. *: new nucleotide site leading to Rif^r was identified in this study but its synonymous mutation had been reported.

From the tables, the following facts can be seen. Among the base substitution mutations induced by 10-keV low-energy N ions, about 75% are the transitions which are the main type of the low-energy N-ion induced mutations, and others are transversions. CG → TA (including GC → AT), AT → GC (including GC → AT), and AT → TA mutations, taking about 93% of the total mutations, are the most preferential mutation types in the base substitution mutations. Among these mutations, CG → TA or GC → AT take the highest portion, about 52% of the total substitution mutations. AT → GC (including GC → AT) takes 24%, AT ↔ TA takes 17%, and GC ↔ CG takes 7%, which is however not found in the ^{60}Co-γ-ray radiation induced mutants. Among the base substitution mutations induced by 10-Gy ^{60}Co-γ-ray radiation, 84% are the transitions which are the main type, and others are transversions. The main mutation types, GC → AT (including GC → AT), AT → GC (including GC → AT) and AT ↔ TA, which are the most preferential mutations in the base substitution mutations, take 98% of the total substitution mutations. Among these mutations, CG → TA or GC → AT takes the highest portion, 63% of the total substitution mutations.

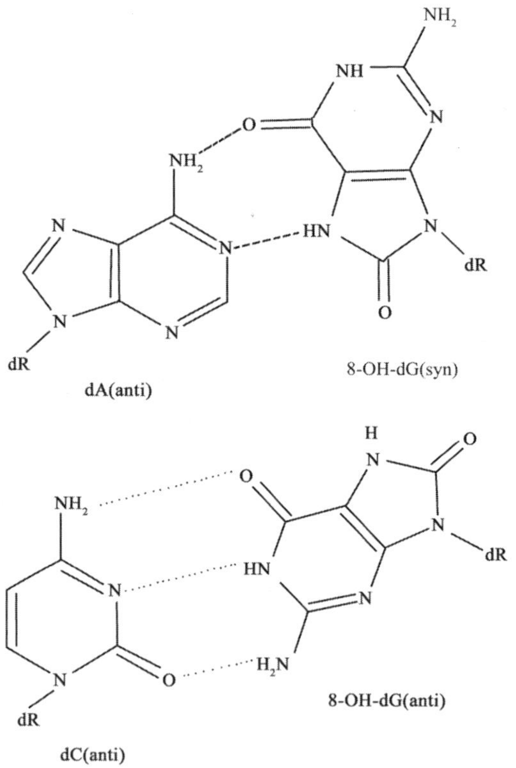

Figure 6.11. Different pairing patterns of 8-OH-dG in DNA molecules. 8-oxo-guanine, the most commonly oxidized adduct, can mispair with an anti conformation of dA when it has a syn-conformation. The occurrences of GC→AT in both mutagens show similar counts and proportion, indicating the oxidized effect raised by both mutagens share similarity.

The experiment of using low-energy N ions to induce mutations of chromosomal DNA in the *E. coli* cell yielded the following results: mutations of CG → TA (including GC → AT) and AT → GC (including GC → AT) had the highest rates, which were in agreement with the result obtained from ion bombardment of the naked plasmid. However, the mutation spectrum of the bacterial chromosomal DNA induced by low-energy N-ion beam was broader, and the mutation types of AT ↔ TA and GC ↔ CG could occur in the cellular DNA but not in the naked plasmid. In brief, compared with naked plasmid DNA, the interaction between the DNA inside the cell and low-energy N ions is complicated, produces broad mutation spectra, and has higher preference to transition mutation. There is water in the living *E. coli* cells. Although the temperature in the target chamber is low and the freedom of the water molecules is relatively low if protected by glycerol, the factor role played by water molecules as an indirect effect cannot be ruled out. Free radicals produced from radiolysis of the water molecules may be one of the mutation factors. For example, DNA adducts such as 8-OH-dG produced by interaction of free radicals with the DNA molecules can have various pairing formations in the DNA molecules, such as pairing with dA in the syn-conformation to produce GC → AT mutation (Figure 6.11)[15]. Experiments have shown that the GC → AT mutation rates induced by both N-ion implantation and ^{60}Co-γ-ray radiation are very close, indicating the mechanisms for producing mutation of both cases in common.

Mutational spectra of low-energy N-ion implantation basically include all mutational types induced by ^{60}Co-γ-ray radiation. This fact also indicates an overlap between the mutation induction mechanisms for both cases. On the other hand, since low-energy N-ion implantation induced base substitution mutation spectra inside the cell are more complicated, its mechanism should be also more complex. Except mutations induced by free radical factors, there are probably other mechanisms. For instance, the mass deposition effect of low-energy N-ion implantation introduces some more complicated N-groups into the cell system, and the interaction leads to formation of some specific DNA adducts, which participate in DNA damage and mutation. This may be the basis of the more complicated mutation mechanisms.

REFERENCES

1. Dickerson, R.E., *et al.*, The Anatomy of A-, B- and Z-DNA, *Science*, **216**(1982)475-483.
2. Barinaga, M., Forging a Path to Cell Death, *Science*, **273**(1996)735-737.
3. Voet, D. and Voet, J.G., *Biochemistry* (2nd ed., John Wiley & Sons, New York, 1995).
4. Yang, J.B., Wu, L.J., Li, L., Yu, Z.L. and Xu, Z.H., Sequence Analysis of *lacZ* Mutations Induced by Ion Beam Irradiation in Double-Stranded M13mp18DNA, *Science in China* (Series C), **40**(1997)107-112.
5. Cairns, J., Overbaugh, J., Miller, S., The Origin of Mutants, *Nature*, **335**(1988)142-145.
6. Cairns, J. and Foster, P.L., Adaptive Reversion of a Frameshift Mutation in *Echerichia Coli, Genetics,* **128**(1991)670-695.
7. Rosenberg, S.M., Evolving Responsively: Adaptive Mutation, *Nat. Rev. Genet.*, **2**(2001)504-515.
8. Hendrickson, H., Slechta, E.S., Bergthorsson, U., Andersson, D.I., Roth, J.R.,

Amplification-Mutagenesis: Evidence that Directed Adaptive Mutation and General Hypermutability Result from Growth with a Selected Gene Amplification. *Proc. Natl. Acad. Sci. U.S.A.*, **99**(2002)2164-2169.

9. Xie, C.X., Guo, J.H., Yu, Z.L., *et al.*, Evidence for Base Substitutions and Repair of DNA Mismatch Damage Induced by Low Energy N^+ Beam Implantation in *E. Coli, High Technology Letters* (English version), **9**(2)(2003)1-6.

10. Jin, D.J., Cashel, M., Friedman, D.I., Nakamura, Y., Walter, W.A. and Gross, C.A., Effects of Rifampicin Resistant *rpoB* Mutations on Antitermination and Interaction with NusA in *Escherichia Coli, J. Mol. Biol.,* **204**(1988)247-261.

11. Jin, D.J. and Gross, C.A., Mapping and Sequencing of Mutations in the *Escherichia Coli rpoB* Gene that Lead to Rifampicin Resistance, *J. Mol. Biol.*, **202**(1988)45-58.

12. Matic, I., Radman, M., Taddei, F., Picard, B., Doit, C., Bingen, E., Denamur, E. and Elion, J., Highly Variable Mutation Rates in Commensal and Pathogenic *Escherichia Coli, Science,* **277**(1997)1833-1834.

13. Miller, J.H., Funchain, P., Clendenin, W., Huang, T., Nguyen, A., Wolff, E., Yeung, A., Chiang, J.H., Garibyan L., Slupska M.M. and Yang H.J., *Escherichia Coli* Strains (ndk) Lacking Nucleoside Diphosphate Kinase Are Powerful Mutators for Base Substitutions and Frameshifts in Mismatch-Repair-Deficient Strains, *Genetics,* **162**(2002)5-13.

14. Severinov, K., Soushko, M., Goldfarb, A. and Nikiforov, V., Rifampicin Region Revisited New Rifampicin-Resistant and Streptolydigin-Resistant Mutants in the Beta Subunit of *Escherichia Coli* RNA Polymerase, *J. Biol. Chem.*, **268**(1993)14820-14825.

15. Wang, D., Kreutzer, D.A. and Essigmann, J.M., Mutagenecity and Repair of Oxidative DNA Damage: Insight from Studies Using Defined Lesions, *Mutation Res.,* **400**(1998)99-115.

FURTHER READING

1. Dickerson, R.E., *et al.*, The Anatomy of A-, B- and Z-DNA, *Science* 216(1982)475-483.

2. Voet, D. and Voet, J.G., *Biochemistry* (2nd ed., John Wiley & Sons, New York, 1995).

CELL DAMAGE EFFECTS DUE TO
ION IMPLANTATION

Up to now we have emphatically discussed the damage to organism at every structural level *in vitro* caused by ion implantation. We have gained some understanding of the primary interaction processes between the ion beam and the biological organism. However, the *in vitro* study of structures can yield only static and isolated concepts, and does not help with interpretation of the effects of implanted ions on life activities. A cell, being the basic unit of biological structure and life activity, from the moment of bombardment by ions to its final division into two cells, can experience a series of abnormal changes. Limited by lack of knowledge concerning cell life activities and modern research methods, a complete analysis of these abnormal cellular activities is not possible. Nevertheless we can study abnormal cell activities induced by these physical effects at various levels, such as cell morphology, subcellular structure, chromosome, and so on. This is the primary topic addressed in the present chapter; we will form a picture or model of the effects of energetic ions at various stages of the cell life process.

7.1. CHANGES IN CELL MORPHOLOGY

The cell is a simple living object constructed from membrane systems with phospholipid double molecular layers as the basic component, and with microtubes, microfilaments, and medium fibers forming the cell backbone and spatial organization in the living object [1]. For a plant cell, there is an additional outer cell wall layer composed of polysaccharides. It is now known that this steric space is far from empty. Various functional proteins, genetic substances and subcellular structures are distributed in an orderly way within the space, forming a unified and coherent mechanical system. Substance transport, energy transformation, information delivery, and metabolism regulation are all carried out in this closed system. What are needed from the outside depend on the inner structure. Cell life activities rely on these structures and morphology, which restrict random changes. Any changes beyond these structural and morphological restrictions would inevitably destroy cell functions and even cause the death of the cell. Thus changes in structure and morphology of a normal cell are always matched with changes in substance and energy in the system. The changes follow a prescribed order, and random variations are not allowed. When the cell structure and morphology undergo improper changes, due to mechanical effects of subsystems, additional forces cause either elongation, or compression, or twisting, resulting in damage to the genetic substance. In the case of cell division, the distribution of genetic substance can be affected, finally causing malformation or aberration of the daughter cells.

It is the general case that ion implantation induces changes in cell morphology. No mater what plant cells and from which organs the cells come, they all undergo changes in morphology when subjected to ion implantation. These changes feature thinning, perforating and cutting of the cell wall at those regions that face the incident ion beam. These changes in morphology are related to the implanted ion species, total dose, and dose rate [2,3]. The cell steric space can even collapse under short-pulse and high-dose-rate ion implantation [3,4].

Prior to ion implantation, the cell is in a vacuum environment, acted on by a tension force directed from the inside of the cell to the outside (Figure 7.1). When a directional beam of ions impinges the cell, an additional force acts on the cell. Let us suppose that ions moving with velocity v impinge vertically on an area element ΔA of the cell; then the impulse I at the area element is

$$I = P\,\Delta A\,\Delta t, \tag{7.1.1}$$

where P is the additional directional pressure due to the ion beam on the area ΔA. Each ion with mass m strikes ΔA with momentum mv. If the ion density is n, then the number of the ions striking ΔA in time Δt is the number of ions in a column with a bottom area of ΔA and a height of $v\Delta t$, i.e., $nv\,\Delta A\,\Delta t$. Hence the total momentum delivered to ΔA by the ions is

$$I = nmv^2\,\Delta A\,\Delta t. \tag{7.1.2}$$

From Eqs. (7.1.1) and (7.1.2) we have

$$P = nmv^2. \tag{7.1.3}$$

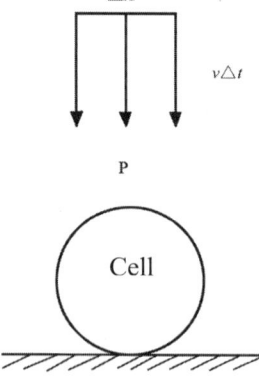

Figure 7.1.   Additional pressure on the cell from implanted ions.

Thus P can be estimated from Eq. (7.1.3). Let us suppose the following experimental conditions. The implanted ion is Ar^+, the implantation time is 2 ms with an interval of 1 min, the ion beam current density is 100 mA/cm^2, and the ion energy is 30 keV. Thus ion density in the beam is about 1×10^{10}/cm^3 and the pressure P is about 100 Pa. This pressure is very small compared to the tension acting on the cell (about an atmosphere). But this additional directional pressure disturbs the balance in the cell tensile force. When the mechanical systems (cell wall, cell framework, etc.) that support the cell structure are injured, this unbalanced interaction repeatedly occurs, together with localized thermal effects, resulting in irreversible changes in the cell morphology, such as elongation and twist as shown in Figure 4.4.

Ion beam sputtering effects (chemical sputtering, ion sputtering, and electron sputtering) described in Chapter 4 perform an ultra-fine processing on the cell surface, like a surgical knife. The cell wall is the first to be processed. The plant cell wall is a complete net formed by linking polysaccharides in specific manners, packing the plant protoplast. The plant cell wall can be plastic or rigid, permeable or non-permeable, and infiltrated with plastic substances or covered by sticky substances. These substances cohere together layer by layer to form fibers or be fused at some point locations to form pores. When such a complex surface of the plant cell wall is subjected to ion bombardment, the etching rates at different locations are different. It has been pointed out in Chapter 3 that the ion beam sputtering yield of different materials depends on the ion (species, energy and incident angle) and the target material surface properties (composition and structure). In the cell wall there are crystalline structures, high-strength cellulose microfibrils, which are filled or cohered with pectin. For similar ion beam parameters, the etching rates are very different for these different structures. The microfibrils may become thinner, the pectin may form molten pits due to sputtering and evaporation, and originally existing pores may be enlarged. Thus it is not surprising that under electron microscopy one can observe etching traces and small-porosities on the cell wall.

Figure 7.2 shows microphotographs of the surface morphology of cultured rice cells implanted with various N-ion doses [5]. For the unimplanted cells, the surface is smooth and flat and the steric space exhibits a certain pear shape. After 30 keV 1.5×10^{14} N^+/cm^2 implantation, the cell wall surface in the direction to the ion beam exhibits increasing roughness and etched traces distributed on the top hemisphere. It should be noted that the distribution of these traces is not homogeneous over the cell surface; the closer to the "top" of the sphere, the denser the traces. This implies a linear dependence of sputtering effect on implantation dose. In the ion implanted system, the ion number captured by the cell surface is proportional to the projected area on a surface perpendicular to the ion beam direction. For the same area, at the sphere top the captured ion number is the greatest, whereas near the equator the captured number of ions is the least. On the hemisphere at the other side of the surface where ions do not bombard, etching does not take place. When the ion dose is increased to 5×10^{14} /cm^2, the etched traces on the cell surface become larger. The ion beam most easily sputters substances that fill in the net structure, leaving pores. When ion beam etching breaks the net framework, the cell wall collapses at that location to form large holes [5].

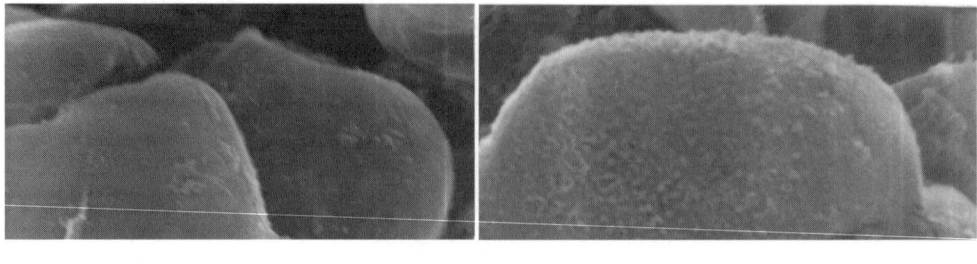

(a)   **(b)**

Figure 7.2. SEM photograph of ion-beam-etched cultured rice cells. (a) Vacuum control, and (b) 30-keV, 1.5×10^{14} ions/cm^2 N-ion-beam etched cell.

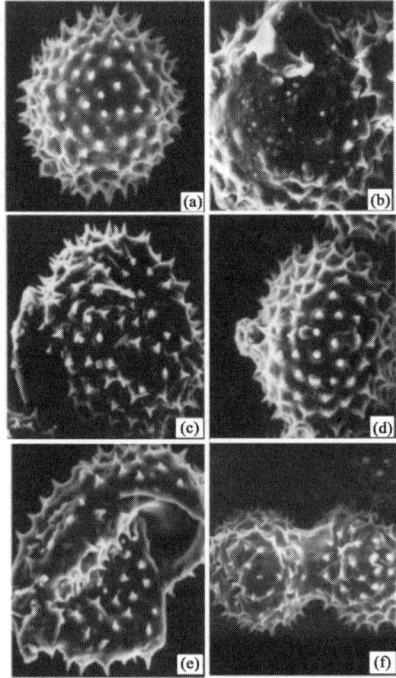

Figure 7.3. N-ion-beam bombardment induced morphological damage to cotton pollen. (a) Vacuum control, (b) 5×10^{15} ions/cm^2, (c) 10×10^{15} ions/cm^2, (d) 15×10^{15} ions/cm^2, (e) 20×10^{15} ions/cm^2, and (f) 25×10^{15} ions/cm^2.

Cotton pollen has a hard shell and thorns. After ion implantation, obvious changes in morphology can be observed under optical and scanning electron microscopy. The degree of change is proportionally related to the ion implantation dose [6] (Figure 7.3). On that part of the pollen particle in the ion implantation direction, the thorns are peeled off, the surface becomes flat, the pollen particle wall is thinned (Figure 7.3b), and parts of the pollen wall are significantly stripped (Figure 7.3c). As the dose increases, "bubbles" form at the stripped regions due to the higher pressure inside. When the dose is further increased, the cell wall ruptures, the inner substance overflows, and the pollen wrinkles (Figure 7.3e). Additionally, the pollen grains may be fused and pollen complexes can be formed.

The ion beam interacts with the cell progressively from the outer surface to the inside. After the cell wall is sputtered and etched, subsequent ions directly interact with the cell membrane, the cell framework and the cell organelle and cause damage to the inner structure of the cell. In fact, in the case of low-dose ion implantation, even though no apparent damage at the cell surface can be observed, the cell inner structure may already be changed. Figure 7.4 shows a fluorescence-marked laser-microscopic image of the microfilament framework of mature pollen of *Lilium davidii* Duch after N-ion implantation at 100 keV with a dose of 10^{13} ions/cm^2 [7]. It can be seen from the figure that after ion implantation (Fig. 7.4a) the pollen actin cytoskeleton exhibits long and dense parallel bundles or or ring structure, while that of control (Fig. 7.4b) appear to be random short and fine network structure. The cell framework is an extremely complicated network composed of microtubes, microfilaments, medium fibers and microbeams. It is not only the supporting structure for a living cell, shaping and strengthening the cell, but also the unitary basis for organizing the internal space and arranging various cell organelles and macromolecules. Changes in the cell framework deeply affect living activities of the cell. Furthermore, with certain doses, an ion beam can damage the nuclei and even smaller organelles such as chloroplasts and mitochondria. The kinds of mechanical damage commonly seen in nuclei include breaking of the nucleus membrane, swelling of the perinuclear cavity, and loss and lagging of chromosomes. More seriously, the nucleus may be eroded into holes. The damage to chloroplasts includes breakage or deletion of the chloroplast membrane, and decrease in the number of basal granules and basal granule lamellas [8] (Table 7.1).

A chloroplast is a genetically semiautonomous cell organelle, containing the genetic substance, DNA. More strictly speaking, the formation of a chloroplast is the result of interaction between chloroplast genetic groups and nucleic genetic groups. Morphological changes in the chloroplast structure lead to definite mutation of the chloroplast DNA (see Chapter 8).

Table 7.1. Statistical data concerning the number of basal granule lamellas of damaged chloroplasts of rice leaf cells.

Chlorop-last	Number of basal granule lamellas							Σ nN
	4 2-6	8 6-10	12 10-14	16 14-18	20 18-22	24 22-26	28 26-30	
Damaged	842* 57.6**	584* 40.0**	28* 1.92**	7* 0.48**	* **	* **	* **	1461
Normal	275* 20.15**	501* 36.7**	320* 23.44**	183* 13.41**	49* 3.95**	21* 1.54**	16* 1.17**	1365

* 100 nN   ** 100 nN/Σ nN

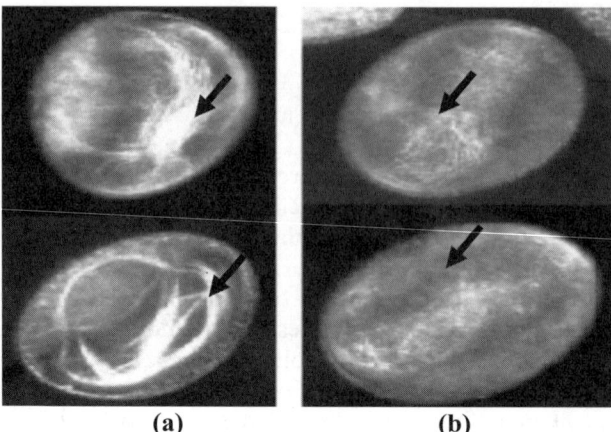

(a)   **(b)**

Figure 7.4. Fluorescence-marked laser-microscopic image of the actin filament framework of hydrated mature pollen grains of *Lilium davidii* Duch (a) after N-ion implantation at 100 keV with a dose of 10^{13} ions/cm^2, compared with (b) control. Arrows point the network of actin filaments.

7.2. CHANGES IN CELL ELECTRICAL CHARACTERISTICS

Cells that have been studied up to now have all been electrically negative. The electrical property of the cell surface is absolutely necessary for maintaining life activities. If the typical potential drop across the cell membrane is 60 mV and the electric capacitance of the membrane is less than 1 μF, the cell surface charge density is 6×10^{-8} C/cm^2.

According to the result calculated from Eq. (4.3.8), when the number of the singly charged positive ions reaches 5×10^{14} /cm^2, the increased charge on the cell surface is 6.4×10^{-5} C/cm^2 — assuming that no accumulated charge leaks away. If the original negative charge on the cell surface is totally neutralized, the net positive charge remaining on the cell surface is 6.394×10^{-5} C/cm^2. This indicates that ion implantation at this dose not only changes the polarity of the electric potential across the cell membrane, but also increases the membrane potential by three orders of magnitude! But actually this is not possible. The cell is different from a tyrosine crystal (described in Chapter 4), which can change its electrical polarity when the ion dose reaches a critical value. A possible reason is that the charge reservoir (about 10^{-3} C/cm^2) existing in the cell balances the cell membrane potential. On the other hand, due to ion sputtering and electron sputtering, ionized groups on the cell surface are continuously released so that the accumulated charge on the cell surface is dynamically balanced. Figure 7.5 shows electrophoresis mobility of cultured rice cells as a function of ion dose.

Ion implantation differs from ionizing radiation in that ion beam irradiation is localized. The ion dose is highest at the top of the cell sphere that faces the ion beam direction, but near the equator the dose is lowest. When the ion dose is very low, the lower hemisphere of the cell is basically not affected. In Figure 7.6, the ion implantation dose near point M is

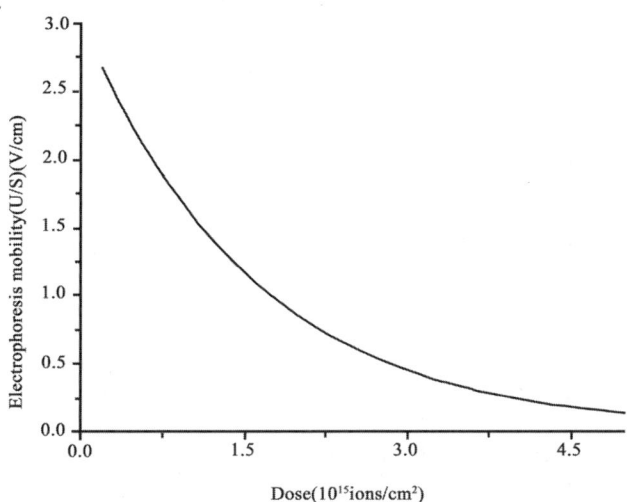

Figure 7.5. Electrophoresis mobility of rice cell as a function of implanted N-ion dose.

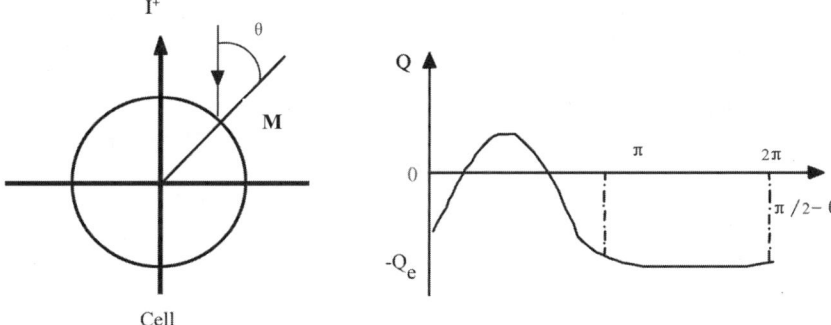

Figure 7.6. Charge distribution on the cell surface as a function of angle θ after ion implantation.

$$D_M = D \cos \theta. \tag{7.2.1}$$

From Eq. (4.3.8) we know that the implanted ion charge accumulated from charge exchange per unit area of the cell surface is proportional to the dose,

$$Q_e = k_e D_M, \tag{7.2.2}$$

where $k_e = 1.6 \times 10^{-19} F_0$, and F_0 is the neutralization efficiency.

If the charge density on the cell surface is Q_b before ion implantation, then the charge on the cell surface after ion implantation is

$$Q = Q_b + \Delta Q_t, \tag{7.2.3}$$

where ΔQ_t is the decrease in negative charge brought about by ion implantation:

$$\Delta Q_t = \Delta Q_E + \Delta Q_e, \tag{7.2.4}$$

where ΔQ_E is the charge change caused by energy deposition, and ΔQ_e is the charge change caused by charge exchange. Both are related to the ion implantation dose, and for low dose the relationship is linear.

After ion implantation the charge distribution on the cell surface can be expressed by

$$Q = Q_b + (k_E + k_e)D \cos (\pi/2 - \theta) \quad (0 \leq \pi/2-\theta \leq \pi), \tag{7.2.5}$$

$$Q = Q_b \quad (\pi \leq \pi/2-\theta \leq 2\pi), \tag{7.2.6}$$

where k_e is determined from Eq. (7.2.2), and k_E is the proportionality coefficient of the negative charge decrease due to ion energy deposition against ion dose. From Figure 7.5, we see that at high dose the cell negative electrical characteristic slowly decreases with increasing dose and eventually reaches a dynamic balance.

For ionizing radiation, at the balance the cell negative electrical characteristic decreases by about 1/4 to 1/3. For ion implantation, at the balance the negative electrical characteristic decreases to 1/4 of the original, or decreases by 3/4. We can say that, approximately, the change in the negative electrical characteristic caused by implanted ion charge exchange is about twice that caused by energy deposition at the same dose. This is true on the average. In fact, the charge exchange over the cell surface is inhomogeneous due to geometric factors. For simplification, the cell is divided into two hemispheres, one implanted and the other unimplanted. Let us suppose that before ion implantation the total charge of the ionized groups on the cell surface is Q and on each hemisphere is $1/2Q$. After ion implantation, the unimplanted hemisphere still has $1/2Q$. It is assumed that the total charge of the cell surface after ion implantation is 1/4 of that before ion implantation, namely,

$$Q' = \tfrac{1}{4}Q, \tag{7.2.7}$$

and it is also known that

$$Q' = \tfrac{1}{2}Q + Q_e, \tag{7.2.8}$$

where Q_e is the charge on the surface of the ion-implanted hemisphere of the cell. From Eqs. (7.2.7) and (7.2.8) we have

$$\tfrac{1}{4}Q = \tfrac{1}{2}Q + Q_e, \tag{7.2.9}$$

or

$$Q_e = -\tfrac{1}{4}Q. \tag{7.2.10}$$

From Eq. (7.2.10) the charge on the surface of the ion implanted hemisphere changes polarity, changing from negative to positive. This change can automatically orient the cell in an external electric field, i.e. the irradiated sphere cell-top is oriented to the negative electrode of the external field, while the unirradiated hemisphere faces the positive electrode. Although the cell moves toward the positive electrode in electrophoresis, the cell orientation is already controlled by the electric field [9].

Changes in the distribution of the ionized groups on the cell surface will substantially affect the potential distribution of the cell membrane. They disturb the equilibrium concentration of the ions in the cell and so influence the normal metabolic cell process [10].

7.3. CHANGES IN CELL ACTIVITY

Active motion is one of the most outstanding characteristics of all life activities. The cell, as the basic unit of life, has a variety of physiological activities, such as maintenance and change of cell morphology, transport of substances and energy within the cell, absorption inward and excretion outward, immunizing behaviors, cell division, etc. All of these activities involve various kinds of motion, or they are themselves a sort of motion in a broad sense. These activities are controlled by DNA. The observed organizations and directivities are determined by the cellular and subcellular structure. Once these structures are damaged, the formation and direction of the cell life motion will be changed. In fact, the cell structure and function are continuously subjected to destructions in physics, chemistry and biology from various hostile factors such as physical factors, toxic substances and viruses. The repair systems in the cell can continuously repair various kinds of damage to maintain normal life activities of the cell. For example, DNA repair enzymes related to radiation-induced DNA breakage can be activated, toxicant-removing reactions related to harmful chemical substances in liver can be enhanced, and immune response related to invasion of bacteria can be induced.

Damage and repair are a couple, coordinating each other in maintaining cell life activities (Chapter 6). Damage may be considered the origin while repair is the intrinsic emergent reaction. In low energy ion beam biological studies, from the moment the first energetic ion interacts with the cell, the cell is subjected to damage on a microscopic scale along the ion trajectory and the life activities of the cell are disturbed. With increasing number of incident ions, the cell suffers from progressively more serious damage. In particular, damage to the genetic substances, DNA and RNA, causes mistakes in the transcription process. Often these mistakes are possibly repaired; however, in case unrepairable mistakes occur, they can accumulate and cause degeneration of cell function and even abnormal death or mutation.

The abnormal death of the cell induced by ion implantation has complicated factors, which include both environmental effects (vacuum) and implanted ion effects. Cotton pollen cells have a survival rate of 94.7% after immersion in saturated water vapor at 40°C for 20 minutes, a rate of 96% after a 25-min time in vacuum, but a rate of only 20.1% after a 4-hour residence time in vacuum. In studying the cell damage effects of ion implantation, experiments can normally be completed in about 25 minutes and hence the effects of vacuum are often neglected.

In the case of ion implantation, the relationship between the cell survival rate and the ion implantation dose does not follow strictly the exponential law of inactivation, but instead, the survival curve presents a flat region (Figure 7.7). From the point of view of life activity processes, cell death is an irreversible stopping of life activities. For mutagen-induced death, either serious damage to the cell structure or damage to genetic substances can cause the cell death. In principle, the degree of physical damage is proportional to the dose. But for the survival curve, when the dose increases to a certain value, a flat region appears and seems to have an increasing trend. This implies that the damage is relieved. This kind of relief is only produced during the ion implantation process or in later life activities, or in other words, it is the result of effects of primary repair and biological repair mechanisms.

We saw in Chapter 6 that the displacement and recombination of single-strand DNA atoms or groups on the double helix column surface induced by momentum exchange between implanted ions and DNA molecules are an important cause of induced plasmid DNA mutation. Breach of single-strand DNA, as an induction signal, can cause the emergent repair system to work (SOS repair), leading to tendency errors of the DNA damage repair. Compared with error-avoiding repairs including light repair, excision repair and recombination repair, tendency-error repair has a very different effect. It increases the survival rate but at the expense of producing a large quantity of mutations.

In normal situations, SOS-reaction-related genes are repressed by the LexA protein. Acted upon by induction signals, RecA protein is combined with the induction signals and excited. The excited RecA protein is dissociated into the repressor LexA protein to result in expressions of the SOS-reaction-related genes. Although many mutagens may induce the SOS reaction, the probabilities and intensities of producing induction signals are different. Due to low energy ion bombardment, atoms or groups of genetic substances are displaced due to cascade collisions (Chapter 3), and vacancies remain. These vacancies probably appear in the form of single-strand DNA breaches. In later life activities, single-strand DNA breaches induce the SOS emergent repair system, leading to repairing of tendency-error DNA damage. The consequence is then a coexistence of increase in cell survival and production of a large quantity of mutations.

Based on the above analysis, the survival-dose curve shown in Figure 7.7 can be explained as follows. Since genetic substances occupy only an extremely small portion of either mass or space of the cell, few energetic ions arrive at the target volume of the genetic substance when ions homogeneously bombard *E. coli* at low dose. The damage to *E. coli* induced by energetic ions mainly includes etching of the cell wall, perforation of the membrane, destruction of the cell framework, and so on [10]. The degree of this damage to cell activity increases with increasing dose. When the dose increases to a certain value, the collision cascade results in a large number of vacancies in the genetic substance. Some of the vacancies in single-strand DNA breaches induce an SOS reaction and increase the survival rate of damaged cells. When the dose further increases, the cells are subject to serious damages, leading to a large quantity of double-strand breaks, which accumulate to an unrecoverable level, and the cell survival rate decreases again.

As described in Chapters 5 and 6, damage to the cell caused by low energy ions results from the combined effects of mass, energy and charge, and consequently stimulates a physicochemical repairing function which exists in the cell before biochemical and biophysical reactions. For example, the competitive reaction of deposited nitrogen ions with neighboring free radicals clears the radicals as the positive ions capture electrons to decrease the probability of producing OH° free radicals from excited water molecules; implanted ions and displaced atoms replace the removed matrix atoms to relieve the cell damage. We may ask if there is any unknown damage repair model responsible for repairing the initial damage in this complicated process. Whether the answer is positive or negative, it is worth continuing further research on mechanisms of abnormal phenomena of activity change in ion-implanted cells.

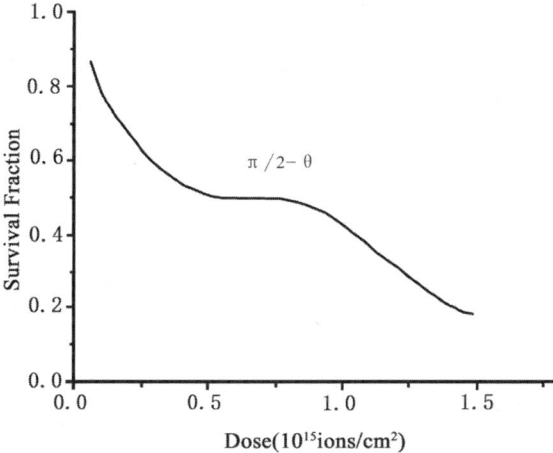

Figure 7.7. Survival fraction of ion-implanted *E. coli* as a function of ion dose.

Implanted ions have effects not only on cell activity but also on the cell cycle. The cell cycle is controlled by a gene to proceed in an orderly way. The number of genes of the cell division cycle (C_dC gene) of yeast bacteria is about 32, and some of them are already cloned. Selection of yeast bacteria to study effects of ion implantation on the cell cycle helps understanding relevant problems at the gene level. The working points of the yeast C_dC_1, C_dC_2 and C_dC_3 mutants are approximately in the beginning and germinating periods of the cell cycle, and the final point depends on the mutant. C_dC_1 stops its functioning just after germination, C_dC_2 stops at nuclear division, and C_dC_3 is continuously divided but not the cytoplasm, thus producing multi-nuclei cells with extended branches. C_dC_4 is the mutant starting DNA combination, while C_dC_8 is the mutant controlling the process of DNA combination. Under normal conditions it takes about 2 hours and 10 minutes for yeast to complete the process from germination to forming two divided cells. Microscopic observation is able to follow and document the entire process of the division activity. For ion implanted yeast cells, continuous germination and growth can be observed but not the division of cytoplasm. After more than 4 hours, the filial generation cell has the same volume as that of the matricyte, but the two are still tightly connected. This may be due to mutation of the C_dC_3 gene. Some mutants are found with restricted germination but expanded cell volumes. This implies that the nuclei do not divide or the DNA combination is stopped, namely C_dC_2 or C_dC_8 gene is damaged.

REFERENCES

1. Lodish, H., Berk, A., *et al.*, *Molecular Cell Biology* (4th ed., W. H. Freeman & Co., New York, 2003).
2. Yu, Z.L., *et al.*, Preliminary Studies on the Biological Samples by Ion Beam Etching, *Journal of Anhui Agriculture University* (in Chinese with an English abstract), **21**(3)(1994)260-264.

3. Song, D.J., *et al.*, Studies on Etching and Damage Action of Low Energy Ion Beam on Microbial Cells, *Acta Biochemical et Biophysical Sinica*, **30**(6)(1998)570-574.
4. Song, D.J., *et al.*, Study on the Direct and Indirect Action of N^+ Ion Implantation on D. *Radiodurans* and *E. Coli*, *High Technology Letters*, **2**(1)(1991)47-50.
5. Yang, J.B., Wu, Y.J., *et al.*, Low Energy Ion Beam-Mediated DNA Delivery into Rice Cell, *Journal of Anhui Agriculture University*(in Chinese with an English abstract), **21**(3)(1994)330-335.
6. Cheng, B.J., *et al.*, Effects of Nitrogen Ion Implantation on Morphology, Viability and Fertility of Cotton Pollen, *Acta Bot. Boreal*, **14**(2)(1994)85-89.
7. Ren, H.Y., Huang, Z.L., *et al.*, Effects of Nitrogen Ion Implantation on Lily Pollen Germination and the Distribution of the Actin Cytoskeleton during Pollen Germination, *Chinese Science Bulletin*, **45**(18)(2000)1677-1680.
8. Yu, Z.L., Ion Beam and Life Science — A New Research Field, *Physics* (Chinese), **26**(6)(1997)333-338.
9. Song, D.J. and Yu, Z.L., Ion Implantation Biology: A New Interdisciplinary Subject Emerging in China, *Science* (Chinese), **000**(011)(1996)49-51.
10. Song, D.J., Yao, J.M., *et al.*, The Etching of Cells and Damage and Repair of DNA in Deinococcus Radiations by N^+ Implantation, *Hereditas* (Chinese), **21**(4)(1999)37-40.

8

BIOLOGICAL EFFECTS OF ION IMPLANTATION

Biological effects due to physical factors are part of our fundamental knowledge of life sciences. When a biological organism is acted on by various physical factors, initial physical or biophysical processes can start within a few minutes or seconds or even immediately. This direct interaction is called the preliminary effect. Later on relatively stable biochemical changes may occur, leading to primary biological effects. Other biological effects can be further induced, eventually bringing about changes in the structure or function at a certain level in the organism, and this is called the biological end-point effect. Even so, forms expressing the biological effects are limited, such as protein denaturation, changes in enzyme activity and structure, cell death, genetic mutation, etc.

In previous chapters we have examined the primary effects of implanted ions on the biological organism. The primary effect is identified as a mechanism leading to the observed biological effect. Interpretation of the mechanisms of implanted-ion-induced biological effects calls for a phenomenological description of our knowledge of biological effects. Although there are many coexisting ion implantation factors affecting the biological effects, making it difficult to distinguish between direct and indirect effects and factors, here we outline a physical picture of the interaction between implanted ions and the biological organism, as well as primary effects. This is helpful for our understanding of the nature of biological effects due to ion implantation. Comprehension of this and subsequent chapters calls for a familiarity with the fundamental principles discussed in previous chapters. This chapter introduces the idea that the induction of biological effects requires time and then summarizes and reviews previous chapters to facilitate a deeper understanding of these effects, particularly the unique characteristics of the effects. Following this, we consider biological effects on physiology, biochemistry, and genetics which are induced by ion implantation. Occasionally we will refer γ-ray irradiation as a convenient reference and comparison.

8.1. PROGRESSION OF ION-IMPLANTATION-INDUCED BIOLOGICAL EFFECTS

From the first moment when a biological molecule is impacted by an energetic ion to the production of final effects, it moves across a broad space-time continuum of reaction.

Temporally, changes can proceed over many years; spatially, the effects can expand from molecules to cells and finally to macroscopic effects. Some changes may disappear instantly with no measurable manifestation in the short or long term and some may persist for a very long time.

Based on current understanding of ion implantation biological effects, the progression of biological effects can be roughly divided into the following stages.

8.1.1. Physical Stage

In this stage the effects are mainly absorption of ion energy, ionization and/or excitation, charge exchange, and ion-molecule reactions in the body of the biological organism. If these effects occur at key biological molecules, direct effects can be induced; and if they occur elsewhere, indirect effects are induced. The duration of this stage is about 10^{-13} seconds.

8.1.2. Physicochemical Stage

In this stage the collision cascade occurs. Atoms are displaced, slowed ions deposit and diffuse, displaced atoms, deposited ions, and matrix atoms are rearranged and chemically combined to form new groups. Excited or ionized molecules produced in the physical stage spontaneously collide with neighboring molecules to produce secondary products that have high reaction energy. The duration of this stage is about 10^{-9} seconds.

8.1.3. Chemical Stage

The products from the preceding stage continue to interact with each other or to react with other molecules in the cell environment, to form biomolecular radicals until the latter form stable molecules through secondary reactions. In this stage, DNA and RNA damage is initiated; enzyme activation and inactivation occur; the content of sulfhydryl groups in the cell decreases; lipid peroxidation begins and many normal biochemical reactions are disrupted; and DNA repair processes proceed. This stage lasts about 10^{-3} seconds.

The effects of these three stages all occur at the molecular level. Damage at the molecular level is microscopic and basic. Before this damage develops into macroscopic biological effects, there is an irreversible biological magnification process, which is the biological stage.

8.1.4. Biological Stage

The biological stage starts upon completion of molecular level changes in the organism. In this stage, as reactions of secondary-reaction-formed $O_2^{\bullet}$ continue, damage to biological macromolecules brings about disorder in the energy metabolism so that supplies of precursors for biological combination are insufficient. Many important biochemical reactions are disturbed, cytoplasm membranes and nuclear membranes are broken, and cell mitosis is delayed. In this stage the metabolic process plays an important role in modifying damage development. Measurements of the biological effects finally displayed determine to a great extent whether the primary molecular damage can be repaired by metabolic processes or persist. The timescale of the biological stage lasts very long, from seconds to years.

The biological effects continuously change and are very difficult to separate clearly. Ion implantation simultaneously inputs energy, ions and charge into the irradiated organism, and thus the primary processes (the first three stages) become extremely complicated. For instance, the rearrangement and recombination of the displaced atoms, deposited ions and matrix atoms are dependent on not only the ion energy, mass number, and charge number, but also the chemical activity of the ion. These dependencies strongly affect changes in the biological stage.

From a macroscopic point of view, primary biological effects are manifested as excitation and damage. It has been commonly thought that low-dose ion implantation normally leads to excitation effects whereas high-dose ion implantation brings about suppression effects. We will see in following sections that for ion-implanted dry crop seeds, the effect is manifested by either decrease or increase in survival rate as well as either suppressed or enhanced growth and development of the resulting plants. These effects are only intrinsic responses to the ion implantation induced damage, and are not necessarily exhibited as heredity. Some of the primary biological effects are repaired during growth, while some are transmitted to future generations through sexual processes. Through screening of many generations, a small part of the characteristic changes can be fixed. At this point changes resulting from ion implantation biological effects end.

8.2. MAIN FACTORS INFLUENCING BIOLOGICAL EFFECTS

The factors that influence biological effects mainly depend on the properties of the implanted ions and the biological objects.

8.2.1. Implanted Ions

Radiation is the propagation of energy in space. When a biological organism absorbs energy radiation over a critical value, biological effects result. Thus, a major parameter in the study of radiation effects is energy absorption, namely, the absorbed dose normally used in radiobiology. Factors related to ion implantation induced biological effects, as distinguished from those due to radiation, include not only energy deposition (and momentum transfer) but also mass deposition and charge exchange. In brief, in the case of ion implantation the influencing factors are threefold – the simultaneous input of energy, substance and charge to the biological organism. Since there are intrinsic differences between ion implantation and radiation, it is reasonable that the responses of the biological organism to the treatments are different.

8.2.1.1. Energy Transfer

An ion carries energy, and when it is implanted into a biological organism it distributes its energy steadily along its path. This is similar to the absorption of radiation. The point is that the energy range considered here is a low energy regime, and the energy transfer between ion and target atoms is dominated by elastic collisions. When an incident ion collides with an atom in the organism, the atom makes a potential jump so as to move from its original position to become a displaced atom if the energy transfer to the atom exceeds a critical value (the bond-breaking energy plus the energy needed to overcome any other binding the potential). If the displaced atom gains sufficient kinetic energy in the primary collision, it will continue to collide with other target atoms to produce new displaced atoms. A repetition of this process can displace many atoms in the neighborhood of the implanted ion trajectory. Thus, as distinct from radiation, implanted ions break biomolecular bonds without following the principle that "breaking starts at the weakest bond"; instead, ion-implantation induced breaking of the bonds relates to the direction of the ion (including recoil atoms) momentum transfer. A 100-keV ion can produce several hundreds of displaced atoms during the implantation process. If the ion energy is high enough, an instantaneous Coulomb-explosion can also produce a large number of displaced atoms along the incident ion trajectory. These displaced atoms are rearranged to new locations, or recombine with neighboring atoms, or become interstitials.

Note that as the incident ion energy is deposited, a small part of the energy reaches the surface to cause secondary particle emission from the biological organism surface, i.e. sputtering. Sputtered particles can be atoms, molecules or ion debris. The incident ion can transfer its momentum directly to surface atoms in a single collision, which can also produce high-energy secondary ions. Furthermore, electron sputtering and chemical sputtering can also release molecular debris from the biological organism surface. All of these phenomena cause mechanical damage to the biological organism surface. The surface is an extremely heterogeneous multi-layered structure, showing thinning or perforation of the cell wall which can be seen by means of electron microscopy.

8.2.1.2. Mass Deposition

Implanted ions input matter to particular locations within the biological organism, while displaced atoms import or export matter to other locations. Doping or removing this matter, which may be either biological structural material or not, changes the molecular structure, composition and chemical environment in specific micro-zones of the organism. If the implanted ions are active, after slowing down they may bond with neighboring displaced atoms and matrix atoms to form new molecular groups. Ions of all elements in the Periodic Table as well as some molecular ions can be accelerated and implanted. Thus, a description of the mass deposition includes the ion mass number. Implantation of ions of different masses yields different results. Note that implanted ions are deposited with a depth profile that is approximately Gaussian, and thus there is a peak mass deposition concentration as a function of depth below the surface.

8.2.1.3. Charge Exchange

Ions carry charge that may be positive or negative. In the energy range of relevance here, the implanted ions deposit their charge in a thin layer of the biological organism surface and hence alter the electrical characteristics of the surface. After ion implantation into a tyrosine crystal, it can be observed in electrophoresis that an accumulation of positive charge occurs. The crystal moves to the positive electrode with decreasing speed and then stops; when the charge is further increased, the direction of motion can be changed from moving to the positive electrode to moving to the negative electrode. This is a direct demonstration of ion implantation induced charge exchange affecting the electrical characteristics of biological organisms. Generally, since the cell membrane is a poor electrical conductor, accumulation of charge on the surface leads to malformed changes to the physical field across the membrane and affects the exchange of matter and information between the outside and inside of the cell. If the base receives additional charge so as to result in a change in the shape of the double potential well, genetic mutation can be induced.

These three functioning factors of ion implantation biological effects can be distinguished in space. If an ion beam is focused to a spot of dimension less than a micrometer, damage to the target biological organism can be observed in 3-dimensional space. This is different from the bulk damage induced by radiation and the deep penetration effects caused by high-energy ions.

8.2.2. Biological Objects

Biological organisms consist of extremely heterogeneous multi-layer structures containing a large amount of water or a large number of micro-pores. In order to explore quantitatively the biological effects of ion implantation, we established in Chapter 4 a simple model of the structure of dry crop seed in the direction of ion beam implantation equivalent to a periodic layered structure. In reality a biological system is extremely complicated, and cannot be replaced by any simple model. On the other hand, the versatility of biological systems is determined by the genotypes. Different genotypes have different sensitivities to ion implantation and thus the induced biological effects are also different. Of more importance is that only when the ion energy, mass and charge are absorbed in a localized area of the biological organism and reach critical values, can significant biological effects be produced. Thus ion-implantation biological effects are determined by the properties of both the implanted ions and the biological organism.

8.3. PHYSIOLOGICAL EFFECTS

After ion implantation of crop seeds, the sprouting period, germination rate, root length, and seedling height can all be affected. They may be manifested as stimulated effects such as increase in the germination rate and seedling sturdiness, or suppression effects such as decrease in the germination rate and late development. The organism usually recovers after a period of time, and the time needed is affected by external factors of self-adjustment, or recovery of the organism functions that are disabled by damage to normal functioning, or by DNA repair. This is called the biological delay time. The biological delay time may be a few seconds or several months or even years. Thus there is adequate time to study induced biological effects caused by ion implantation.

8.3.1. Physicochemical Effects

After a biological organism is ion implanted, the local electrical properties, optical properties, structure, composition and so on may all be changed. The consequences of these changes are that the substance metabolism, expiration cycle, function state, and nutrient reach new levels. Seen from the organism itself, the new levels mean either growth stimulation or growth suppression.

Figure 8.1 shows changes in superweak luminescence during the germinating period of rice 31111s and 1016 after ion implantation and γ-ray irradiation. Biological luminescence is a process by means of which the system effectively converts chemical energy to optical energy. All living organisms possess low-level luminescence characteristics. This luminescence is related to processes such as oxidation metabolism, detoxification, cell division, cell death, growth regulation, and so on. It can be seen from Figure 8.1 that ion implanted rice has a superweak luminescence that is greater than that for unimplanted rice 48 hours before germination, but not so pronounced as that of γ-ray irradiated rice. After 54 hours some changes occur. The luminescence of the ion implanted 31111s-seeds is obviously stronger than that of the control, whereas that of the γ-ray irradiated seed is less than the control. After 54 hours, although the luminescence of the 1016-seeds is lower than that of the control, the luminescence of the ion implanted seeds is still greater than that of the γ-ray irradiated seed.

Experimental evidence shows that cell mitosis luminescence is a prerequisite condition for cell mitosis, as well as a condition for stimulating the activity of cell mitosis, producing reversible morphological changes in the cell and changing the cell permeability. The changes in ultra weak luminescence of rice seeds in the germination process shown in Figure 8.1 may provide information about root meristematic-cell mitosis. Ion implantation of various biological organisms may either stimulate cell division or suppress cell division, but not strongly so, indicating less damage to rice seeds by ion implantation. However γ-ray irradiation causes whole-body damage to the seeds, and thus after 54 hours cell division is always suppressed.

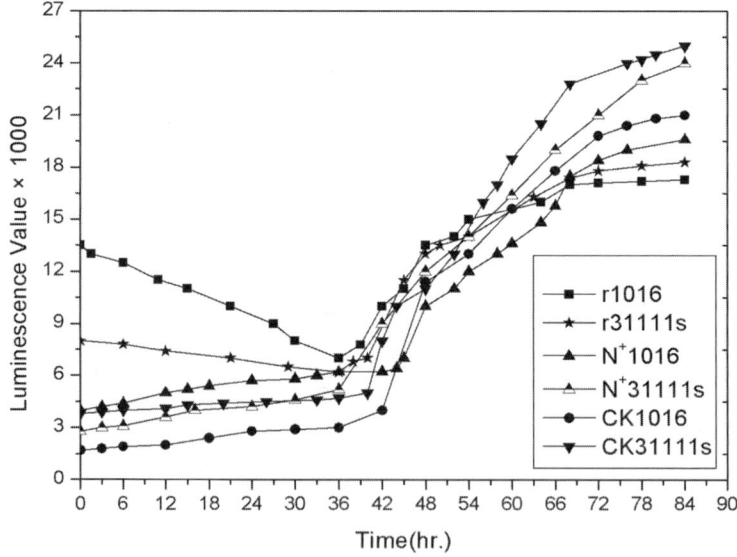

Figure 8.1. Changes in ultra weak luminescence of rice seeds in the germination process after N-ion implantation and γ-ray irradiation, compared with those of the control. In the legends, "1016" and "31111s" are the rice varieties, "r" represents γ-ray irradiation, "N$^+$" represents N-ion implantation, and "CK" represents the control.

8.3.2. Stimulation and Suppression Effects

Stimulation and suppression effects are a kind of artificial classification of biological effects induced by ion implantation [1]. The stimulation effect enhances plant growth activity after ion implantation, whereas the suppression effect decreases the activity or even causes death. In fact, for the biological organism itself, both are responses to ion implantation. Normally stimulation and suppression effects are dependent on ion dose. Low dose generally leads to stimulation, and high dose generally results in suppression.

In one experiment, the target material is wheat (*T. aestivum*) or rye (*Secale cereale*), implanted with 30 keV nitrogen ions to doses of 1–4×10^{16} ions/cm^2 [2]. The fraction of seeds which germinate after 3 days is used as an index of germination potential, and the fraction after 7 days is taken as the germination rate. The results are shown in Table 8.1.

It can be seen from Table 8.1 that the germination potentials of all of the tested materials at either high or low ion doses are lower than that of the control. After 7 days, the germination rates of all of the tested material for low dose (1×10^{16} N ions/cm^2) are either the same as or greater than for the control. From a comparison of the germination abilities of the tested samples between the first 3 days and the last 4 days, it is seen that the germination abilities in the last 4 days at all doses except the highest one, 4×10^{16} N ions/cm^2, partly recover. This may be due to partial repair of the physiological damage induced by ion implantation after a certain time period. But it is generally true that ion implantation has a suppression effect on the germination ability of wheat or rye.

If the ion-implantation-induced physiological damage in the seeds cannot be completely repaired, the germinated seeds may not survive, particularly at high doses. From a mutation breeding point of view, we are more concerned about the relationship between survival rate and ion dose. Table 8.2 shows the effects of N ion implantation on the survival rate of some wheat varieties; the survival rate as a function of ion dose is shown.

Table 8.1. The germination potential and rate for wheat and rye affected by N-ion implantation at various doses.

Germination	Materials	Ion dose (10^{16} N ions/cm^2, at 30 keV)			
		1	2	3	4
Potential (%)	Nongda 139 (wheat)	81.4	77.1	77.1	60.0
	Premebi (wheat)	81.2	78.1	72.0	60.4
	AR72 (rye)	95.2	76.2	71.5	52.3
	AR1 (rye)	94.6	77.3	72.7	56.1
Rate (%)	Nongda 139 (wheat)	99.5	84.5	87.7	58.5
	Premebi (wheat)	100.3	89.4	88.3	52.1
	AR72 (rye)	100.4	85.6	85.6	51.7
	AR1 (rye)	98.1	86.7	83.3	48.0

It can be seen from Table 8.2 and Figure 8.2 that after ion implantation the survival rate clearly depends on ion dose. As distinct from well-known radiation effects, now the survival rate does not logarithmically decrease with increasing dose and there appears a plateau (for Premebi and Youmangbai 4), or the curves even show a tendency of first decreasing and then increasing. The mechanisms involved include not only a fairly strong primary physicochemical repair induced by implanted ions but also some new repair model(s) existing in the organisms, which are to be investigated.

For low dose, N ion implantation has a stimulation effect on seedling growth. Table 8.3 shows seedling heights of the tested samples after one month of turning green. After ion implantation at 1×10^{16} N ions/cm^2, the seedling height for all test samples is 1.1 – 6.2 cm greater than for the control. With increasing dose, the ion implantation suppression effect on the seedling growth becomes increasingly stronger. But when the dose is 4×10^{16} N ions/cm^2, the seedling height generally decreases by 7.6–17.1 cm.

Table 8.2. Effect of ion implantation on the survival rate (%) of wheat seeds

Material	Control	Ion dose (10^{16} N ions/cm^2, at 30 keV)			
		1	2	3	4
Nongda 139	82.2	81.7	73.3	76.7	46.7
Premebi	83.5	83.3	71.7	70.0	43.3
Youmangbai 4	85.0	86.7	76.7	76.7	50.0
Hard-grain Bom	80.0	80.0	63.3	66.7	33.3
8555	53.1	53.0	40.7	50.4	40.3

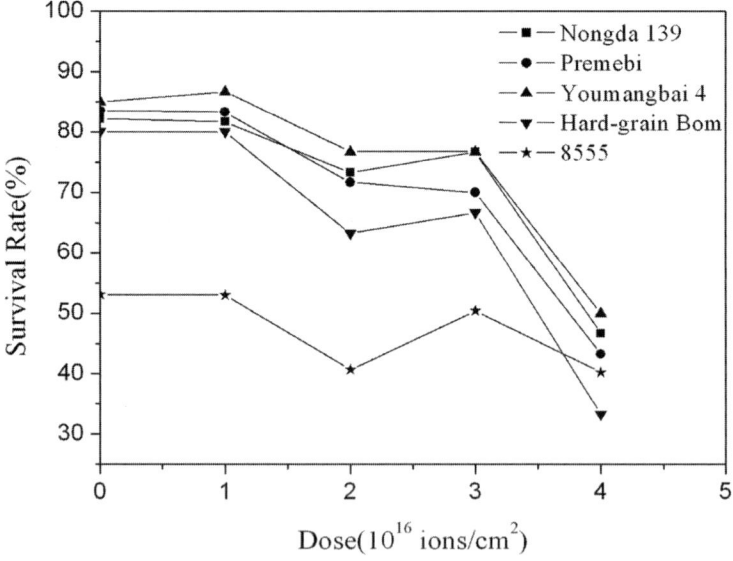

Figure 8.2. Survival rates as a function of ion dose for various N-ion implanted wheat.

Table 8.3. Ion implantation effect on wheat seedling height after a month of turning green. Dose units: N ions/cm^2.

Material	Nongda 139 (wheat)	Premebi (wheat)	Hard-grain Bom (wheat)	AR1 (rye)	Tetraploid rye
Control	50.5	52.4	46.5	75.6	70.4
1×10^{16}	56.4	58.5	47.6	78.4	73.5
2×10^{16}	51.6	52.8	39.8	75.1	70.3
3×10^{16}	47.8	47.2	35.1	69.8	66.2
4×10^{16}	42.9	43.5	29.4	63.4	60.5

8.3.3. Specificity Effects

Specificity as used here refers to properties of the interaction between substances. When a factor selects only a certain specific object to act, this is called the factor specificity. If an object only produces a certain specific effect under the actions of different factors, this is called the object specificity. It is generally known that radiation effects do not possess these specificities. However, for ion implantation, when it acts on some particular objects, certain specificities can be exhibited. For example, it has been found that for crops such as tobacco (*N. tabacum*) and sweet-leaf chrysanthemum, whose seeds are small and have hard shells but leaves are utilizable, after ion implantation of the present generation, leaf production is increased, and the intrinsic composition of the leaf is improved for utilization (Tables 8.4 and 8.5).

Table 8.4. N-ion implantation effect on the intrinsic chemical composition of tobacco (percentage increase or decrease) [3]. N-ion energy: 13 keV, implantation mode: pulse; analysis of intrinsic composition and quality: standard.

Variety	Holo-nitrogen	Holo-phosphorus	Holo-potassium	Protein	Cl	K/Cl
S79-1 (Feng-yang)	-6.9	14.3	17.7	-7.9	-27.4	62.2
K399 (USA)	-19.6	31.4	43.0	-24.1	-8.72	65.6

Table 8.5. N-ion implantation effect on the characteristics of sweet-leaf chrysanthemum (data from Shu Shizhen, Chinese Agricultural Academy) [4].

Germi-nation rate	Seed-ling rate	Leaf produc-tion	Total sugar	Sweet chrysan-themum sugar	High-quality Rebau-dioside A	Total Glycosi-des	A Glycosi-des
+41%	+500%	+41%	+2.8°	+1°	+1.8°	+83%	+162%

There are eight components of special interest in sweet-leaf chrysanthemum, among which are Stevioside and Rebaudioside A. The sweetness of the former is 200 times that of sugar but with a bad taste; the sweetness of the latter is 450 times that of sugar with a sugar-like good taste. Relationships between these substances are indicated below.

> Stevioside
> ↓   dehydrated (OH)
> Steviolbioside   B+H base
> ↓   ferment-dehydrated
> Rebaudioside A

In order to increase the Rebaudioside A content, enzymes should be continuously added to Stevioside for dehydration. Whether ion implantation promotes the generation of Rebaudioside A via auto-metabolism in sweet-leaf chrysanthemum is not yet clear. The mechanisms involved are not understood.

For seeds such as tobacco and sweet-leaf chrysanthemum, the above-mentioned present-generation effects can be repeated to show factor specificities and object specificities when the ion species and dose are appropriately chosen. The biological effects of physical factors are very complicated. The specificities discussed here are only examples. Qualitatively speaking, this kind of specificity is probably related to the expression of some genes that are activated, and thus the kinds, structures and activities of the enzyme proteins determined by these expressions are subsequently changed. The specificity of the enzymatic reaction results in plants evolving in the growth and development process by following a certain change route. If each occasion of ion implantation can activate those genes, the eventual biological effects should be repeatable.

8.4. BIOCHEMICAL EFFECTS

Primary effects produced by ion implantation in organisms induce a series of biochemical processes and cause biological effects. Biological effects can be determined by physical, chemical or biological methods, as well as biochemical analysis.

An amino acid array of any kind of zymoprotein molecules in the organism is determined by a certain sequence of its nucleotide chains (DNA). The regulator sequence on the DNA molecule affects the protein synthesis process, determining the number, space, location and time of protein synthesis. Therefore the activity and spectral bands of the zymoproteins are examined to infer whether ion implantation can cause effects at the genetic level. In this kind of analysis, the isoenzyme is usually used as a biochemical index.

Twelve different patterns were obtained using isoenzyme band analysis of peroxide of second true leaves from 238 present-generation plants of two tested cotton varieties [Xuzhou 553 and CCI (Chinese Cotton Institute) No. 10] implanted with various doses of ions (Figure 8.3) [5]. Maximum 5 bands and minimum 1 band can be seen. The bands from the cathode to the anode are marked as 1A, 2A, 3A, 4A, and 5A, where 4A is the band shared by all individual plants. Aberrations of the isoenzyme bands mainly occur at the cathode region while the activities are mainly exhibited on the 4A band. The enzyme pattern of the control of the two tested varieties is XI. Very few plants are of VII type.

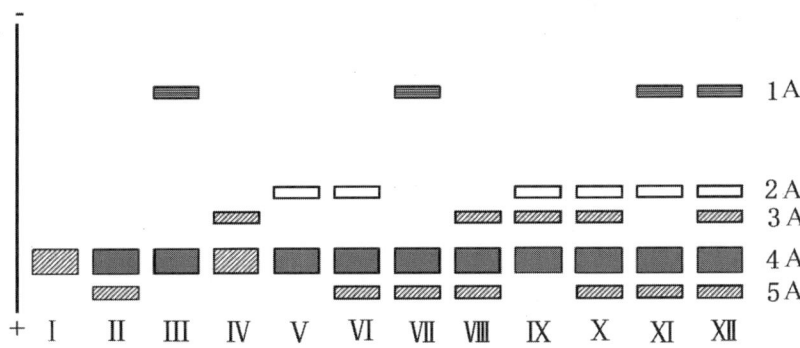

Figure 8.3. Electrophoresis analysis obtained change types of isoenzyme of the second true leaf of M $_1$-generation cotton seeds after N-ion implantation.

The effects of ion implantation dose on the occurrence rate of isoenzyme bands of the second true leaf of M $_1$-generation from ion-implanted cotton seeds are shown in Table 8.6. It can be seen from the table that 1A, 2A, 4A and 5A are the tested variety bands. After N ion implantation there appears a weak band of 3A for about 41% of individuals. Changes in the patterns of all the bands with dose are very different. The occurrence rate of the 4A band does not change with dose and it can be shown as the specific band of the tested samples. The 3A band is a new band. Its occurrence rate increases with increasing implantation dose and reaches a maximum at a dose of 4×10^{16} N ions/cm^2 and then decreases. The occurrence rate of the 5A-band first decreases and then increases with increasing dose. This behavior is similar to the changing tendency of the survival-dose curve given in Chapter 7 (Figure 7.7). Both 3A and 5A bands exhibit specificity of dose effect. The occurrence rate of the 1A and 2A bands decreases as the dose is increased. The relationship between occurrence rate and dose for all bands of ion-implanted cotton seeds of both CCI No. 10 and Xuzhou 553 follow the same pattern.

The electrophoresis patterns of the lipase isoenzyme of ion implanted wheat Premebi and rye AR1 seeds after 48 hours of absorption-swelling are shown in Figure 8.4. It can be divided into E_1 and E_2 zones from the anode to the cathode. In the E_1 zone, ion implantation has no effect on the number of bands and in the general case activity increases with increasing dose. In the E_2 zone, the lipase isoenzyme bands and activities of wheat and rye are related to the implantation dose. It can be seen from the figure that there are eight enzyme bands in the E_2 zone for Premebi and two extra bands, E_2-1a and E_2-1b, at doses of 1, 3, and 4×10^{16} N ions/cm^2. AR1 also has eight bands in the E_2 zone but an extra band of E_2-1a at doses of 1, 3, and 4×10^{16} N ions/cm^2.

The zymoprotein is the direct product of gene expression as well as a catalyst to regulate biochemical reactions in metabolic processes. Ion-implantation-induced changes in this isoenzyme can be considered to be a reflection of DNA structural changes of the genetic substance or phenotypes at the molecular level. These phenotypic changes must produce effects on the formation of organs and growth and development of individuals so as to change the characteristics of the organism.

Table 8.6. The effects of ion implantation dose on the occurrence rate of isoenzyme bands of the second true leaf of M_1-generation from N-ion-implanted cotton seeds. The bands refer to Figure 8.3.

Variety	Ion dose (10^{16} N ions/cm^2)	Occurrence rate of the enzyme band (%)				
		1A	2A	3A	4A	5A
Xuzhou 553	4	68.4	73.7	21.1	100	78.9
	3	76.6	76.5	36.7	100	85.7
	2	83.3	80.0	66.7	100	92.6
	1	93.3	90.0	23.3	100	83.3
	0	100	93.3	0	100	100
CCI No.10	4	78.6	71.4	28.6	100	73.5
	3	80.0	76.0	40.0	100	85.7
	2	83.3	80.0	70.0	100	90.0
	1	90.0	86.7	26.7	100	87.5
	0	100	100	0	100	100
Enzyme band change rate (%)		-17.3	-19.7	40.9	0	-15.4

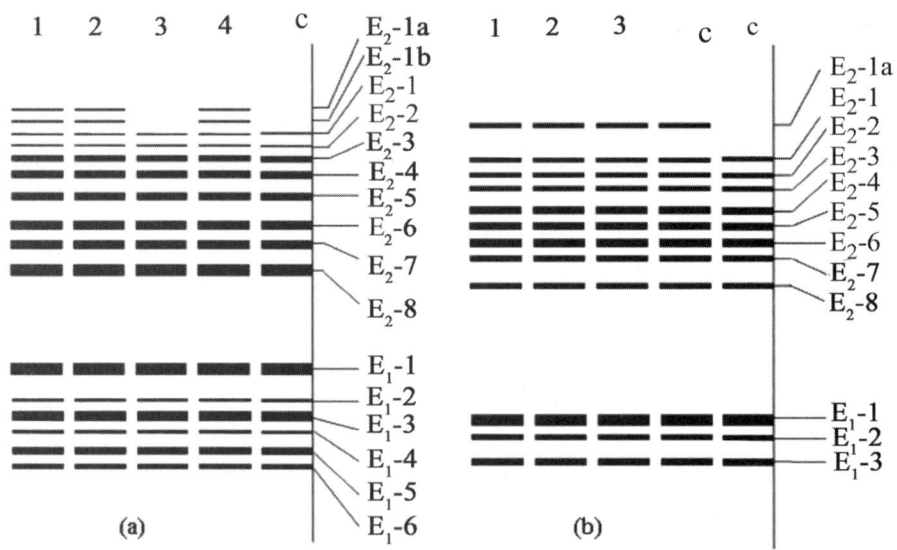

Figure 8.4. Electrophoresis patterns of the lipase isoenzyme of ion implanted (a) wheat Premebi and (b) rye AR1 seeds after 48 hours of absorption-swelling. In both (a) and (b), the right columns marked "C" are the patterns of the control; numbers of 1, 2, 3, and 4 indicate the N-ion doses used in the unit of 10^{16} ions/cm^2.

8.5. GENETIC EFFECTS

Of all the non-lethal effects produced by ion implantation, genetic effects are often of most interest. The aberrations produced by ion implantation are mostly variations which have obvious phenotypes and in the growth and development process exhibit genetic phenomena with visible characteristics such as separation, recombination, linkage and exchange. But the degrees of characteristic variations resulting from these aberrations are normally different. For example, some variations refer to genes which control qualitative character, such as ion-implantation-induced incomplete ventral feet of silkworms, turning the tobacco flower red to white, long grain rice becoming round grain rice, and so on, which are remarkably distinct and are called "great aberrations". Some genes that control the quantitative character, such as ion-implantation-induced changes in the grain red-white color degree, maturation period, and grain quality of wheat, have a continuous spectrum, not easy to distinguish, and must be statistically analyzed for population. This kind of gene aberration is called "micro-aberration" and is controlled by a series of micro-effect polygenes. The effects of these genes are minor but with accumulation they can cause significant change in a trait.

8.5.1. Quality Aberration

Ion-implantation-induced aberrations of the crop qualitative character generally exist, but are fewer than for quantitative character. One of the most typical examples is the temperature-sensitive chlorophyll mutant S_{9040} obtained from ion-implanted present generation of late-mature Zaoxian rice (*Oryza sativa* L. ssp. *indica*, variety: Luwuhong) [6]. This mutant exhibits yellow leaves in the seedling period and chlorophyll synthesis ability increases with increasing temperature. Then the leaf color turns to green and normal in the flowering period. This characteristic aberration continues to the M_{10} generation in which it is still stably inherited. This mutant was crossed and backcrossed with the original variety, and the next generation separations so produced are shown in Table 8.7.

It can be seen from the results in Table 8.7 that the leaf color character of the mutant is controlled by a single recessive gene, and the F_2 leaf color change in green and yellow exhibits a 3:1 simple inheritance. From the results of backcross between the chlorophyll mutant and Luwuhong, it is seen that F_1 exhibits normal green leaves no matter whether from cross or backcross. But for F_2, in the cross combination using the mutant as the parent, the leaf color separation follows the green-to-yellow 1:3 separation proportion, whereas in the backcross combination using Luwuhong as the parent, the leaf color separation does not follow the 1:3 separation proportion. This may be due to inheritance controlled by a nuclear gene. But the behavior of the trait also has certain relations to cytoplasm, attributed to the inheritance of interaction between nucleus and cytoplasm. When the mutant is used as the parent, the nuclear-gene-controlled character can have normal separations which follow the rules. This means that the nuclear genes must be in the mutant cytoplasm background to realize the normal characteristic separation under the coordination of nucleus and cytoplasm. Otherwise, they will exhibit abnormal separations.

Table 8.7. Separation of the etiolation character of the N-ion-implantation-induced S_{9040} mutant in the next hybridized generation.

Combination	P_1	P_2	F_1	B_1			B_2			F_2		
				G (No.)	Y (No.)	X^2 (1:1)	G (No.)	Y (No.)	X^2 (1:1)	G (No.)	Y (No.)	X^2 (3:1)
S_{9040}/ Luwuhong	Y	G	G	33	29	0.143		G		419	117	2.709
S_{9040}/ Zhe15	Y	G	G	23	20	0.209		G		79	21	0.653
Luwuhong/ S_{9040}	G	Y	G		G		26	17	1.49	366	79	12.83
Luwuhong/ Zhe15	G	G	G		G					560		

Note: $X^2_{0.05}1=3.84$, $X^2_{0.01}1=6.63$. G: green, Y: yellow.
P_1: parent 1; P_2: parent 2; F_1: first generation; F_2: hybridized generation from F_1;
$B_1=F_1\times P_1$; $B_2=F_1\times P_2$.

It is commonly thought that aberrations of the allelic genes on the homologous chromosomes in the cell occur independently. The aberration always becomes its allele and produces new differences in the genotype. The leaf-color character of S_{9040} is controlled by a single recessive gene. From the sensitivity dependence of leaf development on temperature, it is seen that the gene controlling this character may be the same as the cold-tolerance gene in the Zaoxian seedling period. They seem to be a pair of single recessive genes. But the critical temperature of the leaf-color turning yellow of the mutant is extended to the normal temperature region. In order to have no separation of this phenotypic character in the offspring, aberrations of pairs of this recessive gene inside several cells in the rice seed embryo must all occur. However, the probability of this happening is extremely low. Ion-implantation-induced phenotypic variations for the present generation also occurred in the mutations of wheat, cotton and tomato. This might be related to interaction between ion beam and dry seeds. There exists a peak in the depth distribution of ions implanted in seeds. Thus the probability of ions simultaneously acting on the same locus in several cells is not negligible.

8.5.2. Variations in Quantitative Character

Studies of the biological effects of ion implantation have been carried out along with plant breeding in synchronism. The studied objects are all economic characters of the crops. For example, the number of ears, plant height, thousand-grain weight and maturation period of rice; the grain weight, protein content and production of wheat; and the fiber length and production of cotton, all are quantitative characters. Therefore, studying variations and genetic rules of quantitative characters is the main way to understand the genetic effects produced by ion implantation. Variations of the quantitative characters have a continuous distribution in the next generation separation. Thus it is impossible to observe clear proportions that may be easily observed for qualitative characters such as the above-mentioned turning yellow and green of the leaves.

In these studies, a large number of individuals are required for statistical analysis, namely, a "population" used as the unit to search for genetic tendencies and rules.

In one experiment, dry seeds of Zaoxian rice (*Oryza sativa* L. ssp. *indica*, variety Zhe-15) were implanted with 30 keV nitrogen ions to doses of 1, 2, 3, and 4×10^{16} ions/cm². The M_1 individuals were densely planted. After ripening, collection was done in a way of three grains per individual mixed in groups. M_2 was individually planted with 2,500 individuals per group. After ripening, 200 individuals per group were collected at random and examined for seeds indoor and the aberration rates were calculated according to the standard of exceeding the critical value of the parent individual (greater or smaller than 1.966%). The results showed that the aberration rates for grain number per ear was 3.11%, for plant height 2.35%, for fruiting rate 1.87%, for earring period 1.05%, for ear length 0.4%, for thousand-grain weight 0.46%, and the total aberration rate was 9.33%.

Plant quantitative characters are controlled by multiple genes. Variations of the quantitative characters are affected not only by the genotypes but also by the environment. Hence in studies of genetic effects, understanding of the fraction of genotype-produced variations in the total variations is very important. If genotype variations dominate, the probability of the next generations exhibiting the parents' characters is high. From a breeding point of view, the reliability of selection based on phenotypes is increased. The measurement of genetic statistic quantity on the relative degree of genotype variations to the total phenotypic variations is called the heritability. The genetic increment obtained by the offspring from the parents via selection is called the genetic progress. The heritability and genetic progress are two important genetic parameters for the study of aberrations of quantitative characters.

The above-described ion implantation of Zhe-15 dry seeds is still taken as an example. In M_3, 35 mutants were picked at random from the total number of mutants. They were individually planted in 77 lines and 10 were used as the control. After ripening, 5 plants were selected at random from each line for seed examination. The heritability in broad sense (h_2), genetic progress (ΔG) and genetic variation coefficient (CVg) of the mutants were estimated as shown in Table 8.8.

Table 8.8. Aberration characters such as heritability, genetic progress and genetic variation coefficient of Zhe-15 dry seeds after 30-keV N-ion implantation. h_2: heritability, CVg: genetic variation coefficient, ΔG: genetic progress, $\Delta G'$: relative genetic progress.

Item	Plant height *	Ear length	Ear number of individual	Grain number per ear	Grain number of fruited ears	Frui-ting rate	Produc-tion of indivi-dual	1000-grain weight
h_2 (%)	74.4	46.2	10.3	52.8	63.5	57.7	23.4	77.2
CVg (%)	6.2	5.8	9.7	13.3	18.5	14.8	21.1	3.9
ΔG (%)	-8.8	1.5	0.3	28.4	26.6	7.4	6.2	2.1
$\Delta G'$ (%)	-11.01	8.12	6.41	19.91	30.37	23.16	21.03	7.06

*: A short stem is used as the selected target and thus negative values occur.

It can be seen from Table 8.8 that in the M_3 generation the mutants exhibit higher or medium heritability in broad sense in the thousand-grain weight (77.2%), plant height (74.4%), grain number of fruited ears (63.5%), fruiting rate (57.7%), grain number per ear (52.8%), and ear length (46.2%), but lower genetic power in the production of individual (23.4%) and ear number of individual (10.3%).

For genetic progress under a selection strength of 5%, the grain number of fruited ears is expected to increase by 26.6 grains, increased by 30.37% compared with that of the original variety. The grain number per ear is expected to increase by 28.4 grains, an increase of 19.91% compared with the control. The fruiting rate is increased by 23.16%, thousand-grain weight increased by 7.06%, and production of individual increased by 21.03% in comparison with those of the control. The plant height is expected to be shortened by 8.8 cm, reduced by 11% compared with that of the control.

The genetic variation coefficient reflects the variation degree induced by ion implantation for each character. The greater character variations include the individual production (21.1%) and grain number of fruited ears (18.5%), while the variations in the plant height, ear length and thousand-grain weight are smaller.

REFERENCES

1. Shao, C.L. and Yu, Z.L., Irritation Effect Model of Low Energy Ions Irradiation, *Nuclear Techniques*, **19**(6) (1996)321-325.
2. Cui, H.R. and Yu, Z.L., Effects of Ions Implantation on Seeds Set in Wild Hybridization of Wheat, *Journal of Anhui Agriculture University* (in Chinese with an English abstract), **21**(3)(1994)303-305.
3. Lin, G.P., A Preliminary Study on Ion Implantation of Tobacco Seeds, *Journal of Anhui Agricultural College* (in Chinese with an English abstract), **18**(4)(1991)317.
4. Shen, M., Wang, C.L., Chen, Q.F., Cytological Effect of Nitrogen Ion Implantation into Stevia, *Acta Agriculturae Nucleatae Sinica*, **11**(3)(1997)141-144.
5. Cheng, B.J., Li, Z., Zhou, L.R., Cytogenetic Effects of Different Ions Implantation in Cotton Seeds, *Journal of Anhui Agriculture University*(in Chinese with an English abstract), **22**(3)(1995)189-195.
6. Liu, F.G., Wu, Y.J., Yu, Z.L., Induction of a Mutant in Leaf Color Sensitive to Temperature, *Journal of Agricultural Sciences,* **24**(1)(1996)16-20.

9

FUNDAMENTALS OF ION IMPLANTATION INDUCED GENETIC VARIATION

Genetics is the study of biological heredity and variation. From a cytogenetics viewpoint, chromosomes are the chief basis of genetics. A chromosome functions according to its basic unit – the gene. Thus, in the final analysis, genetic variations stem from changes in chromosomes and genes, commonly called "mutation".

All living things, including viruses and bacteria, have one or more chromosomes. Genetic substances are situated on the chromosome in a significant sequence. Hence a chromosome is actually a complex higher-class cell organelle, which is the core of cell division activity and provides for growth, development and reproduction with all the necessary information for replication. At the same time, the chromosome is also the product of evolutionary pressure, having high selectivity and adaptability. Any minute variation in the biological organism can be included and transformed only through the carrier of the genetic substance – the chromosome, and then those aberrations that are beneficial to life drive the biological variation, heredity and evolution [1]. Thus the study of ion implantation effects on cytogenetics has great significance to our understanding of the nature of ion-implantation-induced biological effects.

The process of ion implantation producing initial primary biological effects at the molecular level up to the overall final effects on the whole organism crosses a broad range in time and space. If phenotypic effects in the biological organisms are controlled by genetic information, then the chromosome, the carrier of the genetic substance, also experiences changes in time and space. Studies of ion implantation effects on cytogenetics are not limited to the present generation but can be traced to filial generations. These studies focus on root meristematic cells and pollen matricytes as well. This study approach facilitates a comprehensive understanding of the cytogenetic effects of ion implantation for the readers. In the studies, when materials are selected, the homozygous degree should be noted as well as the representative character of the material.

9.1. TYPES OF CHROMOSOME STRUCTURE VARIATION

In a eukaryotic cell, DNA is not independently present, but in a complex form with protein. This material is called chromatin. Chromatin exists in the nucleus, can be stained by alkali dyes, and exhibits various formations along with different cell periods. Under the optical microscope, an interphase chromosome is seen as an irregular reticulate structure. When in the mitosis period, highly agglutinated chromosomes with various characteristics and topographies are formed. Normally, the karyotype of various biological chromosomes is stable. Only when the structural formation and number of chromosomes change will the organism suffer from damage and even death or loss of fertility. A few changes can increase this vigor and fertility and lead to evolution of a new genus.

As pointed out in Chapter 7, gene mutation is point mutation. From a molecular point of view, point mutations are changes occurring inside genes due to increase or decrease or transition or transversion of one to several nucleotides. Aberrations of chromosomes are macro-structural changes, normally meaning changes in the structure, behavior or number of chromosomes. Chromosome aberrations and gene mutations are related each other and cannot be completely separated. Because the chromosome is the gene carrier, any change in chromosome structure or number will be associated with gene changes. On the other hand, a certain gene point on the chromosome changes and then the chromosome structure will be different from the original.

In nature, chromosome aberrations can be caused by effects due to temperature, cosmic radiation or disease. When physical and chemical factors are utilized to treat cells artificially, the rate of chromosome aberration can be greatly increased. As seen from the primary process, after the chromosome suffers damage due to physical and chemical factors, structural changes might result. If the damaged structure recovers, there will be no change in function. But if during the recovery some mistakes occur, variations in characteristics follow. Thus, chromosome damage is a prelude to variations.

Changes in chromosomal structure in the cell can be classified as follows (Figure 9.1).

9.1.1. Deletions

If a small segment is lost from a chromosome in the cell and then disappears during its lifetime, this is called a chromosome deletion. After the deletion occurs, the chromosome segment that contains the centromere can still remain in the newly born cell whereas the segment that does not contain the centromere, which is a so-called fragment, will be lost during cell division. It is frequently seen that after a chromosome bridge is broken, a chromosome with deletions is formed. A chromosome with deletions is shorter than the normal homologous chromosome and thus harmful to the organism. If 1/10 of the length of a chromosome is deleted, non-reproduction of spores is caused and the sporophyte dies.

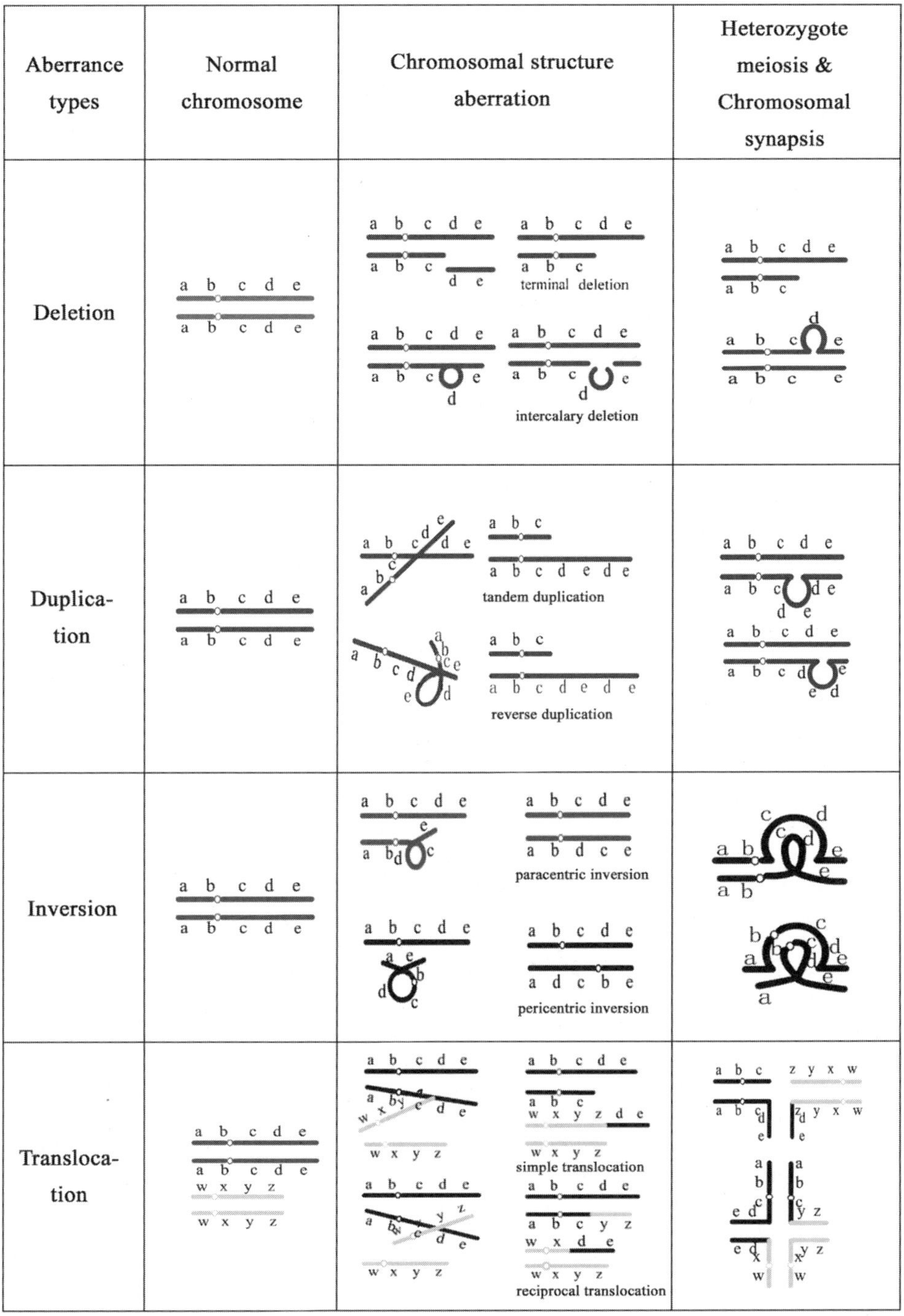

Figure 9.1. Types and illustration of chromosomal structure aberration.

9.1.2. Duplications

If a normal chromosome is increased by a segment with the same sequence, it is called a duplication. If the duplicated segment is linked to the original sequence, it is called tandem duplication; if the segment is linked in the reverse sequence, it is called a reverse duplication. When a pair of homologous chromosomes is broken twice, if asymmetrical exchange links occur, duplication may take place at one of the chromosomes while deletion occurs at another chromosome. Cytologically, looped or nodular structures can be observed in duplicated chromosomes. Duplication also has effects on the organism growth and development as well as the gamete viability, but smaller effects as compared with that of deletion, and duplication that is too large can cause death of the organism.

9.1.3. Inversions

A $180°$-inverted normal sequence of a segment in a chromosome is called an inversion. The inversion can be paracentric or pericentric. An inversion near the centromere occurs inside the arm and the arm length is not changed. In the pachytene stage, the heterotypically inverted chromosomal segments pair each other and thus form inversion loops. If there is an exchange, a dicentric inversion bridge and an acentric piece form and cause the gamete to die. If an inversion of the centromere part occurs between two arms, the arm ratio then changes and can also form an inversion loop. However, even though there are exchanges, they do not form any dicentric inversion bridge and acentric fragments but cause chromosome duplication or deletion and thus the gamete will still die.

9.1.4. Translocations

Deletion, duplication and inversion are all variations of certain segments of the homologous chromosome. But translocation is the transfer of a segment between two pairs of nonhomologs. If two nonhomologs are broken and later on the broken chromosomes and their segments are exchanged and recombined, this is called reciprocal translocation. This is a common type of translocation. If only a segment of a chromosome is transferred to another heterologous chromosome, this is called a simple translocation, which is seldom seen. In the pachytene stage, heterotypically translocated chromosomes are completely paired with each other. If exchanges occur among four arms, these chromosomes are arrayed in loops. Translocations do not normally change the numbers of chromosomes and genes but only alter their original positions. Translocations can cause two normal linkage groups to be reorganized to two new linkage groups.

Chromosomal structure variations can change the characteristics of biological materials. Therefore, in radiobiology, appraisal of chromosome aberration is usually used as a cytological index of whether an organism has mutated. Over more than ten years development of low energy ion beam biology, experiments have demonstrated that variations induced by ion implantation are normally point mutations, namely, gene mutations. Nevertheless, chromosome aberration is still a most active and effective research topic. It has been found that the types of chromosomal structure variations exhibit certain specificity and repeatability, depending on selective implanted ion species and dosage [2,3].

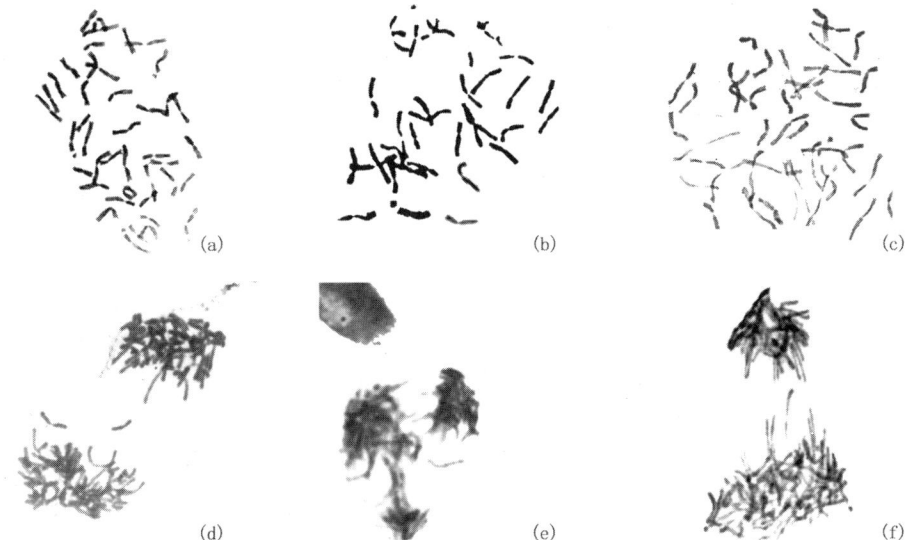

Figure 9.2. Chromosomal aberrations of root meristematic cells of the Premebi M_1 generation. (a) Control, 2n=42, (b) centromere gap, (c) terminal deletion and segment, (d) lagging, (e) 3-pole segregation, and (f) multiple bridge.

9.2. CHROMOSOME ABERRATIONS OF WHEAT PREMEBI

9.2.1. Mitosis of M_1-Generation Root Meristematic Cells

In the metaphase of mitosis of root meristematic cells, the wheat Premebi cell has 42 chromosomes (Figure 9.2). After ion implantation, the number of metaphase chromosomes in the tested cells was not seen to undergo a change, but chromosomal structure variations all existed. These include centromere gaps, chromosomal terminal deletion, one-arm deletion of chromatid and chromosome fragments; chromosome lagging, 3-pole segregation, unequal segregation, and multiple bridge in the anaphase; and multiple bridge, double minute nuclei and micronucleus in the telophase [4]. These change types may occur singularly or in combinative forms with 2 or 3 aberrations, which occur more often than the former. The chromosome aberration rates of the M_1-generation root meristematic cells of N-ion implanted wheat Premebi are shown in Table 9.1.

The centromere gaps, terminal deletion and one-arm deletion are all the results of single breaking of the chromosome. Chromosome fragment may be the result of single breaking (terminal deletion) or double breakings (intercalary deletion). As observed, the locations of chromosome breaking are mainly at the centromeres and the telomere areas. This is because in the metaphase of mitosis, it is difficult to identify from the morphological duplications, inversions and translocations (if the referred chromosome segment is fairly short) formed due to breaks in the middle part of the chromosome arm. The centromere gap is an indicator of primary lesion of the chromosome. It is also a prerequisite for occurrence of breaks at centromere locations, resulting in terminal-centromere chromosome and one-arm deletions of chromatid. At the same time, it can affect the functioning of the centromere and may further cause Robersionian translocations and changes in chromosome number. The micronucleus and minute nuclei are formed from lagging chromosomes or segments, which cause both chromosomal structure and behavioral variations. The 3-pole segregation, unequal segregation and lagging chromosomes are behavior variations caused by lesion at the centromeres and function variations of spindles. The appearance of the bridge reflects the structural variations due to inverted paracentric duplication occurring in sister chromatids. All of these can lead to changes in chromosomal behavior, structure and number in the next cell division.

9.2.2. Meiosis of M_1-Generation Pollen Matricytes

In metaphase I of the meiosis of wheat Premebi, 21 bivalents are arrayed in sequence on the equatorial plate. In the bivalent, closed circular bivalents are in the majority (average 18.6 per cell) and open bar-like bivalents are in the minority (average 2.4 per cell). In anaphase I, homologously paired bivalents are normally separated, and 21 dyads are equally concentrated at two terminals of the dividing cell, without chromosomal lagging behavior. Thus, it can be seen that Premebi is a stable variety and its meiosis follows the normal laws.

After ion implantation, in metaphase I there appear phenomena such as monovalent or even bivalent lagging, and terminal centromere chromosome lagging. The pairing configurations also have changes. The circular bivalents decrease, bar-like bivalents increase, and unpaired monovalents appear. In anaphase I, there are aberrations such as bivalent lagging, unequal segregation and lagging of dyads, and early segregation of the lagged dyads (Figure 9.3). In anaphase and telophase I and anaphase II, chromosome bridges are observed (including mono-bridge, double-bridge and multi-bridge). The aberration rates of the M_1-generation pollen matricytes in meiosis after N-ion implantation are shown in Table 9.2.

Table 9.1. Mitotic chromosome aberrations of M_1-generation root meristematic cells of wheat Premebi after N-ion implantation at various doses.

Dose (10^{16} ions/cm^2)	0	1	2	3	4
Total number of cells observed	368	372	364	382	376
Number of aberrations in metaphase	0	4	8	12	13
Number of aberrations in anaphase and telophase	2	6	11	14	18
Overall rate for aberrations (%)	0.54	2.69	5.22	6.81	8.24

Table 9.2. Chromosomal configuration variations of M_1-generation pollen matricytes in meiosis in metaphase I after N-ion implantation at various doses.

Dose (10^{16} ions/cm^2)	0	1	2	3	4
Aberration rate (%)	0.73	3.19	5.88	7.96	9.06

Table 9.3. Chromosome aberration rates of Premebi M_2-generation root meristematic cells in mitosis after N-ion implantation at various doses.

Dose (10^{16} ions/cm^2)	0	1	2	3	4
Aberration rate (%)	0.51	1.83	3.20	3.97	5.22

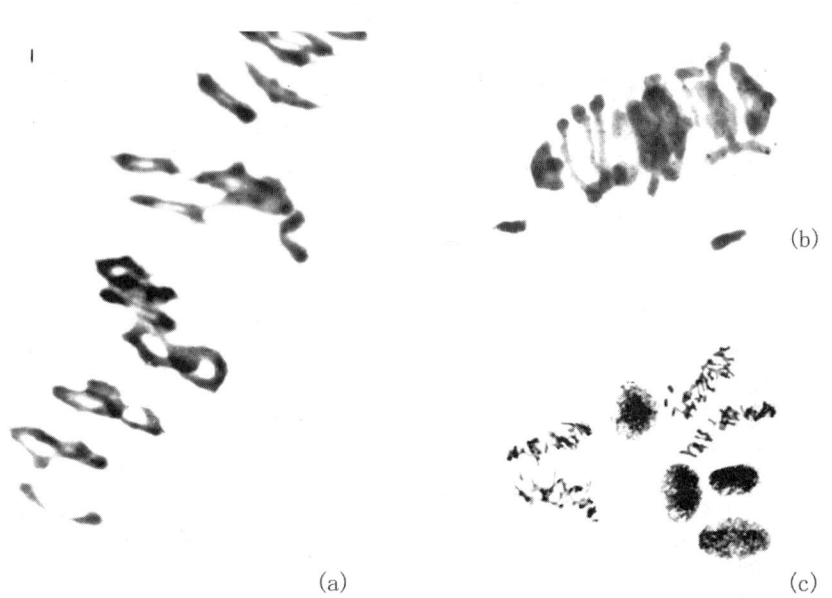

(a) (b) (c)

Figure 9.3. Chromosomal aberrations of wheat Premebi M_1 generation in meiosis. (a) Control, 2n=21, (b) two terminal centromere laggings, and (c) anaphase-II bridge.

9.2.3. Mitosis of M_2-Generation Root Meristematic Cells

After ion implantation, a fraction of the root meristematic cell individuals of the Premebi M_2 generation maintain the normal chromosome number and a few have 41 chromosomes (Figure 9.4) in the metaphase, showing changes in chromosome number. At the same time, in some metaphase divisions, chromosome segment, terminal deletion, chromosome lagging, chromosome bridges and unequal segregation can be observed. The total aberration rate (Table 9.3) increases with increasing ion dose, exhibiting a linear relationship with dose.

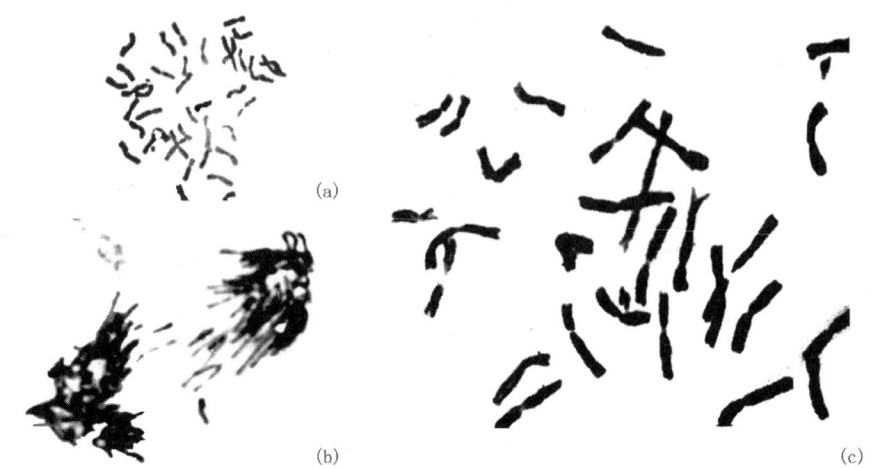

Figure 9.4. Chromosomal aberrations of Premebi M_2 generation in mitosis. (a) Control, 2n=41, (b) bridge and lagging, and (c) terminal deletion and segment.

Table 9.4. Chromosome aberration rates in the anaphase and telophase of Premebi M_2-generation meiosis after N-ion implantation at various doses.

Dose (10^{16} ions/cm^2)	0	1	2	3	4
Number of cells observed	380	370	352	350	330
Number of cells with aberrations	3	11	20	26	28
Aberration rate (%)	0.79	2.97	5.68	7.43	8.79

It can be seen from a comparison of Table 9.1 and Table 9.3 that the chromosome aberration rate of the M_2-generation mitosis is lower than that of the M_1 generation. This indicates that ion-implantation-induced chromosome aberrations can be transmitted in part to next generations in the processes of multiple cell divisions between generations and sexual reproduction, but other parts may be lost or repaired.

9.2.4. Meiosis of M_2 Generation Pollen Matricytes

In meiosis of the M_2-generation pollen matricytes of wheat Premebi, in metaphase I not only changes in chromosome configuration (the circular bivalents decrease as increasing of the dose, but bars increase) are maintained, but also some monovalents appear (about 1.6%) and mostly show lagging behavior. Furthermore, in anaphase I many chromosomes show lagging phenomena. These include bivalent lagging, dyad lagging and early segregation, and dyads and chromatid lagging. In anaphase I as well as telophase I (Figure 9.5) chromosome bridges occur.

The chromosome aberration rates for the anaphase and telophase of the M_2-generation pollen matricyte meiosis are shown in Table 9-4. It can be seen that the chromosome aberration rate is linearly related to the ion dose, and that the rate decreases slightly compared to that for the M_1 generation. This indicates that ion-implantation-induced genetic variations are indeed transferred to the M_2 generation and, on the other hand, the deleted and repaired chromosome aberrations are not noticeable.

9.3. CHROMOSOME ABERRATIONS OF RYE (*SECALE CEREALE* L.) AR1

9.3.1. Mitosis of M_1-Generation Root Meristematic Cells

In the normal mitosis of rye AR1, there are 16 (14R + 2B) or 18 (14R + 4B) chromosomes due to different numbers of B-chromosomes. After ion implantation, the chromosome aberration types are basically the same as those of wheat Premebi but with a few differences such as centromere gaps in the metaphase and chromosome lagging, multi-bridges and unequal segregations in the anaphase and telophase.

(a) (b) (c)

Figure 9.5. Chromosomal aberrations of the wheat Premebi M_2 generation in meiosis. (a) Early segregation, (b) dyads and chromatid lagging, and (c) bridge.

Table 9.5. Chromosome aberration rates of M_1-generation root meristematic cell mitosis of rye AR1 after N-ion implantation at various doses.

Dose (10^{16} ions/cm^2)	0	1	2	3	4
Number of cells observed	314	325	342	346	314
Number of aberrations in metaphase	0	1	4	6	8
Number of aberrations in anaphase and telophase	2	5	10	13	14
Overall rate for aberrations (%)	0.64	1.85	4.09	5.49	6.92

Table 9.6. N-ion implantation effect at various doses on euchromosomal configurations of AR1 in metaphase I.

Dose (10^{16} ions/cm^2)		0	1	2	3	4
Number of cells observed		80	60	50	50	50
Monovalent		0	0.1	0.12	0.12	0.08
Bivalent	per cell	7.0	6.85	6.78	6.70	6.76
Tetravalent		0	0.05	0.08	0.12	0.10

Table 9.7. Effect of B-chromosomes on forms of euchromosomal bivalent after N-ion implantation at various doses. The number of the bivalents is counted per cell.

Dose (10^{16} ions/cm^2)	Number of cells observed	Unpaired B-Chromosome		Paired B-Chromosome		Average number of	
		Number of		Number of		circle II	bar II
		circle II	bar II	circle II	bar II		
0	80	6.30	0.70	6.20	0.80	6.25	0.75
1	50	6.06	0.94	4.92	2.08	5.49	1.51
2	50	6.02	0.98	4.34	2.66	5.18	1.82
3	50	5.88	1.12	3.58	3.42	4.73	2.27
4	50	6.14	0.84	4.76	2.24	5.46	1.54

Chromosome aberration rates in root meristematic cell mitosis of rye AR1 are shown in Table 9.5. Compared to Table 9.1, the aberration rates at various doses are somewhat lower than those for wheat Premebi. But the relationship between rate and ion dose is the same for both materials.

9.3.2. Meiosis of M$_1$-Generation Pollen Matricytes

In normal cases, in metaphase I of AR1, the euchromosomes (R chromosomes) are paired into 7 bivalents, and the ultra chromosomes (B chromosomes) may either form bivalents or appear in the form of monovalents. After ion implantation, the euchromosome configurations are changed (Table 9.6) and monovalents and tetravalents appear at certain rates. The tetraploid is in an "8"-shape which is a typical characteristic of the chromosomal hybrid translocation. Translocations among euchromosomes are observed at four ion doses. The translocation between the euchromosome and the B-chromosome is present only at a dose of 2×10^{16} ions/cm^2. In cells without tetravalents and monovalents present, the formation of euchromosomes paired to form bivalents is also changed. In the cells that contain either one or two pairs of B-chromosomes, only if B-chromosomes are paired, the euchromosomes have bar-bivalents, and the number of euchromosomal bar-bivalents when there are two B-bivalents is more than when there is one B-bivalent. When B chromosomes are not paired, euchromosomal pairs are mostly present in the form of circular bivalents (Table 9.7). This fact reflects the genetic homeostatic effect of organisms, namely, the exchange rate reaches an optimal value through self-regulation. As the ion dose increases, the number of bar-bivalents increases, indicating a decrease in the euchromosomal crossover rate. This is due to, on the one hand ion-implantation-induced direct changes in the chromosomal structure, and on the other hand the influence of B-chromosomes on euchromosomal crossover and recombination. Ion implantation indirectly induces an increase in the number of bar-bivalents through influencing B-chromosomal functions.

In anaphase and telophase I, the chromosomal aberration types are fairly plentiful (Figure 9.6). In euchromosomes, 3-pole segregation, chromosome lagging, early-segregation-formed chromosome bridge and micronucleus occur. B-chromosomes exhibit dyad lagging, chromatid lagging, and go to a pole without segregation; simultaneous two dyads lagging and early segregation, one dyad lagging and early segregation whereas another goes to a pole. This kind of early segregation of B-chromosomes guarantees it to be transmitted to every individual of the filial generation. In anaphase and telophase II, the euchromosomes and B-chromosomes also exhibit abnormalities such as chromosome lagging, chromosomal single and double bridges, and B-chromosomal single bridge. In the diad and quadrant periods, aberration types such as single and double micronuclei, quadrant with micronucleus, and pentant also appear. Chromosome aberration rates of AR1 after N-ion implantation are given in Table 9.8.

9.3.3. Mitosis of M_2 Root Meristematic Cells

In metaphase I of the M_2-generation root meristematic cells of rye AR1, the number of chromosomes changes (15R + 4B) (see Figure 9.7). In the anaphase there is chromosome lagging, 3-pole segregation, and chromosome bridge, and in the telophase there also appears chromosome bridge. The total aberration rates due to N-ion implantation are shown in Table 9.9.

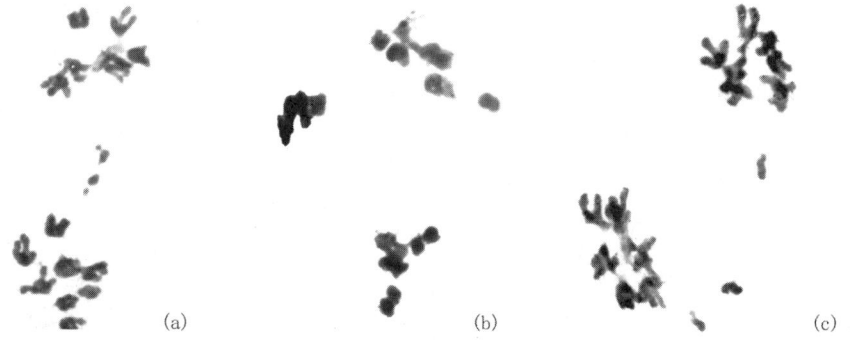

(a) (b) (c)

Figure 9.6. Chromosomal aberrations of the M_1 generation of rye, AR1, in meiosis. (a) One lagging dyad, B-chromosome early segregation with one already going to a pole (anaphase I), (b) 3-pole segregation (anaphase and telophase I), and (c) three B-chromatid lagging.

Table 9.8. Chromosome aberration rates of M_1-generation meiosis of rye, AR1, after N-ion implantation at various doses.

Dose (10^{16} ions/cm^2)	0	1	2	3	4
Rate (%)	0.88	3.03	5.31	7.62	8.71

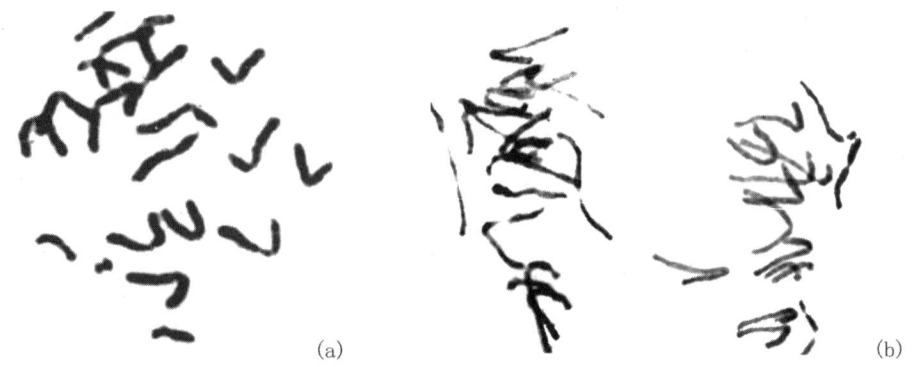

(a)   (b)

Figure 9.7. Mitosis of rye AR1 M_2 root meristematic cells. (a) $2n = 19 = 15(R) + 4(B)$, and (b) lagging chromosome in anaphase.

Table 9.9. Chromosome aberration rates of M_2-generation mitosis of rye, AR1, after N-ion implantation at various doses.

Dose (10^{16} ions/cm^2)	0	1	2	3	4
Rate (%)	0.55	1.54	2.85	3.55	4.72

9.3.4. Meiosis of M_2 Pollen Matricytes

From the diakinesis to metaphase I of meiosis of the M_2-generation pollen matricytes of ion-implanted rye AR1, polyvalents are first produced. The tetravalents have types in the shapes of "+", circle, "8" and chain, and some are hexavalents (Figure 9.8). Secondly, the formation of the bivalent synapsis is changed. As happens to the M_1 generation, circular bivalents decrease with increasing ion dose, while bar bivalents increase. Additionally, in cells without polyvalents, about 3/4 of the cells exhibit the appearance of bar bivalents in the presence of B-chromosomes. This is similar to characteristics controlled by a pair of dominant/recessive genes exhibiting Mendelian segregation in the M_2 generation. For the B-chromosome, since the M_1-generation pollen matricyte has early segregation, which guarantees B-chromosome transmission to every individual of the filial generation, individuals without B-chromosomes are not observed in the M_2 generation. It should be mentioned that B-chromosomal number aberration is observed in M_2 such as individuals having 6 B-chromosomes. When 6 B-chromosomes are paired in each of 3 bivalents, euchromosomes form 6 bar bivalents and 1 circular bivalent. This further shows cross effects of B-chromosomal behavior change on the euchromosomes due to ion implantation.

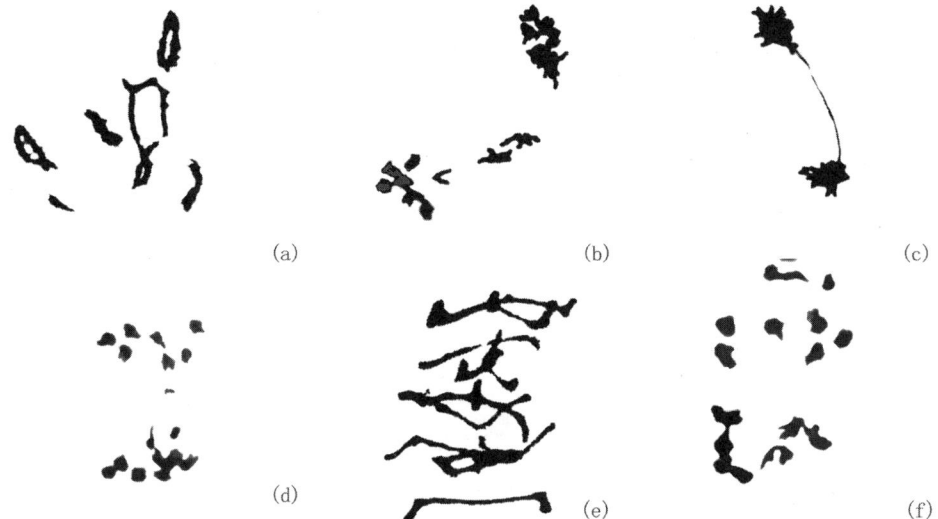

Figure 9.8. Meiosis of rye AR1 M_2 chromosomes. (a) 2n = 1 IV ("8"-shape R) + 4 II (R) + 1 II (B), (b) dyad and chromatid (R) lagging, (c) I bridge in telophase I, (d) B-chromosome bridge (anaphase I), (e) 2n = 1 II (circular R) + 6 II (R) + 3 II (B), and (f) unequal segregation and bridge (R).

In the anaphase and telophase I, there also exist numerous of chromosome variations, as for M_1-generation meiosis. Examples include unequal segregation and R chromosome bridge, B-chromosome going to a pole without segregation, euchromosome dyad lagging and chromatids formed with early segregation, B-chromosomal bridge and chromosome lagging, and lagging chromosome and bridge simultaneously occurring, etc. From anaphase II to telophase II, chromosome aberrations include euchromosome bridge, B-chromosome lagging, and quadrant with single or double micronuclei. The chromosome aberration rates in anaphase and telophase due to N-ion implantation are shown in Table 9.10.

9.4. CHROMOSOME ABERRATIONS OF HOMOLOGOUS TETRAPLOID RYE (*SECALE CEREALE* L.)

9.4.1. Mitosis of M_1-Generation Root Meristematic Cells

Homologous tetraploid rye has 28 chromosomes. In cells observed of the M_1-generation mitosis, no change in the number of chromosomes was found. The structure and topography variations include centromere gaps, chromosome adhesion, terminal deletion, chromosome fragments and circular chromosomes in metaphase; in the anaphase there is chromosome lagging, 3-pole segregation and unequal segregation; and in the telophase, chromosome bridges appear. The chromosome aberration rates are higher than those for wheat Premebi and rye AR1 (Table 9.11), and show a linear relationship with the ion dose.

Table 9.10. Chromosome aberration rates in the anaphase and telophase of M_2-generation meiosis of rye AR1 after N-ion implantation at various doses.

Dose (10^{16} ions/cm^2)	0	1	2	3	4
Rate (%)	0.91	2.53	3.76	4.82	6.13

Table 9.11. Chromosome aberration rates of M_1-generation root meristematic cell mitosis of tetraploid rye after N-ion implantation at various doses.

Dose (10^{16} ions/cm^2)	0	1	2	3	4
Number of cells observed	436	440	427	432	424
Number of metaphase aberrations	0	3	13	16	15
Number of anaphase and telophase aberrations	3	10	15	18	25
Total aberration rate (%)	0.69	2.95	6.56	7.87	9.43

9.4.2. Meiosis of M_1-Generation Pollen Matricytes

Although each chromosome of the homologous tetraploid rye has the same 4 strips, in metaphase I of meiosis of the control, the chromosome configuration has a tendency to form bivalents [5]. After ion implantation, the chromosome configuration has obvious changes (Table 9.12). The tetraploid number increases compared with that of the control, reaching 4-5 in some cells, meanwhile monovalents and quite few hexavalents appear. These changes indicate that the effect of ion implantation on the chromosome synapsis is caused by chromosomal structural changes. In particular, the appearance of hexavalents demonstrates the existence of continuous translocation. In metaphase I, the phenomenon of monovalent lagging also occurs.

In anaphase I to telophase II, there exists a series of chromosome abnormalities such as dyad 15-13 segregation, dyad lagging, and chromosome bridge formation; in telophase, chromosome lagging and double micronuclei; and in anaphase II, chromosome 15-13 segregation. In telophase II, there are chromosomal double bridges, chromatin bridge formation and chromosome lagging. When quadrants form, single and double micronuclei also appear. The chromosome aberration rates are shown in Table 9.13. Comparisons of Table 9.11 and Table 9.13, Table 9.5 and Table 9.8, and Table 9.1 and Table 9.2 show clearly that for all of the tested materials, the chromosome aberration rates of meiosis of M_1-generation pollen matricyte are higher than for mitosis. This is because meiosis is a more complicated and precise process than mitosis. On one hand, meiosis is more sensitive to physical factors such as ion implantation, and on the other hand, it also shows that ion-implantation-induced chromosome aberrations can be contained and transmitted by sex cells.

Table 9.12. The numbers of various chromosome configurations counted per cell in metaphase I of M_1-generation meiosis of tetraploid rye after N-ion implantation compared with that of the control (unimplanted).

Dose (10^{16} ions/cm^2)	Number of cells observed	Monovalent	Bivalent	Tetravalent	Hexavalent
0	60	0	11.55	1.22	0
1	60	0.24	8.40	2.74	0
2	50	0.20	8.06	2.88	0.04
3	40	0.25	7.43	3.15	0.05
4	40	0.20	8.20	2.85	0

Table 9.13. Chromosome aberration rates in the anaphase and telophase of M_1-generation meiosis of tetraploid rye after N ion implantation.

Dose (10^{16} ions/cm^2)	0	1	2	3	4
Rate (%)	1.02	3.79	7.50	9.12	10.91

Table 9.14. Chromosome aberration rates of M_1-generation mitosis of tetraploid rye after N ion implantation.

Dose (10^{16} ions/cm^2)	0	1	2	3	4
Rate (%)	0.56	2.13	3.80	4.86	6.87

9.4.3. Mitosis of M_2 Root Meristematic Cells

In the metaphase of mitosis of M_2-generation root meristematic cells of tetraploid rye, the chromosomal structure and number of some individuals are normal whereas other individuals show both change in the number of chromosomes and differences in structural variations. M_1-generation meiosis normally produces gametes with n = 13 or n = 15 chromosomes. In M_2 root-meristematic-cell mitosis there are individuals having the numbers of chromosomes of 26, 27 and 29 (Figure 9.9). Furthermore, in M_1 the observed centromere gaps, terminal centromere chromosome and chromosome fragments are also transmitted to the M_2 generation. In the anaphase, there also appears to be chromosome lagging, bridge formation, 3-pole segregation, and unequal segregation.

Chromosome aberration rates of mitosis of M_2-generation root meristematic cells of tetraploid rye after N-ion implantation are shown in Table 9.14. These are trends of wheat Premebi and rye AR1, for which the mitotic chromosome aberration rates of the M_2 generation are somewhat lower than those of the M_1 generation.

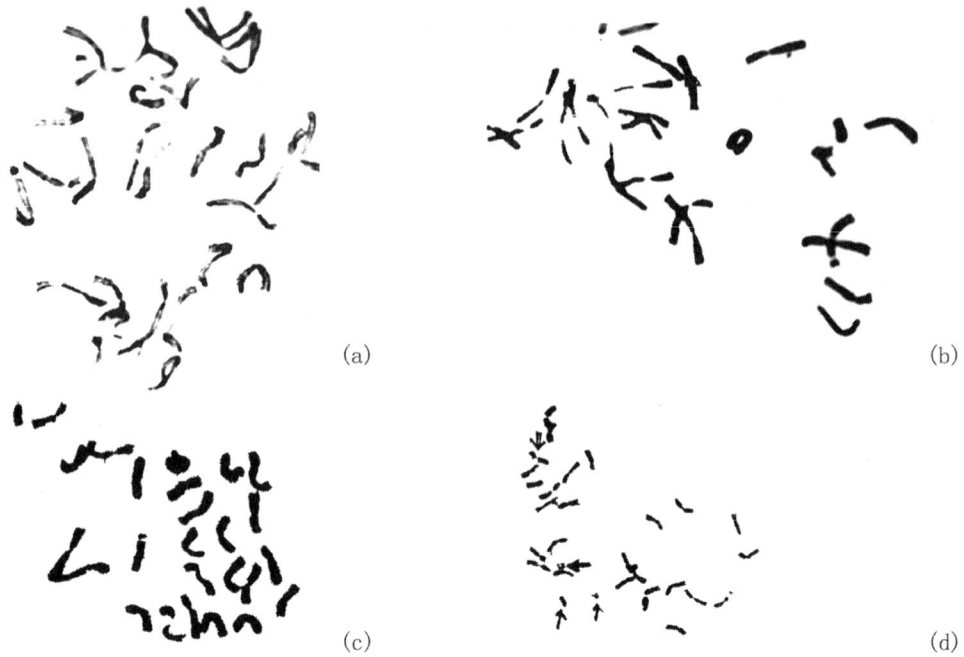

(a)   (b)

(c)   (d)

Figure 9.9.   Change in the number of chromosomes of tetraploid rye after N-ion implantation. (a) 2n = 29, (b) 2n = 26 + 1t (terminal centromere chromosome), (c) 2n = 26 (centromere gap), and (d) 2n = 27 + 1t + 1F (segment).

9.4.4. Meiosis of M_2 Pollen Matricytes

In metaphase I of meiosis of tetraploid rye, the number of chromosomes is still 28, and generally the tetravalents resulting from pairings are fewer. After ion implantation, in metaphase I of M_2 meiosis, the chromosome configuration changes (Table 9.15). Compared with the changes in chromosome configuration of M_1 meiosis, the relationship with ion dose has the same tendency, but is different in that triploids occur in the M_2 generation at three doses of radiation. It should be mentioned that in metaphase I, many chromosomes are strung together, forming a chromosome chain on the equatorial plate (Figure 9.10), similar to Renner's complex particularly existing in Evening Primrose (*Oenothera lamarckiana* Ser) [6]. At the same time, in metaphase I chromosome inversion is observed. As seen from chromosome morphology, it is an intra-arm inversion. In terms of chromosome number, there appears to be terminal centromere chromosomes with numbers of 26, 27, and 27+2, and 29-chromosome individuals. In these individual pollen matricytes, there are monovalents, bivalents, trivalents and tetravalents. The number of the tetravalents is 3 or more, an obvious increase compared with the number of tetravalents of the control.

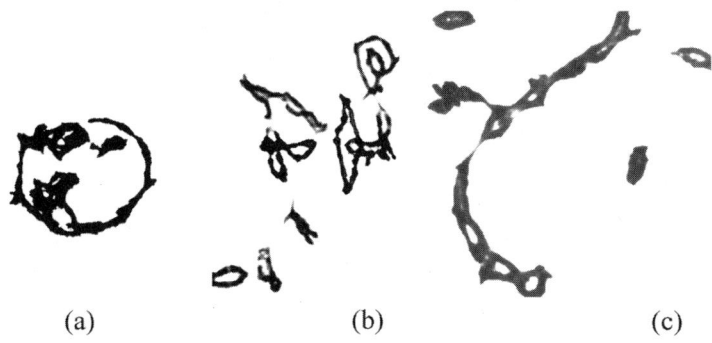

(a) (b) (c)

Figure 9.10. Chromosomal aberrations of the M_2 generation of the tetraploid rye in meiosis. (a) Circular tetravalent, (b) inversion ring, and (c) Renner's complex-like aberration.

Table 9.15. The numbers of various chromosome configurations counted per cell in M_2-generation meiosis of tetraploid rye after N-ion implantation compared with that of the control (unimplanted).

Dose (10^{16} ions/cm^2)	Number of cells observed	Mono-valent	Bivalent	Trivalent	Tetra-valent	Hexa-valent
0	50	0	11.44	0	1.28	0
1	55	0.18	8.20	0	2.86	0
2	75	0.16	7.64	0.05	3.04	0.04
3	62	0.19	7.23	0.03	3.24	0.05
4	50	0.14	7.86	0.02	3.02	0

Table 9.16. Chromosome aberration rates of M_2-generation meiosis of tetraploid rye as a function of N-ion implantation dose.

Dose (10^{16} ions/cm^2)	0	1	2	3	4
Rate (%)	0.98	3.91	7.95	9.64	11.21

In anaphase I to telophase I, chromosome abnormalities are mainly exhibited as dyad lagging and dyad early segregation to form chromatids. Unequal segregations of dyad 15-13 and 16-12 occur, very often accompanied by the appearance of chromosome bridges. There also appears to be 3-pole segregations with lagging chromosomes, bridges with lagging chromosomes, and 3-divided-body accompanied chromosome lagging with fragments and micronuclei. It is apparent that in a cell, there may be more than one type of chromosome abnormality occurring simultaneously.

In anaphase II to telophase II, M_2-generation meiosis of tetraploid rye still gives rise to chromatids, and 16-12 and 15-13 unequal segregations, which can form quadrant spores with the numbers of chromosomes of 13, 15 and 16. Additionally, there also appear variations such as bridge, lagging and minute nucleus. The total aberration rates are shown in Table 9.16.

It can be seen from Table 9.16 that after ion implantation of tetraploid rye, in the anaphase and telophase the chromosome aberration rates of M_2-generation pollen matricyte meiosis are higher than for the M_1 generation. This implies that ion-implantation-induced chromosome damage is not easily repaired [7,8]. Nevertheless, two points are clear. First, M_2 is the segregation generation and thus various chromosome aberrations can all be exhibited, while M_1-generation chromosome aberrations can recombine through the generative propagation process to produce new aberration types. Second, tetravalents and hexavalents in M_2 pollen matricytes may be inharmonious in the segregation process and result in increasing chromosome aberration rates in the M_2 generation.

9.5. TYPES AND CHARACTERISTICS OF CHROMOSOME ABERRATIONS

The previous three chapters showed that N-ion-implantation-induced genetic effects to root meristematic cell mitosis and pollen matricyte meiosis can be treated at the chromosomal level for two continuous generations of typical wheat varieties. The results demonstrate that ion-implantation-induced chromosome aberrations have not only complicated types and generation transmission but also their own unique characteristics.

From the viewpoint of cytogenetics, the types of chromosome aberrations can be classified roughly into variations in the behavior, structure and number of chromosome. However, of course this classification is not absolute. In fact, there are no uncrossable gaps among chromosome variations in behavior, structure and number, but often relationships of causation among them to transform one into another.

Chromosome lagging, unequal segregation and chromatid early-segregation are typical forms of the chromosomal behavior variations. They are all caused by spindle divergence [9]. The occurrence of monovalents from metaphase I of meiosis is caused by ion implantation induced non-coordinate behavior of the chromosome. But trivalents, tetravalents, hexavalents and Renner's complex are combined behavior aberrations transformed from chromosome structure variations. Note that for Renner's complex, since there are no reports from previous studies on ionizing radiation effects on chromosomes. The significance resides in maintaining formation of new life species with the eternal heterozygosity of chromosome structures [10].

As mentioned in the first section of this chapter, chromosomal structure variations can be classified as any of four types, namely, deletion, duplication, inversion and translocation. It can be seen from the results described in the last three sections that N-ion implantation can induce all four types of chromosomal structure variations. The chromosome fragment, centromere gap and terminal centromere chromosome are evidence of chromosome breaks. The circular chromosome is a product of fusion of terminal breaks. Chromosome terminal deletion, chromatid terminal deletion and one-arm deletion are visual exhibitions of N-ion implantation induced chromosome deletion. The increase in tetravalents in meiotic metaphase I of tetraploid rye pollen matricytes after ion implantation indicates the possibility of chromosome translocation, whereas the occurrence of tetravalents in diploid rye is an example of chromosome translocation. The occurrence of hexavalents and Renner's complex in tetraploid rye further demonstrates the existence of chromosomal continuous translocation [5,10]. The occurrence of a single bridge or a double bridge in meiotic anaphase-telophase I or anaphase-telophase II of pollen matricytes is the main indicator of chromosome inversion and duplication. The single bridge in anaphase-telophase I is a product of single exchange or three-line double exchanges occurring in the chromosomal intra-arm inversion circle. The double bridge indicates that there have been four-line double exchanges in the intra-arm inversion circle, or simultaneously single exchanges outside the circle or four-line double exchanges in the circle. A single bridge in anaphase-telophase II is a secondary structure variation of reversed tandem duplication (or inverted parallel duplication). The double bridge is a rare product resulting from four-line exchanges inside and outside the chromosomal in-arm inversion circle [1,5]. In addition, the intra-arm inversion circle observed in tetraploid rye is direct evidence of chromosomes having structural variation, namely, inversion.

Ion-implantation-induced chromosomal number variation has a low rate, nevertheless, it has been found in ion-implanted wheat – Premebi, rye AR1 and tetraploid rye. Chromosomal behavior variations including chromosome lagging and unequal segregation, three-pole segregation and chromatid early-segregation result in either an increase or a decrease in the number of chromosomes in daughter cells. If this case occurs in mitosis, chimeras may be formed; and if this occurs in meiosis, gametes that have a chromosome number greater or less than normal gametes may be produced. After these abnormal gametes participate in the insemination process, aneuploids can be produced in the next generation. Individuals having 41 chromosomes occurring in the M_2 generation of wheat Premebi are the products of combination of the gametes with $n = 20$ and normal gametes ($n = 21$). An excess chromosome ($2n = 19 = 15R + 4B$) in rye AR1 is formed by the insemination of $n = 8R$-gametes and normal gametes ($n = 7R$). In anaphase II of the M_1-generation pollen matricytes of tetraploid rye, chromosomes have the 15-13 segregation phenomenon. This can produce female and male gametes with $n = 13$ and $n = 15$, which, together with normal female and male gametes, participate in the insemination process to form zygotes having the number of chromosomes in the range $26 – 30$. This is why individuals with the numbers of chromosomes of 26, 27 and 29 appeared in the M_2 generation. Ion-implantation-induced chromosomal number variations are significant for selection and breeding of aneuploids.

There is a complete maintenance system in organisms to repair damage caused by external factors. If the damage to the genetic substance (DNA) is repaired, mutations cannot occur. Higher plants have a greater ability to repair ionizing-radiation-induced damage. It has been reported that a γ-ray-induced break of a single strand of carrot DNA can be repaired 50% within 5 minutes [11]. Similar reports [12] are reported for young wheat embryos. In the process of mutagenesis, only those that are unrepaired or repaired with mistakes can produce mutations in DNA replication. Since the mutagenesis occurrence has the feature of a single cell event, the cells compete with normal cells in division and proliferation and some are eliminated. Even though defective cells are not eliminated, this kind of cell only divides into somatic cells in the growth and development process. The somatic cells die and disappear as the individuals senesce. Only when these mutated cells transmit mutations to the male and/or female gametes and bring the mutations into the zygotes through the insemination process to form young embryos, can mutations be maintained and transmitted to form mutated individuals. Practice has demonstrated that ion-implantation-induced variations of the structure, behavior and quantity of chromosome — the genetic substance carrier — are transmitted from the M_1 generation to the M_2 generation according to the above-described process. If time is plotted as the horizontal axis and chromosome variation rate as the vertical axis, the transmission process of wheat chromosome mutation induced by ion implantation can be displayed as shown in Figure 9.11. Whether in the M_1 generation or M_2 generation, the mitotic chromosome aberration rates of root meristematic cells of wheat Premebi, rye AR1 and the tetraploid rye are lower than meiotic chromosome aberration rates of pollen matricytes. A comparison between the M_1 and M_2 generations shows that mitotic chromosome aberration rates of root meristematic cells of the M_2 generation are always smaller than those of the M_1 generation for all three varieties of wheat studied. A main reason is that somatic cells are continuously eliminated or repaired in the division and proliferation process. For the meiotic chromosome aberration rates of pollen matricytes, only that of rye AR1 at the M_2 generation is smaller than that of the M_1 generation, while the rates of two generations of wheat Premebi and the tetraploid rye are close (for the former, M_1 is 3% higher than M_2; for the latter, M_1 is 3% lower than M_2). Ion-implantation-produced initial damage is different from that of ionizing radiation. The latter brings about breaking of single and double strands, whereas the former results in a large number of atoms of the genetic substance being displaced and rearranged as well as reacting with implanted ions. Is this kind of damage difficult to repair or is it easy to have mistakes in the repair process so that the aberration rates in M_2-generation meiosis of some plants are similar to those of the M_1 generation? This is still a question worthy of study.

It is worth mentioning that chromosomal centromere gaps, terminal centromere chromosomes, chromosomal terminal deletions, and chromatid terminal and one-arm deletions all show N-ion-induced chromosome or chromatid breaks mostly occurring at the centromere location. This indicates that an ion beam has effects on heterochromatin. In the meiosis process of bivalent rye AR1 pollen matricytes, the observed translocation between the B chromosome and the euchromosome indicates that the ion beam can induce B chromosome breaks. A B-chromosome bridge occurring in anaphase I and anaphase II is due to structural variations taking place in the B chromosome itself composed of heterochromatin. Many of the triploids and tetraploids formed in meiotic metaphase I of tetraploid rye pollen matricytes are characteristic of terminal cohesion. Collectively, the ion beam has obvious effects on the heterochromatins, affecting both the heterochromatin structures and functions.

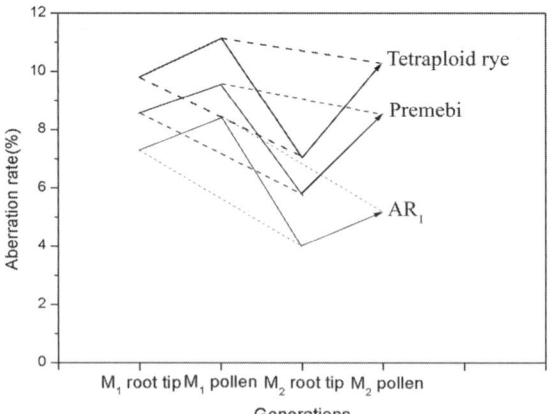

Figure 9.11. Generation transmission of the wheat chromosomal aberrations.

9.6. CYTOLOGICAL EFFECTS OF IMPLANTATION OF VARIOUS ION SPECIES INTO COTTON

In the previous sections, wheat was chosen as the study object and one ion species, nitrogen, was used for implantation. The aberration types and the relationship between the aberration rate and the ion doses were studied from various generations and different chromosome aberration phases. But what are the effects on chromosome aberrations of implantation with different ion species? In one experiment, 30-keV N^+, H^+ and Ar^+ ions were implanted into dry cotton seeds of varieties such as Xuzhou 553, a multi-generation self-crossover tetraploid upland cotton (*Gossypium hirsutum* L.), bivalent Zhongmian cotton, and Jinta cotton (*Gissypuum arboreum* L.) [13,14,15]. The seed shells were peeled off for ion implantation. From each variety, each ion species and each ion dose, 1,000 root meristematic cells were taken to observe cell division, nucleus abnormality and chromosome aberration types. Nucleus abnormalities include nucleus ear, nucleus break, nucleus budding, small nucleus and micronucleus. The aberration types have the bridge, the isolated chromosome, and the lagging chromosome (Figure 9.12).

After ion implantation, root meristematic cell division was affected in both degree and direction (Table 9.17). H-ion implantation had a stimulating effect on root meristematic cell division for Zhongmian cotton and Jinta cotton, and particularly for a dose of 5×10^{16} ions/cm^2, a readily observed stimulating effect on Zhongmian cotton. Ar ion implantation also had a very strong stimulating effect on Xuzhou 553 cotton cell division at high dose levels. But N-ion implantation generally showed a suppressive effect on cell division. The nuclear abnormalities (abnormal nuclei and micronuclei) induced by implantation of all three ion species were pronouncedly more than that of the control. At the low dose levels, the rates of producing cells which have nuclear abnormalities produced by H- and Ar-ion implantations were higher than that for N-ion implantation. At high dose levels (4–5×10^{16} ions/cm^2), the rate of nucleus-abnormalities induced by H-ion implantation was the highest. Generally, the rate of nuclear normalities induced by N-ion implantation was the lowest. Only when the N-ion dose was 2×10^{16} ions/cm^2 for upland cotton (*Gossypium hirsutum* L.) and 3×10^{16} ions/cm^2 for Zhongmian cotton, was the rate of cells with nuclear abnormalities slightly higher than those from H- and Ar-ion implantations.

Within the experimental ion dose range, chromosome aberration rates induced by three ion species are shown in Table 9.18. There are differences in the induced chromosome aberration types (only chromosome lagging and bridge formation were counted). H ions induced the strongest effect on chromosome aberration, and the portion of bridges in these aberration types was higher than those from N and Ar ions. N- and Ar-ion implantations induced more lagging chromosomes than H-ion implantation. This has been verified in cytogenetic studies of ion implanted wheat. For example, for N-ion implanted wheat Premebi, there were many chromosome laggings in root meristematic cell mitosis, and in anaphase I lagging of an entire genome could be found. This phenomenon could normally be found only in meiosis of distant hybrids. This result implies that a certain wheat genome is maybe very sensitive to N-ion implantation. The responding curves of the three ion species induced chromosome aberration rates as a function of ion dose are shown in Figure 9.13. Upland-cotton and Jinta cotton have similar tendencies for implantation by each of the three species of ion used in this study. Chromosome aberration rates increase as the dose increases, but at 5×10^{16} ions/cm^2 they decrease. For bivalent Zhongmian cotton, the mutation effects on chromosomes from the three ion species firstly increase, then decease and finally increase again with the dose.

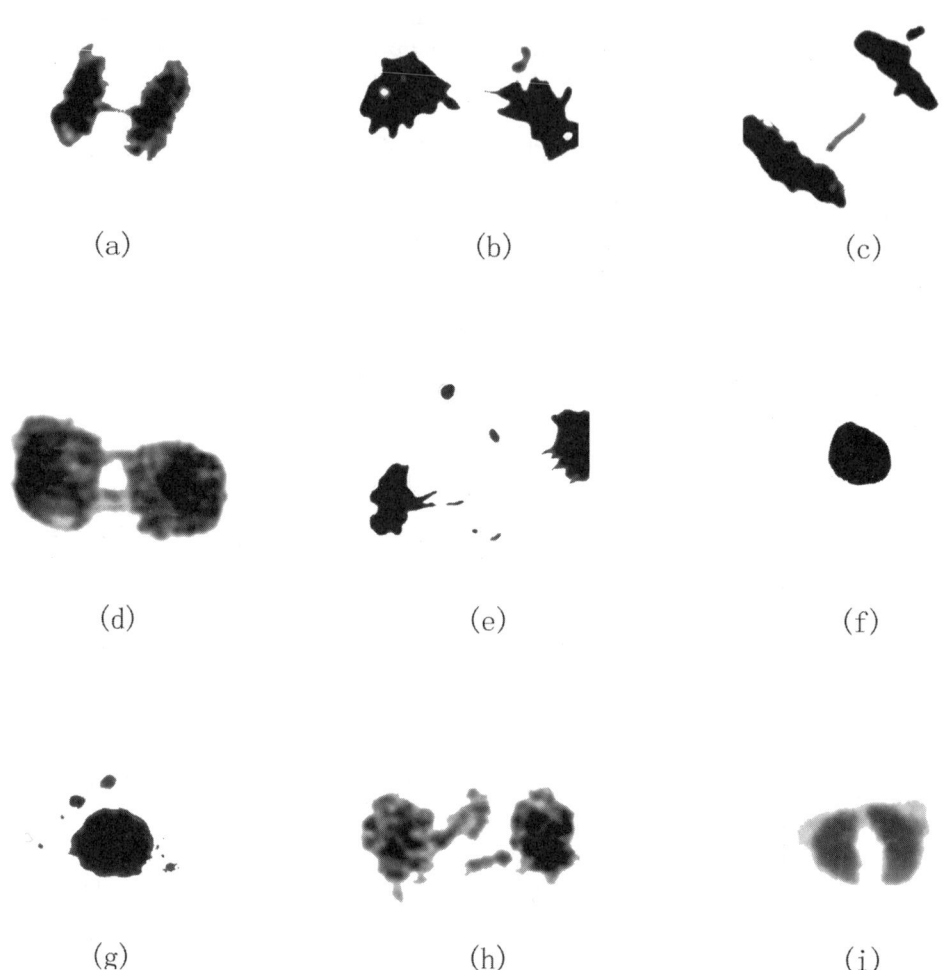

(a)   (b)   (c)

(d)   (e)   (f)

(g)   (h)   (i)

Figure 9.12. Root meristematic cell nuclear and chromosomal aberration types of the M_1-generation cotton seeds implanted with nitrogen ions. (a) Nuclear shrinkage, (b) anaphase bridge formation, lagging and segmentation, (c) micronucleus and anaphase bridge, (d) anaphase micronucleus, (e) anaphase single bridge, (f) anaphase bridge, (g) bridge formation and isolation of chromosome, (h) single and double bridge formation, and (i) chromosome lagging.

Table 9.17. Effects of implantation with different ion species on root meristematic cell division and cell nucleus. C.D.: cell division index. N.A.: nuclear abnormality cell rate.

Dose (10^{16} ions/cm^2)	Type and rate (%)	Upland cotton			Jinta cotton			Zhongmian cotton		
		H$^+$	N$^+$	Ar$^+$	H$^+$	N$^+$	Ar$^+$	H$^+$	N$^+$	Ar$^+$
0	C.D.	6.61			7.01			5.90		
	N.A.	0.36			0.37			0.16		
1	C.D.	5.76	5.97	5.85	7.96	3.43	4.50	6.75	4.90	4.52
	N.A.	4.67	1.68	5.92	4.74	1.17	4.88	4.09	1.28	3.83
2	C.D.	4.82	5.90	7.24	7.49	3.13	2.46	5.81	3.61	6.64
	N.A.	5.39	6.10	4.92	5.94	2.60	6.83	3.62	3.24	4.51
3	C.D.	5.50	4.33	4.05	9.57	4.13	6.36	7.44	4.69	6.14
	N.A.	11.24	4.33	5.34	7.20	2.88	6.80	2.60	3.19	2.77
4	C.D.	6.51	3.84	10.30	5.23	3.93	6.06	4.17	3.24	4.24
	N.A.	12.80	7.27	4.51	8.37	6.46	5.12	6.67	2.27	4.57
5	C.D.	6.02	3.16	8.37	5.15	3.10	5.55	11.97	3.44	5.90
	N.A.	9.58	3.04	6.00	8.48	3.63	3.05	7.55	2.53	3.86

Table 9.18. Effects of implantation with different ion species on chromosome aberration rate of root meristematic-cell mitosis. Lag.: lagging. Brid.: bridge. Total: total aberration rate.

Dose (10^{16} ions/cm^2)	Type and rate (%)	Upland cotton			Jinta cotton			Zhongmian cotton		
		H$^+$	N$^+$	Ar$^+$	H$^+$	N$^+$	Ar$^+$	H$^+$	N$^+$	Ar$^+$
0	Lag.	0.3			0.3			0.2		
	Brid.	0			0			0		
	Total	0.51			0.4			0.3		
1	Lag.	4.33	2.22	3.17	4.49	3.09	3.08	1.92	3.81	4.50
	Brid.	3.92	1.31	1.98	3.89	1.11	5.15	5.36	1.54	2.50
	Total	8.26	4.55	6.15	8.98	5.14	8.33	8.19	5.99	7.50
2	Lag.	4.38	3.22	6.89	5.80	4.50	7.90	1.04	6.13	6.40
	Brid.	4.68	2.42	0.32	5.18	0.90	4.13	5.20	1.43	1.70
	Total	9.86	6.66	9.23	11.90	5.82	12.40	8.95	7.87	8.10
3	Lag.	8.95	4.43	6.88	6.74	1.01	7.36	2.10	4.51	3.22
	Brid.	11.59	4.23	2.53	6.14	5.38	4.50	1.90	1.84	1.31
	Total	21.36	8.67	9.50	13.89	7.30	11.88	6.61	6.36	5.13
4	Lag.	16.73	10.22	5.14	7.48	7.33	6.08	6.47	3.21	6.73
	Brid.	12.30	4.25	2.52	8.80	3.61	3.40	5.48	1.20	0.71
	Total	29.64	14.75	7.87	16.28	10.94	10.38	12.25	4.51	7.44
5	Lag.	9.46	2.63	3.44	9.84	2.89	2.32	8.64	3.66	6.43
	Brid.	8.04	2.83	1.42	5.77	3.49	2.12	5.08	1.83	2.41
	Total	18.41	6.38	6.80	15.66	6.87	5.44	13.82	5.59	7.84

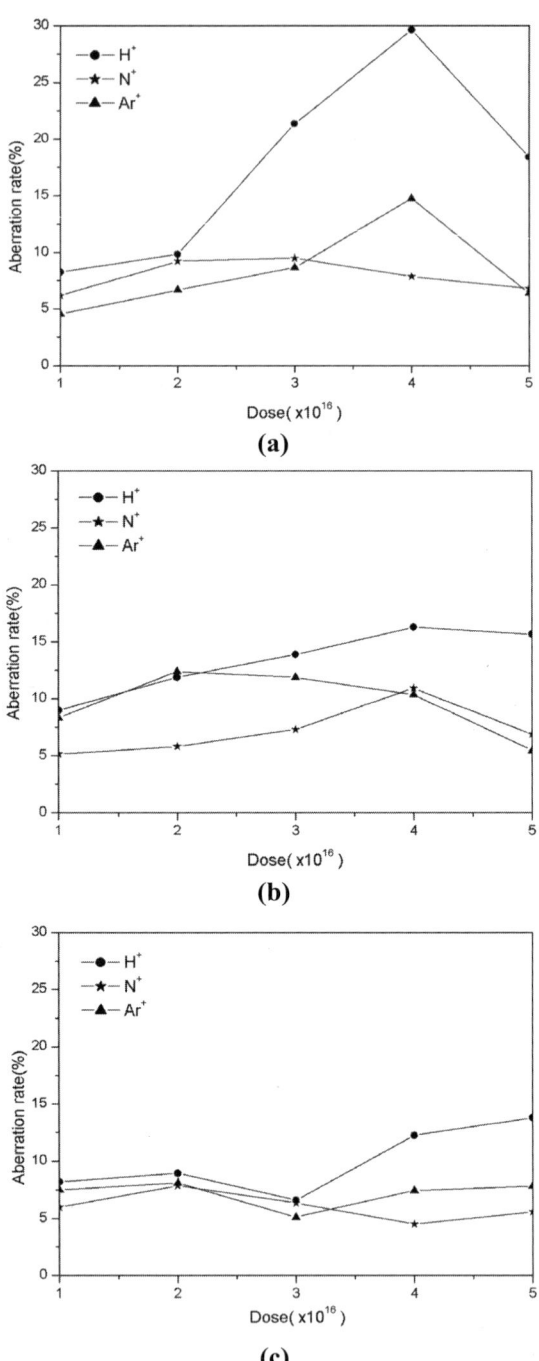

Figure 9.13.   Relationship between the chromosomal aberration rate and the ion implantation dose for various varieties of cotton. (a) Upland cotton, (b) Jinta cotton, and (c) Zhongmian cotton.

Different ions have different atomic masses and also different chemical activities. At the same implantation energy, their effects include sputtering yield, action range, target atom displacement, and combination probability with target molecules. It is seen from Figure 9.13 that at a certain ion dose, the order of effectiveness of the three ion species implantations inducing chromosome aberrations is H > N > Ar, which is just inversely proportional to the order of ion mass numbers. The reason may be that argon, as an inert ion species, produces only a mass deposition effect, whereas H and N ion implantations may produce double damages with respect to both quality and energy of the biological molecules.

REFERENCES

1. Swanson, C.P., et al., Cytogenetics: The Chromosome in Division, Inheritance and Evolution (Prentice Hall, Inc., Englewood Cliffs, New Jersey, 1981).
2. Xie, J.H., Xia, Y.W. and Shu, Q.R., Advances and Prospects in Crop Improvement via Ion Implantation Technology, Nuclear Agricultural Science Bulletin, 15(2)(1994)96-98.
3. Zhu, F.S., Wei, J.Z., Sun, Y.S. and Lu, T., Chromosome Translocation in Rye Induced by Nitrogen Ion Beam Irradiation, Acta Agronomica Sinica, 19(4)(1993)299-303.
4. Wu, L.F. and Yu, Z.L., Radiobiological Effects of Low Energy Ion Beam on Wheat, Radiat. Environ. Biophys., 40(2001)53–57.
5. Schulz-Schaeffer, J., Cytogenetics: Plant, Animal, Humans (Spring-Verlag Press, New York, 1980).
6. Renner, O., Zeit. Bot., 13(1921)609-621.
7. Chen, Z.X., Nuclear Agricultural Science Bulletin, 11(2)(1990)87-88.
8. Qiu, G.Y., Journal of Wuhan University (Natural Science Edition, in Chinese) (1993)9-12.
9. Clark, F.J., Amer. J. Bot., 27(1940)547-549.
10. Moore, D.M., Plant Cytogenetics (Chapman and Hall Press, London, 1976).
11. Howland, G.P., et al., Mutation Res., 27(1975)81-87.
12. Tano, S., et al., Ibid., 42(1977)71-78.
13. Cheng, B.J., Li, Z., Tian, Q.Y., Wang, G.M., Yu, Z.L. and He, J.J., Study on Cytological Effects of Ion Implanted Cotton, Journal of Anhui Agricultural College (in Chinese with an English abstract), 18(4)(1991)329-331.
14. Cheng, B.J., Li, Z., Zhou, L.R. and Yu, Z.L., Cytogenetic Effects of Different Ions Implantation in Cotton Seeds, Journal of Anhui Agricultural University (in Chinese with an English abstract), 22(3)(1995)189-195.
15. Cheng, B.J., Li, Z., Wang, G.M. and Li, Y., The Mutagenic Effects of Nitrogen Ion Implantation in Cotton Seed, Acta Agriculturae Nucleatae Sinica, 7(2)(1993)73-80.

FURTHER READING

1. Li, G.Z., *Chromosome and Its Research Methods* (in Chinese, Science Press, Beijing, 1985) pp108-144.
2. Li, M.X. and Zhang, Z.L., *Crop Chromosome and Research Technology* (in Chinese, Chinese Agriculture Press, Beijing, 1996) pp97-110, pp121-129.

10

ION BEAM MUTATION BREEDING OF CROPS

Mutations are essential for plant breeding. In plant material used for breeding plants, most of the useful genetic variations have evolved naturally and have been selected because of some previous gene mutation or recombination. Crops for human use have been affected by planting practices and artificial selection by human interference. However, for most plant varieties, the applications explored have not approached theoretical production limits. Future plant breeding is facing a difficult challenge and major improvements in breeding methods are needed.

Modern breeding methods, whether originating from physics, chemistry or biotechnology, can be summarized as having two general steps. The first is the creation of genetic variations and the second is screening the mutations for beneficial varieties. Artificially controlled hybridization is the current method of using germ plasm from different crops to obtain variations via redistribution and recombination of genetic material. The problem is that the natural gene reservoir is limited and the natural mutation rate is only about 10^{-7}. To increase germ plasm resources for hybridization breeding, the artificial induction of new genetic variations is most important. An important fact in this aspect is that genetic variation can be induced in modern species that have good adaptability.

The most important application of artificially induced mutations is direct mutation breeding, which uses mutagens to treat the plant sexual or asexual offspring to induce ideal genotypes. Actually direct mutation breeding does not rule out using any useful mutants as germ plasm resources. In principle, induction of excellent germ plasm resources and special genotypic materials can breed groups of excellent varieties. Traditionally, so-called mutation breeding refers to direct applications of mutants but little to indirect applications.

At present there are a variety of physical and chemical mutagens. Since the biological effects of ion implantation were discovered, the ion beam as a new mutagen has received progressively more attention. One of the important reasons is that ion implantation mutation integrates the factors of mass, energy and charge, and the induced damage to the biological materials (including genetic substances) mainly makes the biological molecules and atoms displaced, recombined and compounded. It is these characteristics that make biotechnological applications of ion beams not only suitable for mutation breeding, but also for cell processing and gene transfer.

This chapter mainly introduces applications of ion beams to mutation breeding. Considering that readers in the field of plant breeding may be fairly familiar with γ-ray mutation, this chapter will emphasize those features and technical issues which should be noticed in the processes of sample preparation, ion implantation and screening of mutants from ion beam mutation.

10.1. GENERAL PRINCIPLES OF MUTATION BREEDING

As mutation methods are continuously improved and new mutagens continuously emerge, the quality of mutation breeding has been greatly increased. In many cases, mutation breeding is considered as a tool, similar to hybridization breeding, to create genetic variations flexibly and selectively. Certainly, variations created by mutation methods are related to the properties of mutagenic factors and also the characteristics of the biological objects. In determination of which mutation method to be used to create genetic variations, all of the characteristics of the mutagenic factor, properties of the chosen biological object, and correlations between the factors and the objects should necessarily be considered.

10.1.1. Primary Materials

For mutation treatment, the chosen material should be consistent in its genetics, particularly its mutation characteristics. In mutation breeding, the commercial potential of the primary material and the expected commercial value of the mutant should be considered. Thus, it is necessary to work with materials that have high potential. In practice it may be best to choose mutation materials from separately planted good or high-generation stocks. Before conducting any experiments, breeding objectives and screening plans should be clarified. Choice of materials should agree with a specifically designed goal, which would normally overcome a single defect. This means that except for those characteristics to be mutation-improved, other characteristics should already be excellent. Even for further breeding to look for mutators, the chosen material should have high adaptability and potential for increasing production, as this kind of material is more suitable to be used as the parents for further hybridization breeding. Of course, employment of genotypes that are most agreeable with breeding goals to be the material for development is also very important. Sometimes the source of the specially designed variations is of less concern since the purpose is only to improve the crops, and thus it is not necessary to require the mutation material to be pure and consistent. For accelerating recombination of genes of distant hybrid offspring, choice of F_1-generation seeds as the mutation material is recommended.

10.1.2. Mutagens

After determination of the breeding goal, the mutagen chosen and the dose to be used must be estimated. These depend on many factors. For the first use of a new mutagen, the ideal method is to carry out tentative treatments of the sample material and make a statistical study of the relationship between the mutation rate and damage degree (normally indicated by the survival rate) and the dose. The correlation degree between damage rate and mutation rate varies strongly with mutagen. In the case of direct mutation breeding, the selection should be done using populations that exhibit light to medium damage effects so that the problem of multi-characteristics mutations can be avoided. When looking for mutators for further breeding, high-dose treatments which emphasize M_2-generation screening should be chosen. Generally speaking, even when mutation characteristics that satisfy the breeding goal cannot be found in the M_2 generation, the high frequency of appearance of mutations with other characteristics is also an indicator of good treatment. In formal experiments, improvements in the mutation conditions used for the first time can lead to improved results.

10.1.3. Treatment of Populations

The quantity of treated populations is related to the sensitivity of the plant material to the mutagen, that is, to the mutation rate. Generally speaking, failure to obtain an ideal mutation type may be due to either poor mutability of the material or too few of the chosen populations. In the worst case, certain locations of some genotypes refer to large genetic defects and thus mutations may not occur. Some genotypes may easily have genetic defects, inversion or structural translocation mutations. Sensitivity to radiation is a kind of character of the genetic locus structure, possibly affected by the genetic background. Thus the determination of whether a certain character can be obtained by mutation must be done by selecting this character from large populations of treated materials with various genotypes. In brief, the lower the mutability the material has, the larger the treated populations should be; otherwise, populations can be smaller. When the quantity of treated populations is to be determined, the cost and time spent evaluating results must also be estimated. Experience indicates that when seeds are treated for mutation, it is best to treat several thousands seeds each time. But it must be noted that different mutagens have different mutation efficiencies and the size of the population varies accordingly.

10.1.4. Screening

The degree of effectiveness of a mutation method that has a higher efficiency than any other breeding methods depends on the capability to identify useful genetic variations in the M_2 or M_3 generations. If some other variations that escape or are affected by the environment have high mutation rates and also the populations for screening are large, and/or the identifying methods are simple, they may be selected. Usually with unclear breeding goals, mutation breeding methods should not be used. Only screening methods with high efficiencies should be used. Generally, before a mutation breeding method is chosen, the characteristics to be modified and how populations will be screened should be clear in advance. For mutations that occur early, screening in the tasseling period or flowering period is necessary, as the mutations are most easily identified at that time. The screening for half-short stalk mutations can generally be carried out at any time when normal plant heights are apparent. For characteristics that can be visually screened in the field, such as plant height, plant shape and maturation period, etc., in separated populations (normally M_2 generation), thinning out favors visual screening of mutants. For production characteristics such as the number of grains per ear, weight of 1000 grains and appearance quality, etc., screening should be done by examining fruits indoors. The intrinsic quality of seeds should be established via physicochemical and biochemical analyses. For resistance to toxicities of chemicals, screening is carried out after the chemical reagents are sprayed and sprinkled. For resistance to plant disease, the screening can be done via vaccination or in an environment with a high occurrence rate of the disease vectors.

10.2. MUTATION METHODS

Ion beam mutation has its own unique features, determined by the interaction between low energy ions and biological organisms, and thus greatly determining the methods of sample treatment.

10.2.1. Mutation Conditions

The various factors of ion beam mutation are integrated into entities. The energy determines the ion range in the biological organism and excitation and ionization of biological molecules, leading to possible breaking of DNA single and double strands. A part of the energy is converted to momentum. Momentum exchange causes etching (sputtering) of the biological organism surface to move forward linearly the depth of the ion beam-organism interaction, and also causing displacement and recombination of biological molecules and atoms near the ion trajectory. Mass deposition has a peak in the direction of ion implantation. Most of the elements in the Periodic Table and some molecular ions can be accelerated and implanted into a biological body. If ions of active substances are implanted, they will combine with neighboring molecules to form new molecular groups after being slowed down. Charge exchange affects the biological electric characteristics. The number of implanted charges is related to the number of implanted ions and the ion valence state. One or two or even all electrons outside the nucleus can be removed to leave a naked atomic nucleus. Charge accumulation can cause a Coulomb explosion and the release of biological macromolecules from the surface.

The properties of the ion beam-biological organism interaction determine the features of ion implantation mutation methods. Mastering of these features can achieve the desired mutation effects.

1). Ion implantation is basically conducted linearly. For the ion energy and dose regime discussed in this book, the ion traveling distance is between several micrometers and several hundred micrometers. Only when ions are implanted to locations that can directly or indirectly determine growth and development of the characteristics of the biological organisms and influence building of the formations (e.g. the growing tips of seed embryos and twig buds), can effective mutations be induced. Table 10.1 shows the glutinousness mutation rate at various locations of the ion-beam-irradiated naked dry seeds of medium-mature *japonica* rice (variety: Shao sticky rice 86-87). At the embryo growing tip (when the plumule is directly facing the ion beam direction), the mutation rate is the highest. When the endosperm part is up or the ion beam is blocked, the mutations show no great differences from the control. Thus when an ion beam mutation experiment is carried out in practice, placing the embryo toward the ion beam direction is pivotal. Because the penetration depth of low energy ions in biological organisms is limited, resistance from the seed shell can also affect the mutation rate. For example, for seeds without shells of the M_2-generation early-season rice-213 rice, when the embryos are up for ion bombardment, the chloroplast mutation rate is 0.53%, whereas the rate is only 0.17% for seeds with shells. Similarly for dry seeds of Zhe-15 rice, the M_2-generation chloroplast mutation rates are 0.91% and 0.31% respectively for the seeds without and with shells when they are implanted. Hence, in order to obtain effective mutations, sometimes parts of the seed shell have to be removed. For seeds of some types of plants such as cotton, there is a hard shell and the embryo bud is nearly surrounded by the cotyledon. For ions to reach the expected locations the shell must be removed and then very often the side of the embryo bud is stabbed using a needle to make a hole for the passage of bombarding ions.

2). Ion energy is eventually converted to thermal energy. Water-cooling of the sample holder and pulsing the ion beam limit the maximum sample temperature reached during ion implantation. Intermittent implantation enables the operator to measure the implantation parameters, such as energy and dose, between implantation pulses. However intermittent implantation extends the time for which samples are in the vacuum environment. This is not important for dry crop seeds, but for living organisms, techniques must be adopted to remove free water from the organisms or cells, or to introduce the ion beam outside from the vacuum chamber.

3). To avoid freezing damage to or a detrimental water loss from the sample that highly contains water because of the evaporation of water into the ambient vacuum and the corresponding temperature drop, or affecting ion beam production due to the water evaporation, some effort should be made to isolate those parts of the sample which do not need to be implanted.

4). For microbes or plant cells, the samples are exposed to the environment during ion implantation. Thus, sterile conditions should be maintained while samples are moved in and out of the target chamber. The sterile target chamber at the Institute of Plasma Physics, Chinese Academy of Sciences, was designed to satisfy this requirement.

10.2.2. Sample Preparation

10.2.2.1. Seed-Breeding Plants

Select dry seeds of uniform size and plump ears. Then, after wind-screening, insert the seeds, one by one, into a specially designed sample holder. The seeds should be oriented so that target areas for implantation have maximum exposure to the ion beam. For dry plant seeds such as rice and wheat, each sample holder can hold about 1,000 grains. Normally the number of holders is the same as the number of different ion doses to be used. If the mutation effect of seeds in the germinating state is to be investigated, after germinating the seeds are vacuumed for extra water and wind-screened, then inserted one by one in the sample holder again for the next treatment. For plant seeds with hard shells, such as gingko and cotton, the seed shell part in the beam direction should be removed.

Table 10.1. Effect of ion implantation on glutinousness of glutinous rice pollen from different modes of N-ion implantation of the rice seed embryo.

Ion implantation mode	Implantation in embryo bud	Implantation in endosperm	Blocking of ion beam	Control
Glutinousness mutation (%)	8.7	0.5	0.11	0.15

10.2.2.2. Agamogenetic Plants

For plants which reproduce without sexual differentiation (agamogenetic), bud tips should be implanted. In sample preparation, it is not necessary to take only young buds; a lump of stem or a segment of twig should be included. The lump stem (such as sweet potato) and twig (such as poplar and fruit trees) should be packed with plastic foam and smeared with low melting point wax, so that only the bud tip part that is to be implanted is exposed. The prepared lump stem or twig is fixed on the sample support with bud tips oriented toward the ion beam.

10.2.2.3. Plant Cells and Microbes

For plant cells, pre-culture is necessary before ion beam treatment. The culturing conditions and time are determined by the experiment. For instance, for rice cells, DMSO (Dimethyl Sulfoxide) and sugar solution in certain concentrations are added to the culture for 60 minutes in order to remove free water inside the cells. The cells are cultured using sorbic alcohol or mannitol for a period, and then put into a freezer for rapid cooling so as to simulate the effect of vacuum evaporation freezing during ion implantation on cell survival. For microbes, bacterial solution of suitable concentration is dropped into a culture dish and blow-dried for ion implantation. During the ion implantation, the microbes should be kept in a single layer. For filiform fungus, the microbes should be crushed by glass beads. If colonies are directly implanted, the culture base in the culture dish should be as thin as possible.

10.2.3. Ion Implantation

We consider here the ion beam mutation facility at the Institute of Plasma Physics, Chinese Academy of Sciences, as an example to describe the ion implantation operation.

The prepared sample holders or supports are placed in the target chamber of the ion beam facility. The specially designed target chamber consists of a large chamber and a small sterile micro-environment chamber. There is a rotatable 80-cm revolving disk in the large chamber. To treat dry crop seeds, the big target chamber is usually used. Six to ten sample holders can be simultaneously placed on the circumference of the disk. After ion implantation is completed for one sample holder, a computer control rotates the disk to the second sample holder to be treated, and so on. Control of the ion implantation time can allow the seeds in each sample holder to receive different ion doses. The micro-environment target chamber is designed particularly for treatment of high-water-containing samples, cells or microbes. The volume ratio between the big chamber and the small chamber is 600:1. The small chamber is in a sterile environment. When samples are moved into and out of the small chamber, sterile air is blown in to ensure that the samples are not polluted by foreign bacteria.

After the ion energy and species are chosen, there is still a lot of work to do to ensure the stability and reliability of the ion beam parameters. The first is adjustment and test of the ion source. The cathode temperature, discharge chamber pressure, magnetic field configuration and arc shape all need to be precisely adjusted to obtain optimal plasma density and homogeneity in the ion source discharge chamber. Then, positive or negative high voltage is applied to the extraction system. At this time the energetic ion beam passes through the magnetic mass analyzer to enter the acceleration section where it is further accelerated, and finally the ions are implanted into the samples in the sample holder of the target chamber.

Before the samples are ion implanted, the dose rate should be first measured. Ion implantation of biological samples usually makes use of an intermittent (repetitively pulsed) implantation mode. The number of ions implanted in each implantation cycle is known and thus the total irradiation dose to be received by the samples can be calculated from the total implantation time. For example, samples in a sample holder are implanted once a minute with an actual implantation on-time of 5 seconds and the implantation dose rate is 5×10^{14} ions/cm^2 per implantation cycle of 5 seconds; thus in a total of 50 implantation cycles the total dose received by the samples is $50 \times 5 \times 10^{14}$ ions/cm^2 = 2.5 $\times 10^{16}$ ions/cm^2. The dose rate should be measured not only before implantation but also during ion implantation if a long-duration ion implantation, namely high dose, is required to treat the samples. This is because the ion beam parameters are very sensitive to the discharge gas pressure and the electric field and magnetic field configuration in the beam line, and they may not remain constant throughout a long time run. The measurement is carried out when the beam is interrupted, and thus the time is not counted toward the applied dose.

10.3. PROCEDURE FOR ION BEAM MUTATION BREEDING

As for other breeding methods, mutation breeding should undergo variation, selection and identification. Variation is a prerequisite. Without effective variations, selection loses its meaning. The mutability of biological characteristics is closely related to mutation factors and the biological object. This book has described and discussed the relationships between ion-beam-induced genetic variations and the factors as well as the objects. For plant breeding, it is highly significant to select the needed mutants. Much of the literature gives detailed descriptions and discussions of screening methods and procedures for radiation breeding. Here we only briefly introduce issues related to ion beam mutation breeding. In principle, firstly field planting should be arranged so that mutation states can be sufficiently displayed for easy screening, and at the same time environmental effects on variations should also be considered; secondly, the right time for screening should be controlled. For seed-breeding crops, it is appropriate to select mutants at generational separations. But relations between character variations of various generations and the possibility of occurrence of dominant mutations at the present generation should be considered also.

10.3.1. Plantation and Selection of M₁ Generation

Individual plants grown from seed embryos after ion implantation are called the M_1 generation. Generally the mutated generation should be planted separately. In genetic isolation, a large area devoted to the parent variety is used. The surrounding can be other plant species or other varieties that are obviously different from the M_1 generation. Sometimes, to prevent natural hybridization, it is better to use bag-covering for self-cross. Due to a limited number of ion-implanted samples (seeds), the seeds are individually planted. In the harvest, only the main-stem ears are collected.

Table 10.2. Present generation fruit weight variations of N-ion implanted tomato.

Treatment	Mean single-fruit weight (g)	Variation range (g)	Standard deviation (s)	Variation coefficient c·v(%)	Variation rate (%)
Control (Variety Zaoxia)	65.6	60.0–71.4	3.52	5.36	
Ion implanted (5×10^{15} N/cm^2)	84.0	65.6–134.4	17.57	20.9	8.6

From a genetic point of view, the radiation-bred M_1 generation is a complex mutant chimera. Thus, there is normally no need to do selection in the seed treatment. The M_1 generation, subjected to radiation damage, has malformations occurring in morphology and also abnormal changes in physiology, such as poor growth, slow development, abnormal pregnancy, as well as various malformations in organs. These various changes in physiology and morphology may play certain roles in indicating the occurrence of mutation characteristics of the M_2 generation. Hence careful observations of the M_1 generation are necessary.

In ion beam breeding, the seed embryos are oriented to the implanting ion beam source during treatment. At a given energy, ions are deposited in a thin layer of the growing tip. Owing to atomic displacements and recombination, assisted by ions slowing down, in this thin layer, the probability of mutations occurring simultaneously in one or two or several cells in the growing tip is relatively high. If dominant mutations occur, they can be exhibited normally in the first generation for the diploid plants, but may not be homozygous variants, and instead they may appear together with original characteristics to form chimeras.

Ion-beam-induced heritable present-generation mutations were mentioned in Chapter 8 when great mutations were discussed. Although the mechanisms involved are not clear, the importance is very significant from the breeding point of view. This enables selections to be started from the present generation. Since the population of the present generation is not large, screening is easy. It is more important that the generation for selection is moved up early so that the breeding process is shortened. For example, the hybrid parent tomato, early-mature Daguo-yanfu No.1, was screened from the ion-implanted present generation (Table 10.2). It can be seen from Table 10.2 that the variation coefficient of the single-fruit weight of the present generation is almost four times that of the control. The increase in single-fruit weight obviously results in an increase in the present generation production as well as enhancement of counts of early-mature fruit. The prophase production of the single plant is increased by 39.3% and the total production of the single plant is increased by 34.9%. With the mean single-fruit weight of the M_1-generation control population added with twice the standard deviation as the critical value of the single-fruit weight mutation, the individual plants were selected to plant as the M_2-generation individual-row plot. The result showed that 75% of the mutants which were screened from the M_1 generation in the M_2 generation had higher productions than that of the control, and particularly, 30% exceeded the critical mutation value. The mean single-fruit weight was increased by 28.6%. The variation coefficient became smaller, only twice that of the control. This indicated that the single-fruit weight variation produced in the M_1 generation could still occur stably in the M_2 generation. Plants of the M_3 and M_4 generations were planted based on the plant system for investigation in terms of the individual production, single-fruit weight, and prematurity. Six plant systems were chosen. It was found that increases in production were between 7.1% and 35.1%, except for one system. The single-fruit weights in the six systems were increased by 5.8% to 35.3%. With a self-cross tomato system, 91-1^{nv}, for test cross, in the F_1 generation hybridized with the six mutant systems, five systems showed obvious increases in the prophase single-fruit weight, total single-fruit weight and total production compared with those of the control. One of the systems had increases in the early production, early-period single-fruit weight, total production, and mean single-fruit weight by 23%, 18.3%, 17.3%, and 11.7%, respectively, compared with those of the control. This system passed examination and was planted extensively in large areas in Shandong province. Moreover, favorable mutants were also screened from the M_1 generation of wheat in Anhui and Shandong provinces and ranked at the top in the provincial or regional examinations.

10.3.2. Planting and Screening Methods for M_2-Generation

Mutations occurring in the M_1 generation are transmitted to the gametes through the gametophyte generation to form seeds via insemination. Because the mutators on chromosome are separated and recombined in this process, the M_2 generation is the major generation for separation and selection.

In the M_2 generation, the plants are planted in either ear rows or mixed ways according to the seed collecting system (single ear or mixed collection) of the M_1 generation. In order to enable the mutated plants to display sufficiently or be visually selected, it is necessary to have a certain distance between plants in addition a certain distance between rows.

Most mutations are variations having obvious phenotypes. In the process of growth and development, they exhibit observable mutant character phenomena such as separation, recombination, linkage and exchange. But the degrees of the mutation-caused character variations are normally different. Some variations are found which control quality characteristics, such as color (green or yellow) of rice seedlings. Some mutants affect the color of the wheat grain to be red or white and dark or light, mature early or late, and other aspects of grain quality, for statistical analyses of populations. These variations are controlled by a series of micro-effect polygenes. The characteristics controlled by micro-effect polygenes usually depend on the environment. Thus, environmental factors interfering with the screening of M_2 generations should be considered. Normally the M_2 generation should be planted under typical agronomic conditions so as to facilitate selection of mutants that have various needed characteristics. For example to select barren resistant or soak resistant varieties, the M_2 generation should be planted in an environment that is easy to be made dry or waterlogged. For instance, the ion-beam-mutated rice, Zao-xian series S_{9042}, in the ear-pregnant period in 1991 was subjected to seriously high flooding, overflowed for 4 days and soaked in water for 21 days, but there was still a good harvest; in the period in southern China in 1993 of cold and frequent rain, rice plague was severely spread but this variety exhibited high resistance to the disease. If there had not had these pressures of natural environmental changes, the waterlog resistance and plague resistance of this variety would not have been easily discovered.

Variations of the plant phenotypic characteristics, such as the tall or short stalk, early or late maturation, early or late flowering, big or small grain, color and luster, number of the grain on one ear and thousand-grain weight, can be easily selected via field observation and some indoor investigative planting. Intrinsic qualities of the crop, such as the protein content, nutrition quality, or processing characteristics, should be selected. M_2 is an important generation for protein breeding, hence protein variations should be detected via biochemical analysis. For example, grains of M_2-generation plants of wheat Premebi treated by ion implantation were analyzed for stored protein using electrophoresis. It was found that the A-class electrophoresis spectrum composed of the high-molecular-weight glutelin sub-bases had one more polymolecular glutelin sub-base than the control. The number of high-molecular-weight glutelin sub-bases and their combination forms are related to baking quality. As the number of high-molecular-weight glutelin sub-bases increases, the protein content increases and baking quality is improved. The high-molecular-weight glutelin sub-base is controlled by three loci on the long arm of the first-part homologous-group chromosome, not restricted by environmental conditions. This is exceptional. Generally in identification of content variations of amino acid or protein, notable environmental effects should be taken into account. Hence when protein content changes, the relationship between production and protein content must be considered.

10.3.3. Plantation and Selection of M_3 Generation and Subsequent Generations

Mutants selected from the M_2 generation are threshed plant by plant and then seeds are planted for the M_3 generation. For the M_2-generation plants that are not selected, due to some characteristics probably occurring in M_3 generation, the seeds can be picked in such a way that one or several seeds from each plant are selected for the M_3 generation.

The stability of mutation characteristics in the M_3 generation is explored. The rate of micromutations induced by common physical and chemical factors may be higher than for macromutations. The micromutations involve variations of quantity characteristics which are more important in breeding practice, particularly when the mutants are directly used in plant breeding. Since the quantity-character variations are identified only by statistical analysis of populations, when M_3-generation micromutations are identified, setting up repeated controls is necessary.

The seeds picked from the preliminarily selected plant-system area in the field are examined and finally selected based on field observation and indoor analysis. The seeds of the finally selected plant systems are mixed, some being held in reserve.

From the M_4 generation, stocks can be identified in comparison with controls for spreading good varieties, choosing better ones and removing worse ones. The selected stocks for the next generations are further appraised, tested, and experimented with in various local and provincial places until new improved varieties are finally selected.

10.4 WAYS OF APPLYING ION BEAMS FOR PLANT BREEDING

The effects of ion beam irradiation on biological materials result from multiple factors. The huge number of possible combinations of implanted ion energy, ion species, charge number and particle number leads to rich and diverse biological effects. This makes possible ingenious applications of ion beam to plant breeding. According to the various breeding goals, some ways can take advantage of the stimulation effect of ion implantation on crops, some can use ion implantation induced genetic effects, some may employ ion implantation to create aneuploids of wheat, and so on. However, any effects beneficial to variety improvements can be applied.

10.4.1. Application of Stimulation Effects

For crops such as tobacco and sweet-leaf chrysanthemum, increase in leaf production and improvement in intrinsic plant quality are of economic significance. The grains of these crops have hard shells. Low energy ion sputtering can improve the permeability of the shell surface and facilitate the exchange of external substances (nutrients and water, etc.). Consequently the germination rate is increased, and the growth potential is enhanced and maintained. Laboratory tests indicate that this low dose stimulation effect may not only increase leaf production but also cause intrinsic chemical compositions to improve. When the implanted ion energy and dose are controlled properly, the repeatability of the effects is suitably high. Whether the effects are inherited or not, effects of seed size and regeneration coefficients of tobacco and sweet-leaf chrysanthemum, the stimulation effect can be directly applied to the present generation and its value is significant. This application is to be further explored. For example, airplane sowing of grass seeds onto the deserts and tree seeds on waste-mountains should be done shortly after a rain to increase germination speed and growth potential of the seeds. However, there are few studies concerning the mechanisms of the ion implantation stimulation effect on the present generation, and particularly the mechanisms for improvement of intrinsic quality. This now restricts the use of this application.

10.4.2. Applications of Mutants

In ion beam mutation breeding, the probability of inducing multiple mutations that have favorable characteristics is high. Thus, most breeding workers take ion-beam-induced mutants directly for production as important goals. Indirect applications of the mutants are also increasingly considered as studies of ion beam mutation mechanisms are increased and practical experience is accumulated.

10.4.2.1. Direct Applications of Mutants

In breeding practice, plant breeders seek high production, high quality and high adversity-resistance varieties. In the high-adversity-resistant varieties, resistance to plant diseases and insect pests are particularly stressed. It can be seen from existing results that ion beam induced mutations of desirable characteristics such as high production, high quality and high adversity resistance are easily realized. For example, in rice breeding, in 1988 500 dry seeds of rice Zhe-15 were implanted with 30-keV nitrogen ions to 5×10^{16} ions/cm^2; extensive variations occurred in the M_2 generation and 300 mutated individual plants were selected for further development. Through comparison experiments, mutated early-season rice stock S_{9042} was bred. This stock had an earlier fertilizing period by 4.6 days and a trunk height 10 cm shorter than for the control. By artificial vaccination, the resistance to leaf plague and ear-neck plague reached Class 2.1 and Class 1, respectively. The Institute of Plant Protection, Zhejiang Agriculture Academy, used 20 bacterial series for vaccination and achieved high or medium resistance to 14 bacterial series. In 1990 the variety joined an early-season-rice preparation experiment. Its production in unit area (667 m^2) was 437.0 kg, the top of eight participants, increased by 9.6% compared with that of the control and 6.6% compared with that of original variety Zhe-15. In 1991, this stock was planted in a 2-hectare field. After earring, the plants suffered specially huge flooding, overflowed for 4 days and soaked in water for 21 days, but production in the unit area of 667 m^2 was up to 320.5 kg, an increase of over 38% compared with the locally planted variety early-season-rice-213, showing excellent resistance to water-logging. In 1993, as the weather in southern China was cool and there was frequent rain, rice plague of early-season-rice varieties was a serious problem. Some fields had nothing to gather and 60% of the fields had decreased production. However, in an area of 3333 hectares, S_{9042} highly resisted rice plague disease and the production in an area of 667 m^2 was up to 384.4 kg. In the same year, the government office of seeds organized production tests and S_{9042} achieved the unit-area production of 515.3 kg in a 2.5-hectare field. In 1994, this variety passed appraisal and was rapidly spread in the Yangtze-river region. To 1995 the planting area was up to 200,000 hectares and the 667-m^2-unit-area production was 450 kg.

A new stock D_{9055} of high quality and high production late-rice was produced from ion implantation mutation of variety Eryi-105. Compared with the original, this stock had high quality rice grains, strong growth potential, high stem and big ears, but a 2-day postponed growth and development period. This variety participated in a preparation test of double-harvest late-rice in Anhui Province in 1990 and achieved an average production of 473.1 kg per 667 m^2, increased by 8.4% compared to Eryi-105. In 1991, the variety was tentatively planted in Huaining County and achieved an average unit-production of 452.2 kg, an increase of more than 20% compared with that of the local rice. In 1992, the variety was tentatively planted in Liuan County and achieved an average unit-production of 481.5 kg, an increase to 133 kg per 667 m^2 or 38.25% compared with those of the local varieties. In 1993, in an experiment of the ton-of-rice field organized by the Anhui Provincial Science Commission, variety D_{9055} was planted in association with S_{9042} and had a net harvest with a unit-production of 511.2 kg from an area of 2.5 hectares. The major indices of the rice quality of this stock achieved the Ministry-standard Class 1. In 1993, in the appraisal of high quality varieties of rice from all over the province organized by the Anhui provincial administration, D_{9055} was top-three among 27 sample varieties and appraised as the new high quality variety of Anhui Province. In 1994, this variety passed official appraisals and had been distributed to be planted in 207,000 hectares in Anhui and Hubei Provinces.

Overall, applications of ion beam mutation breeding techniques have achieved 256 new stocks. Among them are a new wheat stock (ranked top in provincial tests) featured with increased protein content and improved appearance quality, a new cotton stock featured with the resistance to the cotton boll pest and red boll pest, and a new high production and high quality sweet potato stock (ranked top in northern China tests). Nineteen improved plant stocks participated in the 1–2-year national or provincial tests. Most of these new stocks were suitable for immediate production.

10.4.2.2. Indirect Applications of Mutants

As a new mutagen, ion beams can create abundant variations. However, few of these variations are directly applied in production. For example, according to a paper-survey at the end of 1994, the author's Institute carried out ion implantation of plant seeds for more than 60 projects from all over the country, found a total of 11,831 mutants and obtained 1,382 mutated materials. There were actually only 21 new stocks, 1.5% mutated materials, including 3 plants directly bred from mutants and some new stocks that had passed the regional 1–2-year tests. Most of the mutated materials would be indirectly applied. From the current situation, it is seen that indirect applications of aberrations may follow different ways.

1). Mutants are hybridized with other parent material to produce hybrid next generations. From these generations, new stocks that have improved characteristics or new materials that have special applications are selected. For example, as mentioned above, rice yellowing mutants were hybridized with photo-sensitive sterile stocks to transfer the yellowing gene into photo-sensitive sterile stocks W6154s, 8912s, 8902s and 7001s producing a group of new stocks which have good breeding, spawn-separation and disease resistance characteristics. These new stocks had the yellowing marker. The self-cross of sterile stocks produced yellowish leaves while the hybrid F_1 had normal green leaves. Thus, it was very easy to remove false hybrids during the field-seedling period. Even if the yellowish seedlings were not artificially removed, due to their weak growth potential they would be self-eliminated during the F_1 growth period. This is very important in the practice of hybrid rice production.

2). Ion beams can be used to modify some bad characteristics of hybrid parents to explore useful genes and breed good hybrid combinations. For example ion implantation was carried out on tomato stock "Zaoxia (Morning Glow)" to produce seven new stocks characterized by early maturation and large fruit. These were bred into hybrid tomato Yanfu-1 which had early-maturation and high-production characteristics. Farmers planted this variety and gained incomes increased by 2,100 RMB Yuan per 667 m^2. In another example, rice Guang-qin-he stock 02428 was treated with ion implantation, and from subsequent generations mutant series that had flowering 1–2 hours earlier. This was selected and bred to overcome the physiological obstacles which caused the Guang-qin-he series and Xian-rice not to meet each other in the hybridizing flowering period. The selected and bred mutant series that had long stem section at the high part facilitated the use of the wind to spread pollens to short-stem sterile mothers. The applications of these mutant series are notably significant in hybrid rice study and production.

3). Mutants can be used for studying gene regulation and control. The basic clue is to look first at a certain function of the biological organism and then induce and identify mutants which cannot realize this function. Genes normally code various characteristics and these characteristics can be clearly expressed from experimental biological organisms. If mutants can be obtained and they also have long enough survival time for analysis, then important inferences on transmitting coding can be derived from them. Therefore, utilization of ion beams to produce rich variations is not only of important significance to production but is also of important value in scientific research.

10.4.3. Applications of Distant Hybridization

Distant hybridization is an important way to create new biospecies and new resources and transfer the allo-genetic substances. But distant hybridization has difficulties in hybrid fruiting and hybrid sterility and thus rarely produces hybrid offspring. Ion implantation of either fathers or mothers or hybridized generations can have effects on the hybridization fruiting rate, hybrid fruiting rate and hereditary process of the offspring.

10.4.3.1. Effects on Hybridization Fruiting Rate

Ion implanted common wheat Premebi as the mother was hybridized with rye AR1. The fruiting rates under three ion doses were all higher than the control (0.64%) with a mean rate of 4.03% which was 6.3 times the control. The ion implanted tetraploid rye as the father was hybridized with octaploid rye and produced an obvious effect on present generation fruiting rate. Generally the fruiting rate was increased by $1.5 - 3$ times. For treatment of either father or mother using a 30-keV nitrogen ion beam, the best effect was produced when the dose was 2×10^{16} ions/cm^2.

10.4.3.2. Effects on Hybrid Fruiting Rate

In an experiment, the hybrid F_1 of the controls of octaploid rye DDIB and tetraploid rye had an ear-fruiting rate of 25% and mean fruiting of 0.38 grains per ear; and the hybrid F_1 of the controls of octaploid rye DD7A and tetraploid rye had an ear-fruiting rate of 37.5% and mean fruiting of 0.56 grains per ear. After ion beam treatment, the ear-fruiting rates and mean fruiting grain numbers per ear of both hybrids F_1 all increased. With a dose of 2×10^{16} ions/cm^2, the mean ear fruiting rate was increased by 32.15% and the mean fruiting grain number per ear was increased by 2.58 for both hybrids.

10.4.3.3. Effects on the Hereditary Process of Hybrid Offspring

In the offspring of distant hybrids, due to gene partition and recombination, segregation starts from the second generation. For example for the hybrid of sea-island cotton and land cotton, generational segregation and selection can obtain stably hereditary cotton plants after more than ten generations. But the characteristics of the cotton are of either the sea-island cotton or the land cotton. It is very difficult to obtain new varieties which have both characteristics such as high quality from the sea-island cotton and high-productivity from the land cotton. However after ion implantation treatment of the normal hybrid or reciprocal hybrid F_1 seeds, the variation coefficient became smaller and smaller, namely, the character segregation degree continuously decreased, and tended to be stable as the number of generations increased. Up to the F_4M_4 generation, the variation degree (the ratio between the total variation coefficient and the F_2-generation total variation coefficient) was only 10% whereas the variation degree of the F_4 generation (i.e. non ion-implanted hybrid F_4) was 30%. This indicated that ion implantation could noticeably accelerate hereditary stability in the sea-land cotton hybrid. The reasons might be that N-ion implantation induced high-rate lagging chromosomes were easy to eliminate when forming gametes or/and ion implantation

Table 10.3. Types and rates of aneuploid wheat obtained by N-ion implantation.

Tested material	Number of Chromosome control	Ion dose $(10^{16}$ cm^{-2})	Number of individual examined	Number of chromosome	Number of types of aneuploids		Rate (%)
Premebi	42	3	38	41	2n-1	2	5.26
		4	22	41	2n-1	1	4.55
AR1	18	4	23	19	2n-1	1	4.53
	(14R + 4B)			(15R+ 4B)			
		1	27	27	2n-1	2	7.41
		2	29	29	2n-1+2t	1	3.45
Tetra-		2	29	27	2n-1	2	6.89
ploid rye		3	36	27	2n-2+1t	1	2.78
	28	3	36	27	2n-1	2	5.56
		3	36	26	2n-2	1	2.78
		4	33	26	2n-2	2	6.06
		4	33	27	2n-1	1	3.03
		4	33	29	2n+1	1	3.03

induced chromosomal displacement. These could accelerate the exchange of hereditary substance between sea and land hybrids and thereby speed up hereditary stability.

10.4.4. Creation of Aneuploids

Aneuploids are the basic materials of plant genetic breeding, playing an important role in genetic analysis and crop improvement. As described above, in ion implanted wheat, tetraploid wheat and AR1 rye, lagging, unequal segregation, 3-pole segregation and early segregation were all observed. A large amount of chromosomal lagging can definitely induce changes in the chromosome number and thus possibly result in new idioplasms of aneuploid wheat.

Table 10.3 shows the types and rates of aneuploid wheat obtained by ion implantation. Aneuploid wheat was obtained from all M_2 generations for three tested materials. With N-ion implantation doses of D3 (3×10^{16} ions/cm^2) and D4 (4×10^{16} ions/cm^2), monomers with the chromosome number 41 appeared in wheat Premebi. In rye AR1 with dose D4, monomers with the number of chromosomes of 19 appeared, 15 of them being normal chromosomes and 4 as B chromosomes which were obviously a trisome. In tetraploid rye, the aneuploids all appeared with four doses. There were not only the numbers of chromosomes as 26, 27 and 29, but also telomeres. The appearance rate of aneuploids was between 3.03% and 7.41%. Compared with the control, the appearance of aneuploids was unchanged and the fruiting rate was slightly decreased. In chromosome engineering, it is ideal that inside the cell a section of the gene vector (chromosome) that controls bad characteristics is cut and replaced with a section of chromosome that controls good characteristics. Currently this kind of operation inside the cell is not easy to do. But the appearance of wheat of the aneuploids such as monomers and nullisomes provides wheat chromosome engineering with easy and cheap original material. In order to create wheat aneuploid material, the normal way should be first having original variety resources such as the China Spring (CS) monomer, and through hybridization, the transmitting breeding procedure of the aneuploids can then be accomplished. After hybridization, identification, selection and breeding continuously for several generations are carried out, ion implantation induced variation of chromosome number provides breeding and developing aneuploid wheat with a new route.

REFERENCES

1. Lodish, H., Berk, A., *et al.*, *Molecular Cell Biology* (W. H. Freeman & Co., New York, 2003).
2. Yu, Z.l., *et al.*, Preliminary Studies on the Biological Samples by Ion Beam Etching, *Journal of Anhui Agriculture University* (in Chinese with an English abstract), **21**(3)(1994)260-264.
3. Song, D.J., *et al.*, *Acta Biochemical et Biophysical Sinica*, **30**(6)(1998)570-574.
4. Yang, J.B., Wu, Y.J., *et al.*, Low Energy Ion Beam-Mediated DNA Delivery into Rice Cells, *Journal of Anhui Agriculture University* (in Chinese with an English abstract), **21**(3)(1994)330-335.
5. Cheng, B.J., *et al.*, *Acta Bot. Boreal*, **14**(2)(1994)85-89.
6. Ren, H.Y., Huang, Z.L., *et al.*, *Chinese Science Bulletin*, **45** (18)(2000)1677-1680.
7. Song, D.J., *et al.*, *High Technology Letters*, **2**(1)(1991)47-50.
8. Yu, Z.L., *Physics* (Chinese), **26**(6)(1997)333-338.
9. Song, D.J., Yu Z.L., *Science* (Chinese), **219**(1996)49-51.
10. Song, D.J., Yao, J.M., *et al.*, *Hereditas* (Chinese), **21**(4)(1999)37-40.

11

ION IMPLANTATION MUTATION BREEDING OF MICROBES

Mutation is the fountainhead of genetic variation. Although hybridization induced gene recombination is one route to variation, mutation is more important. Without different mutation-induced allelomorphic genes to enrich the gene pool, it would be impossible to talk about gene recombination. Inducing artificial mutation is an important means of accelerating mutation and the artificial rate can be thousands of times greater than the natural rate of mutation [1]. It is natural that artificial-mutation-based microbial breeding monopolizes germ breeding in the fermentation industry. At present, most of the various production strains of microbes used in the international and domestic fermentation industries are mutated strains [2]; in particular, almost all of the strains used in the antibiotic industry are mutant strains. Although genetic engineering is well developed for the improvement of strain species and there are numerous examples of good achievements, applications in production are not yet very extensive.

Applications of ion beam technology to plant breeding provide microbial breeders with a reference. However, microbes, as distinct from dormant dry seeds, are growing and have high water content during the implantation process. Hence they are very sensitive to the environment in the equipment and to screening patterns. This chapter first makes an estimation of the efficiency of ion-implantation-induced microbial mutation by identifying auxotroph mutations, then describes the procedure for microbe modification by ion implantation, and finally introduces production practices of ion beam microbial breeding.

11.1. EFFECTS OF ION IMPLANTATION ON MICROBIAL MUTATION

After energetic ion implantation, microbes suffer from various kinds of damage: damage to the morphology of large cells, changes in various subcellular structures, and modification of the biological macromolecules that compose the cells. This damage affects the life activities of the microbe. When the genetic substance DNA is damaged, corresponding DNA repair enzymes are activated to induce repair systems to function. No matter which type of DNA damage is to be repaired, it is inevitable that some errors occur, and these can lead to genetic mutations.

In the study of the effects of energetic ion implantation on microbial mutation, those types of microbes that have clear genetic backgrounds are the best objects to be chosen. A phenotypic mutation easy to screen and identify is the auxotroph mutation. The auxotroph mutation is a type of biochemical mutation. Microbial cells having this mutation are unable to synthesize components such as amino acids, nucleotides, vitamins and some fatty acids by themselves, so the mutation can grow only in culture media that contain these relevant components [3]. For instance, the bacterial solution after mutation treatment is smeared on a complete culture medium and grows to form colonies; each colony is inoculated to various basic culture media; those strains that cannot grow in the basic culture media are selected to be inoculated to various culture media for determining which essential component is missing. These culture media include complete culture media, basic culture media, culture media with various amino acids added, and basic culture media with various vitamins added. If a mutant strain can grow only in the complete culture media and vitamin-containing culture media, it is then known that this is a vitamin-absent strain, and then which kind of vitamin is needed can be determined.

In an experiment, bacteria of *E. coli* were implanted with 30-keV nitrogen ions to a dose of 1.5×10^{15} ions/cm^2. Bacteria that had the tryptophan-absent mutant type were screened. Statistical results showed that about 22.1% of the bacteria were affected with this defect and 1.8‰ of them were determined to be in the tryptophan-absent mutant type. It is known that the tryptophan operon has a total of five genes. They are linked next to each other, coding three enzymes and catalyzing the synthesis reactions from chorismate to tryptophan. It is obvious that in the tryptophan operon, mutations occurring in the regulation gene or structure gene can interrupt the tryptophan synthesis process.

The doses of both ion beam and γ-ray radiation are adjusted for the same survival rates, and the auxotroph mutation rates and the mutation efficiencies of the *Salmonella typhimurium* induced by γ-ray radiation and energetic ion implantation are compared. It can be seen from the results (Table 11.1) that in most cases the mutation rate from ion implantation is higher than that from γ-ray radiation, and the mutation efficiency (the ratio between the mutation rate and the biological damage rate) from ion implantation is also higher. As described above, energetic ion implantation has not only the characteristic of energy deposition, as does also γ-ray radiation, but also momentum-exchange-induced cascade damage. The latter leads to atomic displacements and rearrangement or gene deletion of the genetic substance. Furthermore, ion-implantation-induced damage includes chemical damage resulting from recombination reactions between the slowed ions, displaced atoms and matrix atoms, and damage caused by transfer of electrons in the biomolecules due to charge exchange. In other words, ion beam mutation is in a sense something like combination mutation. Experience in mutation practices has revealed that combination mutation is more effective than single mutation. Thus, it is not difficult to understand why in most cases ion implantation of microbes results in higher mutation rates and efficiency than γ-ray radiation.

It can be seen from Table 11.1 that ion implantation causes less damage to the bacteria than γ-irradiation. The proliferative rate 2 hours after ion implantation is 1 to 7 times greater on average than for γ-ray radiation. For a dose of 20×10^{15} N$^+$/cm^2, the mutation rate can be as high as 32×10^{-5}, which is more than 30 times greater than for spontaneous mutation, and 30% higher than the highest mutation rate from γ-ray radiation. This example demonstrates that ion implantation of microbes, as for crop seeds, is able to achieve higher mutation rates while only inflicting light damage in the target materials.

Table 11.1. Survival rate, growth and mutation of *Salmonella typhimurium* for treatments by N-ion implantation and by ^{60}Co-γ-ray radiation.

Mutagen	Dose (ions/cm^2)	Mean lethal rate (%)	Relative proliferative rate in 2 hours (%)	Mutation rate ($\times 10^{-5}$)	Mutation efficiency ($\times 10^{-5}$)
	0	0	100	1.01	-
	0.75×10^{15}	66.9	93.5	1.60	2.4
Ion	1.0×10^{15}	77.4	65.8	4.00	5.1
implanta-	7.0×10^{15}	85.0	31.5	8.90	10.4
tion	10×10^{15}	90.7	27.3	13.80	15.3
	20×10^{15}	96.9	24.7	32.00	33.0
	25×10^{15}	99.7	14.4	23.30	23.4
	0 Gy	0	100	0.94	-
	25 Gy	68.5	93.8	1.10	1.6
	50 Gy	76.6	27.3	3.20	4.1
^{60}Co-γ-ray	100 Gy	83.8	25.4	7.50	8.9
	200 Gy	91.4	11.2	24.60	26.2
	400 Gy	96.7	3.2	12.80	13.2
	600 Gy	99.9	2.1	17.30	17.3

The auxotroph mutation is notably a mutation type easily identified. In fact, energetic ion implantation induced mutation types in unicellular lives are rather wide-ranged. For instance, in a study of the cell cycle of Saccharomyces, ion implantation can induce a variety of types of genetic mutations, such as continuous germination but no division or germination suppression (see Chapter 7). Investigation of the effects of ion implantation on these single-gene-controlled mutations of quality characteristics provides a basis for ion implantation modification of industrial microbes. The production of the metabolic products or the accumulated products in the cell is generally controlled by micro-effect polygenes. The mutations of these micro-effect polygenes have a cumulative action, namely, the forward mutation of every micro-effect gene contributing to the production of microbe products. Therefore, one can take advantage of the mutation effects of the combined parameters (ion species, energy, dose, charge number, etc.) of ion implantation on the micro-effect polygenes to obtain eventually mutant strains which meet breeding goals.

11.2. MICROBE BREEDING PROCEDURES

Ion implantation induced mutation of microbes can be applied to production processes. Because of the special features of factors and objects, the breeding procedure of ion implantation of microbes should not be simply a copy of those for dry crop seeds. In the treatment of dry crop seeds, the embryo of each seed must be facing the ion beam incident direction. But microbes are sufficiently small that in whatever direction of ion implantation there is no problem for ions to enter the cells. The real problem is that in the ion energy range discussed in this book, ions cannot penetrate the glass containers that usually house the bacteria. Therefore, for microbe breeding, the microbes to be irradiated must be totally exposed to the ion beam. How can we guarantee that the samples are not infected by other bacteria because of this total exposure? If the bacteria are placed in a culture medium to be ion implanted, how can we guarantee that water evaporation from the culture medium will not affect the target chamber vacuum? These are some of the problems of ion implantation of microbes that need to be solved first.

In chapter 4 we have described how samples that contain a high water content can suffer a severe drop in surface temperature in vacuum because of surface evaporation. When a two-phase equilibrium is reached, the temperature of the interface between the two phases is below 0°C and the sample surface is frozen to an ice shell. If the pumping system of the target chamber has a speed high enough, the vacuum will be maintained (i.e. the vapor pressure is constant).

The following describes the ion-implantation microbial breeding procedure used at the bioengineering ion beam facility in the Institute of Plasma Physics, Chinese Academy of Sciences.

Ion implantation of microbes is carried out inside a micro-environmental sterile target chamber (see Chapter 2). The volume of this small target chamber is 1/600 that of the large target chamber. There is a gate valve to separate the two chambers. After a petri dish holding the culture medium is put into the small chamber, a high-speed mechanical rotary pump is used first. After 30 seconds, the pressure in the small chamber reaches about 1×10^3 Pa. Then the gate valve is opened and the pressure in the whole system rapidly reaches about 2 Pa (the initial pressure in the big chamber is 7×10^{-3} Pa). At this pressure the diffusion pump in the big chamber can operate efficiently. About one minute later, ion implantation can be started.

A ventilator is designed in the small target chamber to admit sterile air. Surroundings of the small chamber, equipped with ultraviolet light, are equivalent to a sterile operation stage. When the door to the operation stage is opened, a sterile air stream is automatically blown so as to guarantee sterile conditions for transporting the sample into and out of the chamber.

11.2.1. Pre-Treatment of Samples

A small quantity of single-cell microbes or spores is selected from the slant (or plate), diluted and shaken in sterile water, and 0.1 ml taken to smear homogeneously on a petri dish. Some dishes have culture media and some not. With culture medium, the medium should not be spread too thickly, normally about 2 mm. For filiform fungi that do not produce spores, the bacteria should be mashed with glass beads beforehand and then followed by the above-mentioned steps. If a petri dish is used for growing bacterial colonies, it can be simply placed inside the chamber for treatment but the culture medium must not be too thick.

11.2.2. Ion Implantation

Since the bacterial varieties used in production have already been pretreated, they easily resist mutagens normally used. However, the ion beam is a new type of mutagen in which the primary process mainly results in displacement, addition and recombination of genetic substance atoms. But this does not imply that the ion species to be implanted do not need to be selected. When the ion implantation effect is inefficient, a change to another ion species is recommended.

Inside the micro-environment target chamber, the lid of the petri dish to be treated is removed, the small target chamber is opened, and the dish is placed at right angles to the direction of the ion beam. The mechanical rotary pump is started and operated for about 30 seconds, and then the separation gate-valve is opened. The ion beam is already prepared and the dose rate measured beforehand. After about a minute, ion implantation starts. The implantation dose is normally an order in magnitude lower than that used for the implantation of dry seeds. The optimal dose should be determined by experiment, namely, a plot of survival rate of the treated microbes as a function of ion dose should be obtained. A dose that results in survival rate in the range 10% – 30% should be used. When the ion dose reaches a preset value, ion implantation is stopped by closing the separation gate-valve and then blowing sterile air into the small target chamber. The small chamber is now opened, the sample is removed, the dish lid is closed, and post-treatment can be performed in the laboratory room.

11.2.3. Post-Treatment of Samples

If a bacterial solution is directly smeared onto the petri dish beforehand, the sample can be washed with sterile water, diluted to a certain concentration, and smeared with culture medium. Attention should be paid to the time sequence of the colonies which appear on the culture medium so as to determine which colonies should be selected for further screening. If the bacterial suspension solution is smeared with culture medium beforehand, a direct culture is possible. After the colonies grow, some colonies should be selected for natural segregation. If ions are implanted into grown colonies, after ion implantation thalluses are scraped and taken from the surface of each colony, diluted, and recultured.

11.2.4. Screening of Mutated Strains

Screening of mutated microbial strains is not quite the same as for mutated agricultural crop plants, but the procedural steps are basically similar, including primary and secondary screenings and identification of generation transmission [4]. Primary screening roughly determines the extent of the mutated strains. Secondary screening is used to finally decide whether the screened strains meet the breeding objectives. The generation transmission study mainly investigates the genetic stability of the chosen strains. In these three steps, the key is setting up the sieves of the primary screening, such as the inhibitory zones and the azure reaction zones of amylase, etc. But primary screening of some strains may not be so easy. For example, in the author's laboratory, for modification of the vitamin-C macrocin and microsin using a two-step fermentation method, setting up the screening method for primary screening took 3 months.

The principles of screening for microbe breeding can be summarized as follows. 1). Attention must be paid to establishment and development of a fast, accurate and convenient method of analysis. This has a decisive influence on the success or failure of screening work. Different methods are used at different steps in the screening process. For instance, as there are many samples in primary screening, rough and simple detection methods can be used; in secondary screening, precise determination methods should be adopted.

2). Attention is paid to observation of the morphology of the implanted colonies. High production mutations have a linkage relation with morphology mutations. Before selection, high-production and low-production colonies should be studied. Thus, in the primary screening, those low-production strains can confidently be bypassed. This is about the same as expanding the screening quantity. But note that for implanted colonies with a culture medium, after ion implantation they are cultured for some time and the grown colonies are morphologically irregular, hence special attention should be paid to natural segregation screening, because high-production mutants often hide in the damaged colonies.

3). After new mutants are selected, culture media and conditions should be frequently adjusted so that the selected strains can further display their potential. For the culture media and conditions, several cross experiments can be designed to determine an optimal culture plan.

4). Step-type multiple ion implantations and screenings can generate a small quantity of high-production mutations each time and accumulate results to achieve the breeding goal.

11.3. ION IMPLANTATION MODIFICATION OF PRODUCTION BACTERIA

Fermentation engineering is an important component of biotechnology with an impact on national economics and standard of living. Functional foods, pharmaceuticals and biochemical preparations are almost all produced from microbial fermentation. Besides automatic reactors and modernization of complete facilities, the key of fermentation engineering is based on high efficiency and high quality seeds. Because of this, almost all fermentation production facilities are equipped with professional microbe breeding laboratories. Mutation still plays a major role in plant breeding technology. Frequently used methods include physical and chemical mutation approaches such as ultraviolet, γ-ray and X-ray irradiation, or combinations of these mutagens [5]. Long-term use of a mutagen will induce resistance to it, and thus the exploration of new microbe breeding methods is a goal of microbe breeding. In the past ten years, genetic engineering has emerged as part of microbe breeding. However, for most of production microbes, genetic engineering technology has not yet extensively applied to modification of original production microbes.

The use of ion beams is a new technique for inducing mutation. It has exhibited fairly high mutation rates and mutation efficiencies in experiments. But biochemical mutation is a type of quality-characteristic mutation, whereas the current aim of modification of bacterial species for the fermentation industry is to select and breed high production strains with the ability to synthesize desirable metabolic products. The synthesizing abilities of metabolic products which have been isolated mostly show quantity characteristics, controlled by micro-effects of multiple genes and are easily affected by culture conditions. In addition, the synthesizing ability of a metabolic product forms a continuous spectrum. Therefore, in ion implantation modification of production bacteria the screening procedure should be handled carefully. If it is impossible to achieve the goal in one step, multiple step implantation can be employed to reach the final breeding goal. After each step of improvement in the fermentation level of the strains, hybridization experiments should be done to check the culture medium for its ability to sustain the mutants.

Some examples of modification of bacterial varieties used in production are described below to show the practice of ion implantation for microbe breeding.

11.3.1. Zhijiang Bacterin

The Zhijiang bacterin is a type of new antibiotic screened by Huang Wencai of the Zhejiang Agriculture Academy. This antibiotic is characterized by safety and high effectiveness in curing white dysentery in fowl and livestock. The aim of ion implantation was to increase the antibiotic titer of these production bacteria. It was expected that screening would reveal highly productive mutated strains for the fermentation industry. Nitrogen ions were used for ion implantation at 30 keV with five doses. Single spores were smeared on Gao-1 agar plates. When the ion dose was 5×10^{15} ions/cm^2, the survival rate in each plate was 44.2%, whereas when the dose was increased to 3×10^{16} ions/cm^2, total death resulted. If the colonies formed with a large number of spores were ion-implanted, the survival rates of the tested bacteria were not much affected. Even when the dose was 4×10^{16} ions/cm^2, the bacteria could still survive. The experiment demonstrated that the positive mutation rates of strains with titer 10% higher than the control were 6.75% – 38.48% with an average of 26.78%. This revealed highly productive strains. At a dose of 7.5×10^{15} ions/cm^2, No. 5 to 22 strains were screened and their potency was increased by 75.3% compared to the starting material. At a dose of 1.5×10^{16} ions/cm^2, No. 30 – 38 strains were screened and their potency was increased by 61.5%. These two groups of strains had stable characteristics and were put into experimental production in 3-ton and 5-ton fermentation pots and the potency was increased by more than 60% compared with that of the starting material. The products healed white dysentery and enteritis of chicken and diarrhea of young pigs with an effectiveness higher than 95%.

11.3.2. Rifamycin Production Bacteria

Rifamycin is a frequently used medicine for treating tuberculosis. The domestic fermentation level is normally about 5,500 units/production-pot. The Institute of Plasma Physics, Chinese Academy of Sciences, started to use ion implantation for modification of the rifamycin production bacteria in 1992. The detailed steps are as follows. Mycelium was extracted and mashed with glass beads, and the suspension liquid was smeared on a plate. Due to the small size (1 – 2 mm in diameter) of the rifamycin colonies, primary screening was not convenient. Thus, each colony was positioned on a slant, put in a shaking bottle, and screened by using chemical titer as the standard. The titer of the mutated strains screened in the first stage was increased by 10% compared to the starter. In the second stage, the mutated strains selected from the first stage were ion-implanted again and the titer after the primary screening was increased by 20% – 30%. From the third-stage screening, the titer was as high as 8,600 units and another up to 9,800 units. Over two years, 3,100 shaking bottles were provided and 31 mutated strains were selected with chemical titer increased by 20% – 90%. By segregation and generation transmission, three strains were finally selected for tentative production in 7-ton and 20-ton fermentationors. The highest chemical titer of the first products from the first fermentationor production was up to 6,800 units and 6,100 units on average. The titer was increased by 35% compared with the simultaneous starter. Through a year of preliminary production, the average titer was increased by 11.5%.

11.3.3. Saccharifying Enzyme Production Bacteria

At present, after repeated application of physical and chemical mutagens, the fermentation levels of domestically used saccharifying-enzyme production bacteria mostly remain at about 10,000 units. Ruan Lijuan, Zhejiang Agriculture Academy, chose and used saccharifying enzyme bacteria provided by the Zhejiang Tonglu Enzyme Product Plant as the starting material. Bacterial spores from different growth periods were used as target material and implanted with 30-keV nitrogen ions to doses of 5×10^{15}, 1×10^{16}, 1.5×10^{16} and 2×10^{16} ions/cm^2. Single-spore segregation was carried out for each treatment group to obtain 6,322 strains. Through primary agar plate screening, 1,582 strains were obtained having an increase in production rate of 10%. The positive mutation rate was 25%. The positive mutants were secondarily screened by pot-shaking to obtain three mutants with an increased enzyme activity of 20% compared with the original strain. One of the three had an activity of fermentation enzyme production as high as 26,000 units, and 20,000 units on average for total three strains. The characteristics of the enzyme production were stabilized through fixed-period tests of generation transmission.

11.3.4. Polyunsaturated Fatty Acid Production Bacteria

In all of the examples described above, either using microbial metabolic products or increasing the activities of the microbes themselves was taken as the breeding goal. In this example, *Mortierell alpina* was used as the starter and the breeding goal was an accumulation of endocellular fatty acids. Hence, in the screening of mutated strains, it was required to have both a high collection rate for dry mass and high fat content, and an unsaturated fat component in the thallus. The strains underwent multiple-step combinative implantation with 30-keV – 50-keV H$^+$, Ar$^+$, and N$^+$ ions. The fermentation time was found to decrease from 11 days to 9 days. Through fermentation in a small 20-l pot, the collection rate of the dry thallus was increased from 19 g/l to 30 g/l, the total fat quantity was increased by 20%, and the content of the unsaturated fatty acid was increased from 32% to 50%.

11.3.5. Vitamin C

L-Ascorbic acid (vitamin C), a water-soluble vitamin, has been widely used in industry applications, such as the food and beverage industry, the pharmaceutical industry, and other new fields [6]. The current global production has been estimated at approximately 80,000 metric tons per year with a worldwide market in excess of 600 million US dollars. The vast majority of ascorbic acid is produced from the 70-year-old Reichstein and Grussner method via a single biocatalysis step within a series of chemically based unit operations [7,8]. However, this method is highly energy consuming and relies on the use of a number of environmentally hazardous chemicals. A two-step process of Lascorbic acid manufacture has been established in a commercial scale in China by using a mixed culture of *Gluconobacter oxydans* and *Bacillus megaterium* [9,10]. Although γ-ray, X-ray, and ultraviolet have been used as mutagens to increase the production in the two-step process, the transformation rate of L-sorbose to 2-keto-L-gulonic acid (2-KLG), a key intermediate, has remained at a level of 75–80% for decades [11].

In the Laboratory of Ion Beam Bioengineering, Institute of Plasma Physics, Chinese Academy of Sciences, relevant research was focused on the mutation screening of *Gluconobacter oxydans* (GO29) and *Bacillus megaterium* (BM80) by ion implantation [12]. Ion implantation was carried out with the heavy ion implantation facility at the Institute. 2-KLG in whole culture broth was determined by iodometry. Mutants were screened by single-colony isolation and 2-KLG accumulation in broth. GO29 and BM80 were implanted by either hydrogen ions (H^+) or nitrogen ions (N^+) with various doses, respectively. The average transformation rate of BG112-302 (mixed BM and GO) bred by ion beam in Gram-molecule was increased from 79.3 to 94.5% after eight passages in shaking flasks. Furthermore, in 180-ton fermentationors in Jiangsu Jiangshan Pharmaceutical Co. Ltd, the transformation rate was stable at 92.0%, indicating a producer could get 0.99 kg of gulonic acid from 1.0 kg of sorbose. BG112-302 bred by ion beam implantation dramatically increased the transformation rate by 19.2%, which greatly increased efficiency and reduced the cost of L-ascorbic acid manufacture in the two-step process.

11.3.6. L(+)-lactic Acid

L(+)-lactic acid is mostly used in food industry as a preservative or flavor–enhancing additive. Various lactic acid salts are being used in pharmaceutical industry for their therapeutic functions. Particularly, L(+)-lactic acid can be used to synthesize poly lactic acid (PLA), which is proposed for use in manufacturing a new biodegradable plastic.

Rhizopus oryzae is a common fungus used in industry for production of highly optically pure L(+)-lactic acid. In order to obtain industrial strain with higher L(+)-lactic acid yield, the wild type strain *Rhizopus oryzae* PW352 was mutated by means of nitrogen ion implantation (15 keV, $7.8 \times 10^{14} \sim 2.08 \times 10^{15}$ ions/cm^2) and two mutants, RE3303 and RF9052, were isolated [13]. After 36-hour shake-flask cultivation, the concentration of L(+)-lactic acid reached 131~136 g/L, the conversion rate of glucose was as high as 86%~90% and the productivity was 3.61 g/L·h. It was almost a 75% increase in lactic acid production compared with the wild type strain. Maximum fermentation temperature of RF9052 was increased to 45°C from original 36°C. At the same time, the preferred range of fermentation temperature of RF9052 was broadened compared with PW352.

11.3.7. Antifungal Substance

Bacillus species produce a variety of secondary metabolites with antimetabolic and pharmacological activities, most of which are small peptides that have unusual components and chemical bonds, so are recognized as having a high potentiality for biotechnological and pharmaceutical applications [14]. *Bacillus subtitles* JA that was isolated by the Laboratory of Ion Beam Bioengineering produced a large yield of antifungal substances, which had strong inhibitory activity against various fungi and bacteria which are pathogenic to plants and human being, such as *Rhizoctonia solani, Fusarium graminearum, Pseudomonas solanacearum, Trichophyton rubrum* and so on. A

208

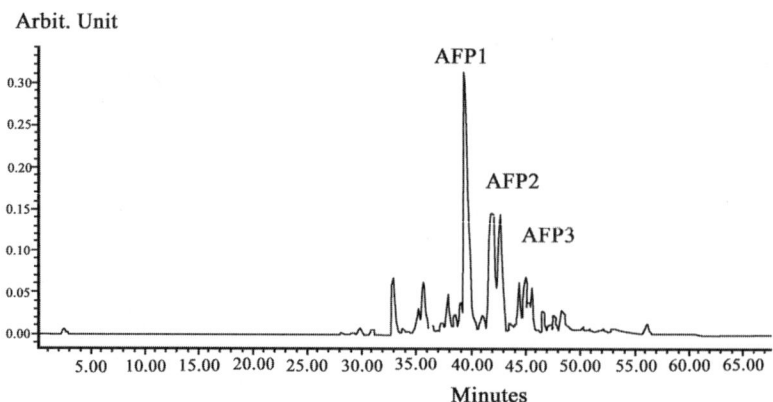

Figure 11.1. Purification spectrum of antimicrobial peptides by means of FPLC (Fast Performance Liquid Chromatography) on RP-C$_{18}$ column using an AKTA^{TM} (Amersham Biosciences) explorer. The peaks stand for different antimicrobial peptides with different molecular weights and properties.

mutant strain of JA-026 resulting from bombardment of the parent strain with nitrogen ions found to have higher antagonistic activity than the original strain [15]. The optimal energy was 20 keV and dose was $3.12\times10^{15} \sim 3.90\times10^{15}$ ions/cm^2. The antifungal substance was purified by AKTA^{TM} explorer (Pharmacia) and three active fractions were analyzed by LC-MS (Finnigan). Their molecular weights were 1462.645D, 1476.390D and 1490.530D, respectively, composed of Thr, Ile, Pro, Val, Glx (Glu or Gln), etc., as demonstrated in Figure 11.1.

The above seven examples show applications of cooperation between the author's Institute and a number of companies concerned with ion implantation modification of industrial microbes. Ion implantation modification of microbes is a relatively new development and is not as common as ion implantation for crop breeding. Nevertheless, as seen from the above examples, success has been achieved in a number of directions. The prospects for further applications of ion implantation for microbe breeding in the fermentation industry are excellent.

REFERENCES

1. Zhu, G.J. and Wang, Z.X., *Experimental Technique Handbook in Industrial Microbiology* (in Chinese, China Light Industry Publishing House, Beijing, 1994)p.384.
2. Zhou, D.Q., *Principle of Microbiology* (in Chinese, High Education Publishing House, Beijing, 1993)p.244.
3. Zhou, D.Q., *Principle of Microbiology* (in Chinese, High Education Publishing House, Beijing, 1993)p.252.
4. Zhou, D.Q., *Principle of Microbiology* (in Chinese, High Education Publishing House, Beijing, 1993)p.251.

5. Zhou, D.Q., *Principle of Microbiology* (in Chinese, High Education Publishing House, Beijing, 1993)p.234-235.
6. Hancock, R.D. and Viola, R., Biotechnological Approaches for Lascorbic Acid Production, *Trends in Biotechnology*, **20**(2002)299–305.
7. Reichstein, T. and Grussner, A., Eine Ergiebige Synthese der 1-Ascorbinsaure (C-Vitamin), *Helvetica Chimica Acta*, **17**(1934)311–315.
8. Holding, T.J., Vitamin C Manufacture. In: *World Biotech Report.* (Pinner, UK: Online Publications, 1988)pp. 221–227..
9. Kieslich, K., Present State of Biotechnological Production of Pharmaceuticals, *Proceedings of The 3rd European Congress on Biotechnology* **IV**(1984)39–72.
10. Lu, S., Huang, S. and Yu, J., Rules of the Supply and Demand of Oxygen and Scale Up for Vitamin C Fermentation, *Chemical Reaction Engineering & Technology* (Chinese), **1**(1985)72–82.
11. Yin, G., Lin, W., Qiao, C. and Ye, Q., Production of Vitamin C Precursor-2-keto-L-gulonic Acid from D-sorbitol by Mixed Culture of Microorganisms, *Journal of Hygienics* (Chinese), **41**(2001)709–715.
12. Xu, A., Yao, J.M., Yu, L., Lü, S.J., Wang, J., Yan, B., Yu, Z.L., Mutation of Gluconobacter Oxydans and Bacillus Megaterium in a Two-step Process of L-Ascorbic Acid Manufacture by Ion Beam, *J. Applied Microbiology*, **96**(6)(2004)1317-1323.
13. Ge, C.M., Gu, S.B., Zhou, X.H., Yao, J.M., Pan, R.R., Yu, Z.L., Breeding of L(+)-Lactic Acid Producing Strain by Low-Energy Ion Implantation, *J. Microbiol. Biotechnol.*, **14**(2)(2004)363-366.
14. Yazgan, A., Ozcengiz, G., Ozcengiz, E., *et al.*, Bacilysin Biosynthesis by a Partally-purified Enzyme Fraction from Bacillus Subtilis, *Enzyme and Microbial*, **29**(2001)400-406.
15. Liu, J., Wang, J., Yao, J.M., Pan, R.R., Yu, Z.L., Properties of the Crude Extract of Bacillus Subtilis and Purification of Antimicrobial Peptides, *Acta Microbiogica Sinica*, 2004, in printing.

12

ION BEAM INDUCED GENE TRANSFER

Plant genetic transfer, or gene transfer, is an important component of genetic engineering. Genetic engineering is a process in which, according to needs, external genes from separation and cloning are assembled into a certain type of vectors to be transferred into a receptor cell using physical or chemical or biological means so that gene expression is realized (at the cell, tissue, and whole body levels), and the new genetic background is passed along. It not only provides for genetic expression, regulation, and heredity studies with an ideal experimental system, but also more importantly provides an effective route for future plant improvement and molecular breeding.

In plant gene transfer, simple and effective means are sought to realize gene transfer into walled cells so that complicated protoplast culturing processes can be avoided. Hence physical methods were developed and good progress has been achieved. These relatively successful methods include electroporation [1], particle gun or microinjectile bombardment [2], microinjection and macroinjection, microlaser [3], and ultrasonication [4]. All of these methods perforate the cell wall, without injuring the cell itself, to provide exogenous genes with a passage to enter the receptor cell.

In the study of the interaction between low energy ion beams and cell surfaces, the author discovered the thinning effect of the ion beam on cell walls and proposed the idea of ion beam induced gene transfer. Attempts for ion beam induced gene transfer have been successful accomplished by independent studies of three Ph.D. student groups in the Institute of Plasma Physics, Chinese Academy of Sciences [5-8], as well as other international research institutions [9-11]. This chapter summarizes the study of this new biotechnology. Section 1 introduces the principle of the ion beam induced gene transfer method. Section 2 deals with factors affecting the transfer rate. Section 3 describes the procedure of the ion beam induction method with examples of gene transfer in rice. Section 4 discusses ion beam induced DNA macromolecule transfer. Readers will gain a basic understanding of the principles, methods and procedures for ion beam induced gene transfer through these expositions, so that they can creatively apply the tools for further development of the method.

12.1. PRINCIPLES OF ION BEAM INDUCED GENE TRANSFER

As described in the previous chapters of this book, three important phenomena can be observed in the study of interactions between implanted ions and biological organisms. (1) Ion implantation causes atoms of the biological substance to be displaced and recombined, and if the ions act on genetic substances, mutation can be induced. (2) The action of low energy ion beams on the biological organism can etch and perforate the surface. (3) Charge exchange between charged particles and the cell surface affects and changes the cell's electric character. These facts or factors constitute the foundations of ion beam genetic manipulation. Mutation breeding described in previous two chapters is the result of comprehensive utilization of these factors.

Ion beams have the ability to process biological organisms or cell surfaces. Can the ion beam parameters be controlled to such a critical extent that localized stripping does not affect the living activities of the cell wall, producing only repairable micropores in the cell membrane so that permeability for exogenous substances to transfer into the cell is increased for the purpose of gene transfer? The answer is positive. An ion with several tens of keV energy has a super strong sputtering action on the surface of the biological organism [12]. The beam is like a pounding driller stripping the cell wall layer by layer. As the organism surface (including the cell wall) has a multi-layer, multi-phase and porous structure, the stripping effect can connect pores that previously were not connected so that subsequent ions etch the organism at deeper layers. This kind of continuous action results in an inevitable perforation of the cell wall [13]. Photographs in Chapter 7 show ion-beam-etched cultured rice cells. The cell surface that was not etched by ions is smooth. With low-dose ion beam action, the hemispherical cap of the cell surface that was oriented toward the ion implantation direction becomes rough, leaving traces of grooves and pores. With increasing dose, the eroded pores on the cell surface become deeper and deeper. When the dose further increases, the cell wall is completely perforated and possibly broken. In one experiment, an Ar-ion beam at 30 keV etched the cuticle of a Malus leaf through a 1-mm aperture. When the dose was $5 \times 10^{15}/cm^2$, diffraction patterns could be observed through a magnifying glass on a single-crystal silicon substrate positioned at the rear of the 20-μm-thick membrane. This indicates that the membrane had been totally etched through. Figure 4.11 shows that when the dose reached $1 \times 10^{17} N^+/cm^2$, micro-holes were created on a 50 μm thick tomato fruit skin through which α particles could freely pass.

Table 12.1. Effect of ion etching on the permeability (ppm) of naked rice seeds in solution. Data are on two N-ion doses.

Treatment	Control	$3 \times 10^{16} N^+/cm^2$	$6 \times 10^{16} N^+/cm^2$
I	3.0	15.0	29.0
II	4.5	20.5	32.0
III	3.0	23.0	32.0
IV	6.0	10.5	23.0
Average	4.3	17.3	29.0

For mature rice embryos (often used as a receptor material in gene transfer), in order to show effects of ion beam etching on the surface permeability, changes in electro-conductivity can be measured for samples before and after ion implantation. If ion etching really makes micro-holes in the mature embryo surface, the cell inner substance will permeate into solution through micro-holes to finally reach a balance. Table 12.1 shows changes in electro-conductivity of non-shelled rice seeds of *Indica* Guanglu-short No.4 after ion beam etching at doses of $3 \times 10^{16} \, N^+/cm^2$ and $6 \times 10^{16} \, N^+/cm^2$.

The mean permeability of four groups of samples that were not ion-beam-etched was 4.3 ppm, and the mean permeability of the samples etched with a dose of $3 \times 10^{16} \, N^+/cm^2$ was 17.3 ppm, more than four times of that of the control. The mean permeability of the rice etched at a dose of $6 \times 10^{16} \, N^+/cm^2$ is 29.0 ppm, 6.7 times of that of the control. This shows that as the ion etching dose increases, the out-permeability of the cell-contained substances increases, indicating that as the permeability of the cell increases, the probability of exogenous DNA entering the cell also increases. But the increase in cell permeability should not affect cell activity. It will be seen in the next section that for exogenous genes to enter the cell and be expressed, the etching dose needs to be in an appropriate range. It is not the case that the greater the permeability, the better.

Low-energy ion-beam etching of the biological organism surface provides pathways for exogenous gene transfer. Due to the ion beam controllability and focusing power, one can in principle precisely control the diameter and depth of the pathway formed on the receptor surface to facilitate gene transfer [13]. One can transfer not only gene segments (including their vectors) but also naked vital DNA macromolecules without affecting the normal life activities of the cell. As technical precision is increased, control of the degree of damage to the receptor chromosomes at the bottom of the pathway becomes possible.

Ion beam processing of biological organisms results from ion sputtering, electron sputtering and chemical sputtering. Sputtering is a kind of cold processing. Compared to the heating that induces evaporation, cold processing has a greater precision and avoids injuring neighboring tissue or cell organelles.

Pathway formation is a prerequisite and basis for gene transfer into walled cells and mature embryos. It is a question worthy of discussion, however, how exogenous genes enter the cell, actively or passively. As described previously, charge exchange between charged particles and the biological organism surface can change the electric polarity of the biological molecule (e.g, positive charge replacing negative one). For the cell, although the change in electric polarity cannot be macroscopically measured, as the cell is a very poor electrical conductor, a local electric polarity may change at the hemispherical cap irradiated by positive ions (Chapter 7). Thus it is possible that due to accumulated positive charge, the ion etched pathway walls may become electrically positive. This causes an electrostatic attractive force to the electrically negative exogenous DNA and facilitates transfer into the cell, so that the associated DNA can be admitted. Even if local polarity does not change at the gene transferring pathways, a decrease in short-distance electrostatic repulsion between two electronegative points can also be beneficial to the entrance of new genes due to decrease in receptor negativity. On the other hand, ion etching takes place in vacuum, and water in the mature embryos must be exhausted first. When dry embryos are put into a DNA-containing solution, the probability of exogenous DNA entering cells via newly formed channels are increased due to the increasing effect of absorption and expansion

Ion implantation induces some long-living free radicals in the biological organisms and at the same time causes damage to the chromosome in the plant cell, inducing and stimulating repair and reconnection of DNA in receptor cells. All of these factors favor inserting and integrating exogenous DNA into the host chromosomes and increase the efficiency of transfer.

The disadvantages of ion-beam-induced gene transfer are also obvious. Since ion beams are necessarily produced in vacuum, loss of water and freezing cause a decrease in survival rates of water-containing receptors in gene transfer (Table 12.2 [6]). A sterilized micro-environment in the target chamber can only partly solve this problem, but relevant technology is very high. This problem is expected to be solved when ion beam technology, particularly the technology of introducing ion beams to atmosphere, is further developed.

12.2.   FACTORS AFFECTING GENE TRANSFER

In an experiment of ion beam bombardment, mature embryos of rice 02428 and an embryo suspension cell line cultured from mature embryos were used as receptors. The DNA donor was plasmid pBI221, which carried the GUS gene encoded for β-glucoronidase introduced by the CaMV35s promoter. After ion beam treatment at a certain dose, the receptors were put into the SSC (mixture solution of NaCl and citric acid) buffer medium that contained the pBI221 plasmid. The exogenous gene either passed through micro-pores or permeated through the thinned cell wall to enter the cells and the GUS gene was expressed. Figure 12.1 [7] shows the instantaneous expression of the exogenous gene entering cells treated by 30-keV N ions at a dose of 2×10^{15} N^+/cm^2. Due to the GUS, the gene transferred cells show blue color. Note that a number of cell clusters show blue and the color distributes uniformly at the side facing the ion beam, but at the other side no blue coloration is seen. This seems to indicate ion-cell interaction. For the cell clusters, at the side irradiated by the ion beam, the exogenous gene could be easily introduced, but at the far side that was not ion-beam-etched, few exogenous genes could enter and be expressed.

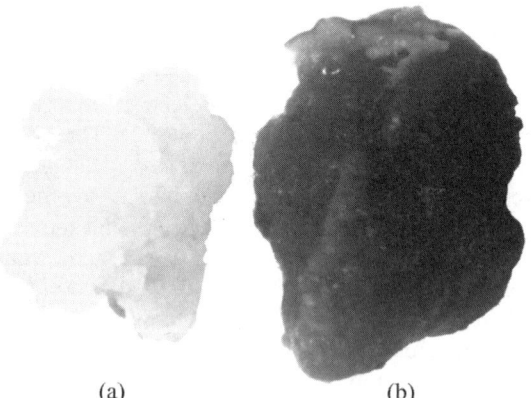

(a)   (b)

Figure 12.1.   Instantaneous expression of the GUS gene entering rice embryo cells treated by 30-keV N ions at a dose of 2×10^{15} N^+/cm^2. (a) Control, and (b) due to GUS, the gene transferred cells show blue (here darker) color. (See also color plate).

Table 12.2. Survival rates as a function of N-ion dose for rice suspension cells measured in author's laboratory.

Dose (N$^+$/cm^2, 30 keV)	Cell survival rate (%)
0	> 90
1×10^{15}	70–75
2×10^{15}	50–55
3×10^{15}	25–30
4×10^{15}	10–15
5×10^{15}	<5

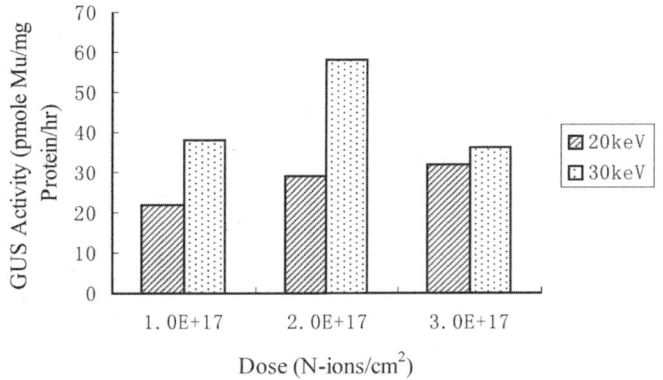

Figure 12.2. Effect of N-ion energy at various ion doses on instantaneous expression of GUS gene.

12.2.1. Ion Energy Effect on Expression of GUS Gene

Treating the receptors with the same implantation dose but different energies has an obvious influence on the expression of the GUS gene transferred into mature rice embryos (Figure 12.2 [7]). Generally speaking, the activity of the GUS gene in receptor cells treated by 30-keV N ions is greater than for treatment at 20 keV. This is more evident at low dose — at high dose the difference between the two cases is not very noticeable. Thus, we may infer that the depth of the etching process in interactions between low energy ions and biological organisms is more dependent on the sputtering rather than the ion range. This is shown by the fact that the difference between the ranges of 20-keV and 30-keV heavy ions in water is far smaller than the cell dimension. The size of ion-beam-created microholes is also related to the degree of etching, depending on the number of ions and the ion mass number. Naturally, different energies have different etching rates. From Figure 12.2 it can be seen that for mature rice embryos, 30-keV N ions have a higher etching rate than 20 keV N ions. Thus, in practical applications, with mature embryo receptors, ion energy should be 30 keV rather than 20 keV.

Table 12.3. Effect of ion etching dose on GUS gene activity expression.

Receptor	Ion dose ($\times 10^{15}$ N$^+$/cm^2)	GUS gene activity (pmole MU /mg· protein/hr)	GUS-gene-activity ratio of ion-treated/control
Cultured rice suspension cells	0 (control)	7.8	
	1.5	36.9	4.7
	2.0	54.8	7.1
	2.5	32.9	4.2
	3.0	27.8	3.5
Mature rice embryos	0 (control)	19.0	
	10	34.9	1.8
	15	38.5	2.0
	20	59.2	3.1
	25	35.7	1.8
	30	32.0	1.7

12.2.2. Ion Dose Effect on Expression of GUS Gene

Ion beam treatment at constant energy (30 keV, N ions) and with variation of ion dose shows a variation of GUS gene activity for exogenous gene transfer in rice cells and expression in mature embryos (Table 12.3 [7]). For suspension cells, the GUS gene expression activity increases with increasing dose. The GUS gene activity is a maximum of 54.8 pmole MU/mg· protein/hr at a dose of 2×10^{15} N$^+$/cm^2, 7.1 times greater than the control. As the dose is further increased, the GUS gene activity weakens. There is a similar trend for mature embryo receptors. Since there is a layer of seed skin protecting the mature embryo, higher energies are needed to peel away this layer. From Figure 12.3 it is seen that for mature embryos, the GUS gene activity is maximum at a dose of 2×10^{16} N$^+$/cm^2, 10 times greater than the optimal dose for suspension cells. The suspension cells and mature embryos have similar GUS gene activity, but the relative intensity (treated/control) of the GUS gene activity is lower for mature embryos than for suspension cells. This may be related to the structure of mature embryos and the state of the mature embryos being measured.

12.2.3. Effect of Transfer Media on Expression of GUS Gene

The transfer medium is an important factor affecting exogenous gene entrance into receptor plant cells and expression. Buffer media commonly used nowadays include SSC, T_1E_{10}, or Tris-EDTA (mixture solution of trismethylaminomethane and ethylene-diamine-tetra-acetic acid), and Kren'sF (one kind of media for exogenous DNA transfer), etc. The experimental results [7] shown in Figure 12.3 indicate that under ion beam induction, SSC buffer medium is obviously better than T_1E_{10} and Kren'sF. Kren'sF contains Ca^{2+}. Ca^{2+} has good effect in increasing stability of the cell membrane and improving membrane permeability. Kren'sF must be added for exogenous gene transfer in receptors of protoplast using the PEG (polyethylene glycol) method or the electroporation method. However, in ion-beam induction, buffers containing Ca^{2+} are not necessary and may result in adverse effects. Now let us consider the case of SSC buffer to which DMSO (dimethyl sulfoxide) is added. DMSO is a substance which increases cell permeability. In experiments on ultrasonic transformation of leaves, addition of DMSO effected the promotion of exogenous DNA absorption. SSC_1 is an SSC medium to which 1% DMSO is added, while SSC_5 contains 5% DMSO. Both do not increase the GUS gene activity in ion-beam-induced rice suspension cells, and even make the GUS gene activity weaker than using SSC buffer alone. Both of these experiments indicate that ion beam etching is able to change the cell permeability to ease the exogenous DNA entering the cell without need of assistance from Ca^{2+} or DMSO.

12.3. PROCEDURES FOR ION-BEAM-INDUCED GENE TRANSFER

Ion beam etching results in formation of micro-holes or passages in the cell surface layer, providing exogenous gene with transfer channels, while charge exchange and vacuum drying also increase the probability for the exogenous DNA to enter the cells. For the same receptor materials, the effect of gene transfer is related to the ion beam energy and dose, but mainly depends on the dose. For plant cells, the dose is lower, whereas for mature embryos, the dose should be one order of magnitude higher. This section considers as an example an experiment in which mature rice embryos were used as receptors and ion-beam-induced gene transfer was used to modify plants, describing thoroughly the operational procedure of this gene transfer method.

The receptor material was *Japanica* rice R8018. The donor was DNA plasmid pBI222, which carried the CaMV35s-promoter-introduced hph gene (encoded hygromycin phosphate transferase), amplified in host bacteria *E. coli* D5a. The introducing medium was SSC solution (15 mmol/L of NaCl, and 1.5 mmol/L of sodium citrate; the pH of the resulting solution is 7.0). The culture medium for inducing callus formation was MS (Murashige and Skoog basal salt mixture) as the basic component, with 2 mg/L of 2.4D, 0.5 mg/L of BA (benzyladenine), and 500 mg/L of rice casein at pH 5.8 added. The subculture medium was MS as the main component along with 1 mg/L of 2.4D at pH 5.8. The differentiation culture medium was N6 (one kind of plant culture medium) with 2 mg/L of NAA (Naphthalene acitic acid) and 2 mg/L of KT (Kinetin, one kind of cytokinins) at pH 5.8 added.

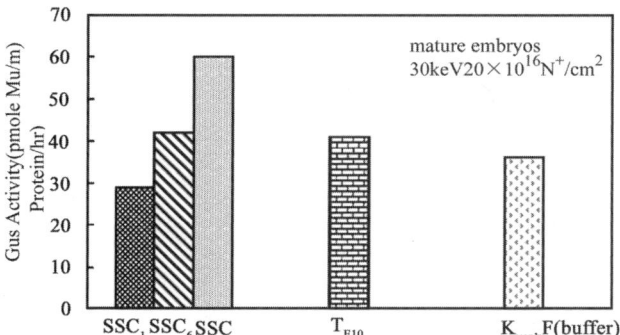

Figure 12.3. Transferred GUS gene activities expressed in mature rice embryos for different buffer media. Ion beam conditions: nitrogen ions, energy at 30 keV, and dose to 2×10^{17} ions/cm².

12.3.1. Ion Beam Bombardment

Mature seeds of rice R8018, with shells removed, were sterilized first with 75%-alcohol for 2~3 min and then 10%-antiformine for 25 min. The seeds were rinsed with sterile water three times and dried with sterile gauze. The sterilized seeds were cut to separate the embryos. A sterile operation stage and surgical knife were used. Embryos were placed uniformly on a plate with the embryos facing upward. The plate was put into the microenvironment chamber that was sterilized beforehand with alcohol. The seed embryos were aligned to the ion beam direction. The chamber was pumped for 30–60 seconds and then the gate-valve opened to allow a 30-keV N-ion beam to bombard the seeds to a total dose of 2×10^{16} N^+/cm^2. After ion bombardment the samples were taken out, immediately soaked in SSC guiding medium that contained DNA (50 μg/mL), and incubated for an hour. The seed embryos were then taken out, rinsed with SSC three times, dried with sterile filter paper, and inoculated into MS callus-induction culture medium (this treatment was marked as T). Two controls were simultaneously set. Control 1 consisted of seed embryos that were put into the sterile microenvironment chamber but not ion beam bombarded, and then soaked in the DNA-containing SSC medium in the same way (marked as CK1). Control 2 was seed embryos that were ion-bombarded to the same dose but incubated in the medium without DNA (marked as CK2).

12.3.2. Screening for Resistant Callus and Plant Regeneration

The treated samples and two controls were incubated in the MS callus-induction culture medium for 2 days, and then transplanted to the same culture medium but now supplemented with hygromycin (40 mg/L) for culturing. Initially, the embryo base parts turned black. But after 5–8 days, about 70% of the embryo shielding pieces started to form calluses that could be seen with the naked eye. The color and luster were gray and dark. Growth was extremely slow. After 20 days the calluses were transplanted to the MS subculture medium that contained hygromycin (50 mg/L). Subculture was done once every two weeks. Now the controls stopped growing, gradually became brown and all died in 30–40 days. Only those treated by ion beam produced bright and fresh callus with light yellow color around the nearly brown tissues. The resistant callus could stably regenerate on hygromycin-containing (50 mg/L) subculture medium, and the regeneration rate was not different from that of normal rice tissue culture. Through subculture screening for 1–2 months, three stable-resistant cell lines were obtained and cultured separately. Finally 32 green seedlings were obtained, and 25 of them were transplanted into pots and grew.

12.3.3. Molecular Detection

Normal leaves from the regenerated plants were cut to get 1-g pieces, extracted and DNA-purified for PCR (polymerase chain reaction) analysis. Two primer sequences were GTTCCAACCACGTCTTCAAAGC and ATTTACCCGCAGGACATATCC. The former was a segment of the complemented oligonucleotide of the CaMV35s promoter site, and the latter was a segment of the complemented oligonucleotide of the hph-gene coding site. The total length of the amplified segment was 350 bp. The PCR was carried out in a 50-μl reaction volume (10-mmol/L, TRIS-HCl, pH8.4, 50-mmol/L KCl,

1.5-mmol/L MgCl$_2$, 200-mmol/L dATP, dTTP, dCTP and dGTP). Tag DNA polymerase in 0.25 U and two synthetic primers in 100 ng each were added. Plant DNA (500 ng) was used as the template. The denaturation temperature was 94°C (1 min), the annealing temperature was 55°C (1 min), and the elongation temperature was 72°C (1.5 min). The number of amplification cycles was 40. After the reaction was completed, 20 µl of the amplified product were taken for gel electrophoresis. From PCR detection of the plant material, a unique amplified band (350 bp) was observed as expected, indicating that the 35s-promoter-introduced hph gene indeed existed in the examined cells. However, in the control sample R8018 which was directly germinated, none of the expected amplified products were detected (Figure 12.4a [8])

The gene-transferred rice DNA in 10 µg was taken for digestion with the XhoI restriction enzyme and then separated by 0.8%-gel electrophoresis. The DNA was transferred to a nylon membrane according to the Southern method. The nick transfer method was used to label the radiation isotopes for hph-coded DNA (the donor plasmid was digested with BamHI to obtain a 1.1 kb segment). Isotopes were used as the probe to hybridize the transferred gene-group DNA as well as for autoradiography analysis. The results obtained were positive (Figure 12.4b [8]). In the digestion of the gene-group DNA of the transferred plant, the single XhoI restriction enzyme (this enzyme did not have an enzyme restriction point on the pBI222 plasmid) was used. The molecular mass of the gene-transferred plant sample was obviously greater than the hybrid band (about 10 kb) of the donor plasmid. This provided direct evidence that the transferred objective gene had already been integrated into the rice gene group.

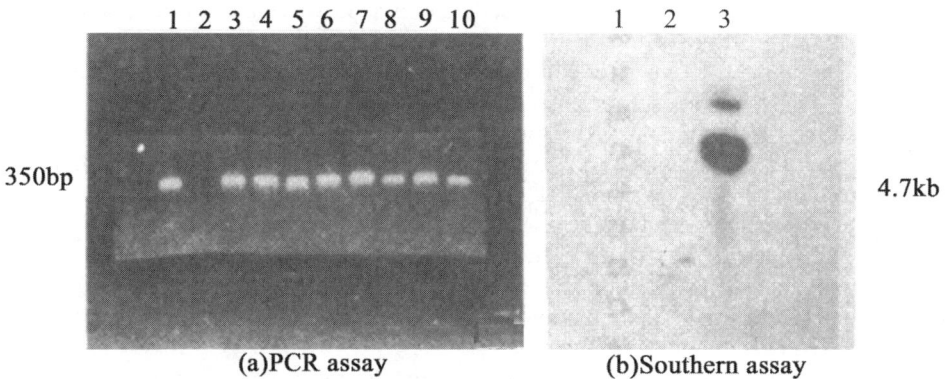

(a)PCR assay　　　　(b)Southern assay

Figure 12.4. Molecular assay of the gene-transferred rice DNA. (a) 1: positive control (pBI222 plasma), 2: negative control (untransgenic rice genome DNA) 3 ~10. The amplified bands of different transferred plants genome DNA are as templates. (b) 1: negative control (untransferred rice genome DNA digested with XhoI), 2: hybridization brand of transferred plant genome DNA segment digested with XhoI (> 5kb), 3: positive control (pBI222 plasma).

12.3.4. Genetic Analysis

Twenty five gene-transferred plants and controls (seedlings generated from the seeds) were planted in the field. The present generation (R_0) grew well, without observable differences from the controls in agronomic characters such as tasselling period, plant height, maturation period, and thousand-grain weight. Eventually all of the gene-transferred plants bore fruit via normal autogamy. The seeds were harvested and threshed, and subsequently seeds from eight randomly chosen gene-transferred plants were cultured for germination on hygromycin-containing (50 mg/L) MSO culture medium. The controls were completely suppressed on the culture medium and quickly turned brown and apparently died. However gene-transferred plants that had derived hygromycin resistance could inherit characters and pass them on to the next generation through the seeds (Table 12.4 [8]).

12.3.5. Characteristics of Ion-Beam-Induced Gene Transfer

From the process of ion-beam-induced gene transfer in rice, it can be seen that this gene transfer method has distinct characteristics. First, there is no risk of endangering the cell's life processes. Table 12.5 [8] shows the callus production rate of ion-etched gene-transferred seed embryos. Compared with the control groups CK1 and CK2, there is no obvious difference under the culturing condition of 40 mg/L for the first 20 days, but after 40 days differences occur. This indicates that the ion-etched mature embryos were not seriously injured and thus callus production rates were not affected. It could be seen from the regenerated plants that no difference in agronomic characters of the present generation was observed. This shows that ion beam etching of the cell surface formed gene transferring channels which did not endanger essential living activities and provided exogenous genes with good pathways and chances. Second, in the resistance screening, from the beginning (2 days after ion-beam-induced gene transfer), high-concentration hygromycin could be used for screening (40–50 mg/L). Under this strict screening condition, almost only mutants could grow. This not only reduced the work load for next generation culturing, but also in the early period of the experiment concerns could be concentrated on those materials that were, though few, highly probably successes, thereby saving time and energy. It is worth mentioning that this strict hygromycin-screening had no noticeable negative effects on the later-period green-seedling split and plant regeneration. Furthermore, using the seed embryos as explants facilitated large scale operation and simplified selection of materials as well, and was not limited by seasonal conditions.

It is seen from Table 12.4 that the absolute transfer rate is 3%. This is not very high. However it should be noted that formation of ion-beam-etching induced gene transfer channels depends not only on the ion species, dose and energy, but also on the ion beam angle of incidence. Due to various shapes of the seed embryos, it is difficult to guarantee that the same locations on every embryo receive the same ion beam irradiation at the same angle during ion bombardment. This problem, together with other internal and external factors that influence the transfer rate increase, should be resolved sooner or later.

Table 12.4. Hm^R (hygromycin resistance) characteristics transmission of filial generations of gene-transferred plants. Hm^S: non-Hm^R characteristics.

Seed sample serial No.	Number of examined seeds	Number of the seedlings expressing Hm^R	Number of the seedlings expressing Hm^S
CK (control)	64	0	64
1-1	31	27	4
1-3	40	34	6
1-6	43	33	10
2-1	46	31	15
2-5	45	33	12
2-6	52	41	11
2-8	42	29	13
3-6	66	50	16

Table 12.5. Callus production rate of ion-beam-induced gene transferred samples.

Treatment group	Number of treated embryos	Callus production rate (40mg/L hygromycin, 20 days)	Resistant callus (pieces) (50mg/L hygromycin, 40 days)	Regenerated plants
T(treated)	100	70.4	3	32
CK(control)1	100	71.4	0	0
CK(control)2	100	72.7	0	0

12.4. ION-BEAM-INDUCED GENE CLUSTER TRANSFER

In the genetic transformation process, normally the gene is separated and extracted (or artificially synthesized). After being cut outside the cell and combined with a vector (such as a plasmid), the plasmid is transferred by various ways into a receptor cell and integrated into the gene group of heterologous cells hopefully to be expressed. Therefore, genetic transformation can be considered as hybridization at the molecular level. Due to restrictions from the vector system, the contained meaningful genetic sequence is not very long. This sort of small gene segment is too small for the plant cell gene group. Even though the gene can be expressed in regenerated plants, it may still be lost in the next generation. Hence, exploration of transferring long segments of DNA molecules is one of the directions of gene transfer technology development.

Transferring naked vital DNA macromolecules is in the same category as transferring gene vectors. Barriers for macromolecules to enter the cell must be overcome. The essential character of ion-beam-induced exogenous gene transfer is ion etching of the cell wall to create micro-holes and charge exchanges to change the local electric characters of the gene transferring pathways of the receptor, so that negative vectors can be actively admitted by electrostatic extraction forces. Both factors, beneficial to transfer, are dependent on the ion etching dose. We know from various experimental results that for 30-keV N ion beam bombardment of cultured rice cells the appropriate dose is $(1.5–2.6) \times 10^{15}$ ions/cm^2, while for mature embryos the optimal dose is $(1.5–2.0) \times 10^{16}$ ions/cm^2. If the dose is too low, the receptor system has poor permeability, which is not beneficial for the exogenous gene to enter the cell. If the dose is too high, although the permeability is increased, the receptor is damaged and the survival rate is decreased and the transfer efficiency is also affected.

In exploration of ion-beam-induced gene-cluster transfer, the systems used are very important. The receptors should be able to accept highly effectively exogenous gene clusters and stably propagate genes to the next generations. The donors should have excellent characters consistent with the breeding goal. In one experiment as described below, naked rice seeds were the receptor while the donor was purple-featured *Maize* DNA. For simplicity, the method of direct *in vivo* transfer without *in vitro* culturing was used. Changes in phenotypes in next generations were identified to determine whether the transfer succeeded.

12.4.1. Extraction of *Maize* DNA

Seeds of *Maize* species with purple color (the marker character) were put in a germination box to promote sprouting. A week later, 100 g of young leaves were taken to freeze in liquid nitrogen, put into a pounding machine, added with 50-ml NaCl-EDTA (ethylene diamine tetra acetic acid)-SDS (sodium dodecyl sulphate) solution, mashed into homogeneous paste, and pressed and filtered with a multilayered gauze. The filtered debris, with quartz sand added, was ground in a mortar. A double-volume chloroform–isoamyl-alcohol water solution was then added and vibrated for 5 min. The solution was extracted and centrifuged at 1000 rpm at 4°C for 20 min. Supernatant was taken and chloroform–isoamyl-alcohol added in the same volume to centrifuge again. The supernatant was added to double-volume cold 95% alcohol and some fiber-like DNA could be sampled using a piece of glass cane. The primary DNA should be purified by removing RNA and protein so that fairly pure extracted DNA-liquid could be derived for experiment.

12.4.2. DNA Transfer

Plump-eared seeds of *Indica*-213 rice were selected and placed on a sample holder plate with the embryo of each seed facing the incident beam direction. The embryos of rice seeds were firstly bombarded with 35-keV Ar ions at a dose of 1.0×10^{16} ions/cm^2. They were then removed from the target chamber, some of the bombarded embryos were smeared with *Maize* DNA using a brush pen, and then bombarded again. There were two purposes in doing this. First, the total *Maize* DNA could be cut by the Ar ion beam into smaller segments; and second, these smaller segments could be pushed by the energetic ions into previously formed micro-holes. For the secondary ion bombardment, the Ar ion beam dose was 5×10^{15} ions/cm^2 and so the total dose was 1.5×10^{16} ions/cm^2. The twice-ion-bombarded seeds were smeared with DNA again and then were soaked in *Maize*-DNA SSC solution with a concentration of 30 μg/ml for DNA transfer. Those seeds that were not smeared with DNA, together with the controls, were put into a non-DNA SSC solution. The three groups of samples were cultured at a slightly elevated temperature for 1 hour. After residual liquid dried, the samples were placed in a

28°C-constant-temperature box for germination and then planted. Each generation was in continuous self-cross using sheath bags, and changes in characters were observed.

12.4.3. Variation Amplitude

The Ar-ion bombarded and implanted seeds, whether they were smeared with the *Maize* DNA or not, suffered damage to various degrees, which were mainly evidenced by a decrease in the seedling height, root length, seedling survival rate and bearing rate compared with the control. The bearing rate of plants from the ion bombarded and *Maize*-DNA-added seeds was the lowest, only 17.7% (Table 12.6). The M_2 generation was planted in sapling rows. Each treatment resulted in 200 rows planted with 25 plants per row. The plants treated with the ion-etching-induced *Maize*-DNA transfer showed definite segregation phenomena. There were numerous variation types and the main mutation characters included growth period, plant height, color and luster, aerial root, leaf tongue, leaf ear, spike shape, and plant form, etc. The total mutation rate was up to 23.6%. The first generation bearing rate was low or even zero. The group to which DNA was not added but ion-bombarded had a total mutation rate of 2.8% and the mutations were mainly evidenced by plant height, growth period and chlorophyll concentration. Most of the mutants of this group became stable in the M_3 generation and a few were also stabilized in the M_4 generation. The group that was treated by adding DNA still had "mad" separations at the M_3 generation. The marker character, the purple color, appeared in some of the separated next generations and was mainly distributed at leaf sheaths, leaf ears, grain husk tips, and post head, etc. The growth period, plant height and leaf shape were also characters that had greater variation. By directional selection, these character separations lasted continuously to the $M_6 - M_7$ generation. Six selected excellent mutant series still had separations till the M_8 generation.

12.4.4. Analysis

The wide mutations and mad separations produced from the ion-bombarded and *Maize*-DNA-added rice (variety 213) were not consequences of ion beam mutation. Ion implantation induced recombination of genetic substance atoms could produce variations but only with small amplitudes. Up to now, the highest mutation rate is 14.8% for ion-implanted rice of high-affinity-series 02428. In this experiment, the *Indica* 213 rice control had a mutation rate of only 2.8%. Mutation-produced aberration characters became stable at the M_3 generation and a few at the M_4 generation, and this was generally

Table 12.6. Damage effects due to different treatments.

Treatment	Seedling height (cm)	Root length (cm)	Seedling survival rate (%)	Bearing rate (%)	Mutation rate (%)
Ion etching + DNA + SSC	5.42	3.66	33.4	17.7	23.6
Ion etching + SSC	5.15	3.75	30.6	44.8	2.8
Control	6.02	5.31	76.2	73.5	0.0

observed. In the experiment, the mutations produced from the ion-implanted 213 rice seeds which were then soaked in SSC solution became stable at the M_3 generation and no longer had separations in subsequent generations. But the experimental group transferred with *Maize* gene clusters not only had wide variation amplitudes but also separations in the main mutation characters lasting up to the M_8 generation. In the genetic background of rice *Indica* 213 there was no purple-color marker character, and the purple color of the mutants transferred with the total *Maize* DNA was not produced by mutation. The air-born root was the specific root system of *Maize*, not mutated from the rice species that had no air-born roots. Therefore the phenomena of wide mutations and mad separations of the gene-transfer experiment group could come only from one possible source, that is the *Maize* DNA which had been transferred into the gene groups of the rice *Indica* 213 cells. In particular the transferred gene could absolutely not be single-character gene segments but very probably a large gene cluster.

Generally speaking, exogenous total DNA contains about 10% structural gene and 80%–90% regulator sequences and reiterated sequences. In the receptor cell, an exogenous DNA segment can be integrated onto the receptor DNA by various means such as homologous recombination, allo-insertion and genetic substance rearrangement. Different integrated locations have different effects on subsequent generations. *Maize*-DNA transfer induces wide variations, which seem to reside not only at the DNA structural gene parts but very probably dominantly at regulator genes. Except for roles in direct and specific regulations of gene expressions, this part of the DNA probably plays many roles in indirect and non-specific regulation of gene expression. This means that recombination of the reiterated sequences and local amplification or reduction result in changes in DNA structure or nuclear types. Jumps and displacements of the mobile components cause gene rearrangement, change gene combinations, switches and expression intensities, so as to lead to phenotypic variations. In the case of the *Maize* gene clusters transferred in rice, either the structural gene or the non-structural gene part can enhance gene rearrangement. Changes in the genetic substance structure and system make it impossible to complete the activities of gene rearrangement within one or two generations. Instead, the activities will continue for a long time until mutation characters are stabilized when the optimal combinations are reached. This may be the molecular basis of the extensive variations and mad separations produced by the total *Maize*-DNA transferred *Indica* 213 rice.

The above discussion uses the reduction-to-absurdity method to demonstrate qualitatively the possibility of naked *Maize* DNA-macromolecule transfer in the gene group of rice 213, but lacks evidence of specific molecular detection. Due to this lack, no proper *Maize*-DNA probe has yet been found. Molecular evidence has not yet been found. Nevertheless, the extensive variations produced from *Maize* DNA transfer in rice provide screening for useful mutants with excellent prospects. Six very good new rice variety series have been selected from these mutants. Their photosynthetic efficiencies are increased by 64% – 80% compared with *Indica* 213. They have shorter and stronger stems, resistance to toppling and drought, and high production (600 kg in a 667-m^2) in experimental fields.

Attempts at ion beam induced naked DNA macromolecule transfer for providing crop breeding with a means for creating extensive variations have been made. The high mutation rates and broad mutation spectra resulting from this method provide field screening with many opportunities. This can reduce work loads and increase breeding objectivity at the same time. By using this transfer method, the total DNAs of Hongma and wild-type Byak cotton can be transferred into Simian-2 cotton by either direct DNA transfer or first DNA transfer into pollen grains and then followed by pollen-insemination. The resulting mutation rate is 5 – 7 times greater than from direct ion-implantation-induced mutation of cotton. A non-gland-character plant series has been screened from the Byak-cotton DNA transferred cotton; and a plant series with withering-resistance and yellow-withering-resistance has been screened from the Hongma DNA transferred cotton. This induction method is equivalent to molecular hybridization, and is not restricted by plant taxonomy. It is simple in operation and does not need *in vitro* culturing. The technique is available for all to use freely.

REFERENCES

1. Fromm, M.E., Taylor, L.P., Walbot, V., Stable Transformation of Maize after Gene Transfer by Electroporation, *Nature*, **319**(1986)791-793.
2. Vasil, V., Castillo, A., Fromm, M.E., Vasil, I., Herbicide Resistant Fertile Transgenic Wheat Plants Obtained by Microprojectile Bombardment of Regenerable Embryogenic Callus, *Biotechnology*, **10**(1992)667-674.
3. Tsukakoshi, M., *et al.*, *Appl. Phys. B.*, **102**(1993)1077-1084.
4. Zhang, L.J., *et al.*, *Chinese Agricultural Science*, **24**(2)(1991)83-89.
5. Yu, Z.L., Yang, J.B., Wu, Y.J., Cheng, B.J., He, J.J. and Huo, Y.P., Transfer Gus Gene into Intact Rice Cells by Low Energy Ion Beam, *Nucl. Instr. Meth.*,, B**80**(1993)1328-1331.
6. Yang, J.B., Wu, Y.J., Wu, L.J., Wu, J.D., Yu, Z.L., Low Energy Ion-Beam-Mediated DNA Delivery into Rice Cells, *Jorunal of Anhui Agricultural University* (in Chinese with an English abstract), **21**(3)(1994)330-350.
7. Li, H., Wu, L.F., Yu, Z.L., *Acta Agricultural Nucleatae Sinica*, **15**(4)(2001)199-206.
8. Yang, J.B., Wu, L.J., Wu, J.D., ,Wu, Y.J., Yu, Z.L. and Xu, Z.H., *Chinese Science Bulletin*, **39**(16)(1994)1530-1534.
9. Hase, Y., Tanaka, A., Narumi, I., Watanabe, H., and Inoue, M., Development of Pollen-Mediated Gene Transfer Technique Using Penetration Controlled Irradiation with Ion Beams, *JAERI Rev.*, **98-016**(1998)81-83.
10. Anuntalabhochai, S., Chandej, R., Phanchaisri, B., Yu, L.D., Vilaithong, T., Brown, I.G., Ion-Beam-Induced Deoxyribose Nucleic Acid Transfer, *Appl. Phys. Lett.*, **78**(2001)2393-2395.
11. Phanchaisri, B., Yu, L.D., Anuntalabhochai, S., Chandej, R., Apavatjrut, P., Vilaithong, T., Brown, I.G., Characteristics of Heavy Ion Beam Bombarded Bacteria *E. Coli* and Induced Direct DNA Transfer, *Surf. and Coat. Technol.*, **158-159**(2002)624-629.
12. Yu, L.D., Phanchaisri, B., Apavatjrut, P., Anuntalabhochai, S., Vilaithong, T., Brown, I.G., Some Investigations of Ion Bombardment Effects on Plant Cell Wall, *Surf. and Coat. Technol.*, **158-159**(2002)146-150.
13. Yu, L.D., Vilaithong, T., Phanchaisri, B., Apavatjrut, P., Anuntalabhochai, S., Evans, P., Brown, I.G., Ion Penetration Depth in the Plant Cell Wall, *Nucl. Instr. Meth.*, B**206**(2003)586-590.

FURTHER READING

1. Vilaithong, T., Yu, L.D., Apavatjrut, P., Phanchaisri, B., Sangyuenyongpipat, S., Anuntalabhochai, S., Brown, I.G., Heavy Ion Induced DNA Transfer in Biological Cells, *Radiation Physics and Chemistry*, **71**(2004)927-935.
2. Voet, D. and Voet J.G., *Biochemistry* (John Wiley & Sons, New York, 1990) pp771-1178.

13

LOW-ENERGY IONS INDUCED SYNTHESIS OF ORGANIC MOLECULES

Water, and the other simple molecules which comprise the atmosphere, do not react with one another under ordinary conditions, but when ionized and supplied with a small amount of extra energy may react to form complex molecules. Even some reactions, such as that between carbon and water, which hardly take place in chemical engineering, can occur. So, it is expected that low-energy ions in nature have an important effect on molecule evolution. How reactions lead to changes of simple molecules to complex ones, inorganic ones to organic ones and further to amino acids and nucleic acids, is a central question in the study of biomolecular chemical genesis. This chapter will concentrate on gas discharge and ion implantation to illustrate the process. The first section introduces principles and details of some interesting experiments, as well as the results of the reactions between the ions of carbon, nitrogen, carbon dioxide, methane etc. with water. The second section discusses the synthesis of the preliminary forms of organic molecules induced by ion implantation. The third section discusses the potential effects of low-energy ions on amino acid synthesis. The last section forecasts the application of low-energy-ions to a possible scenario for the generation of life and formation of interstellar molecules.

13.1. REACTIONS BETWEEN LOW-ENERGY IONS AND WATER

Reactions between gaseous ions and water can take place through ion implantation into either ice or liquid water. The former can be realized using ion implantation apparatus, while the latter can be done by gas discharge in the cathodic potential-drop zone.

First, let us look at features of the gas discharge [1]. The distribution of the gas discharge potential can be divided into three different areas, as shown in Figure 13.1. Around the cathode is the cathodic potential-drop area, which is short ($\leq 10^{-4}$ m) and has a voltage drop of about 10 V. There are many positive space charges in this area. Around the anode is the anodic potential-drop area, which is also short and has negative space charges. Between these two electrode areas is the arc positive column, or the plasma. In this area, the electric field in the axial direction is basically uniform. So, if the sample solution is put in the cathodic area, the ions are accelerated by the cathodic potential drop and are implanted into the solution; while if the solution is in the anodic area, the electrons will be implanted into the solution.

The experimental apparatus, as shown in Figure 13.2 [2], is primarily composed of a needle-like anode A and a table cathode C. D is the discharge area, P is a high-voltage dc power supply, and R is a current-limiting resistance. The sample solution S is contained by a stainless-steel container B, which is placed on C. High-pressure gas is introduced into the discharge chamber H through the gas inlet orifice I and expelled through the gas exit orifice O. The experiment is started by turning on the power supply. The voltage is adjusted to obtain a steady discharge. If the positive and negative poles are swapped, study of the reaction between electrons and the sample solution is also possible.

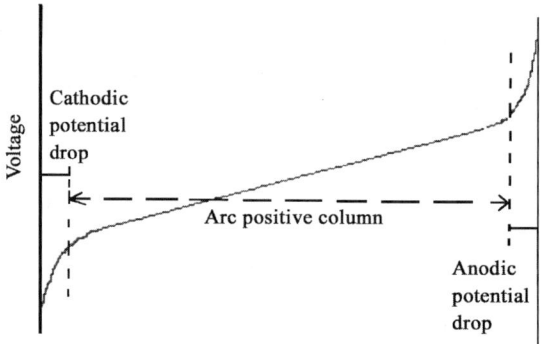

Figure 13.1. The distribution of potential of arc discharge in the axial direction.

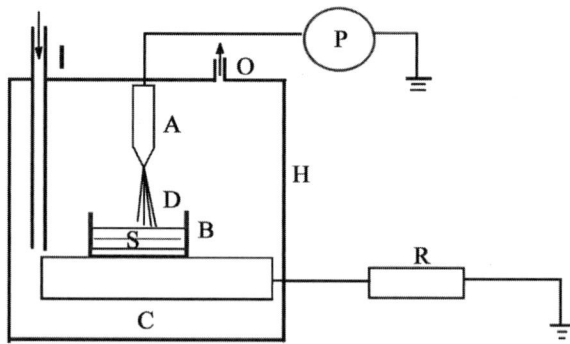

Figure 13.2. Gas discharge apparatus for studying reactions between low-energy ions and solution. P: power supply, A: anode, I: gas inlet orifice, O: gas exit orifice, D: discharge area, B: container, S: sample solution, C: cathode, H: discharge chamber, and R: current-limiting resistance.

COLOR PLATE

(a)   (b)

Instantaneous expression of the GUS gene entering rice embryo cells treated by 30-keV N ions at a dose of 2×10^{15} N$^+$/cm^2. (a) Control, and (b) due to GUS, the gene transferred cells show blue color.

13.1.1. Chemical Reactions of Nitrogen Ions with Water

Pure nitrogen gas is injected into the experiment apparatus. After one hour of discharge the Proton Nuclear Magnetic Resonance (^{1}H-NMR) spectrum of the water solution is examined (Figure 13.3). The highest peak in the spectrum is the proton peak. At 6.95 ppm with the coupling constant 52 Hz, there are triple equal-intensity peaks, which represent the coupling of proton and nitrogen expressed by the ionized form NH_3^+ of amino NH_2 [3,4]. This kind of ion can only be formed in a strongly acidic environment. The pH of the sample as measured by a pH meter shows that the longer the discharge time, the stronger the acidity of the sample. Both diphenylamine and ferrous sulfate demonstrate formation of HNO_3 [5]. The sharp single peak at 1.95 ppm is due to the chemical shifts of amidogen, which is unable to form cation, imino (-NH-) and other groups. From this experiment, it is confirmed that low-energy nitrogen ions can react with water and form amine, etc.

After nitrogen ions enter the water solution, the water molecules are excited and ionized, or bond broken directly, or converted to free radicals. Other active fragments are probably created. The formation of amine may result from nitrogen capturing hydrogen in the water, as illustrated below:

$$H_2O + N^+ \longrightarrow {}^\bullet H, \ {}^\bullet OH, H_2O^+ \dots ,$$
$$N^+ + {}^\bullet H \longrightarrow {}^\bullet NH \longrightarrow {}^\bullet NH_2 \longrightarrow NH_3.$$

13.1.2. Reaction of Methane and Carbon Dioxide with Water

The reaction between methane or carbon ions and water is a little more complicated than that between nitrogen ions and water. ^{1}H-NMR spectra of methane ion and carbon-dioxide ion implanted water are shown in Figure 13.4 and 13.5, respectively. In the former case, carbon ions that are generated by the methane gas discharge react with water and form some organic molecules. The framework of these molecules mainly consists of carbon chains, such as dimethyl ether, ethanol and some other carbonyl compounds, even some complex aromatic cycle, heterocyclic or othercyclic compound [6].

From both Figure 13.4 and 13.5 one can find many common peaks representing products of the reaction between the methane-discharged ions or carbon-dioxide-discharged ions and water. The effect of carbon-dioxide ions on water results in dimethyl ether (single peak at 3.33 ppm) and ethanol (triple peak at 1.15 ppm and quadruple peak at 3.62 ppm). The relatively higher single peak at 8.28 ppm indicates that many aromatic cycles or heterocyclic compounds are formed. The single peak at 2.06 ppm is the contribution of some kinds of methyl combined with carbonyl. In general, the compounds are nearly the same for the reaction of carbon dioxide and water and the reaction of methane and water. This indicates that the element carbon has an important effect on the chemical combination with water, while oxygen or hydrogen seems to have little effect on the production because water itself can provide these two elements to form organic compounds.

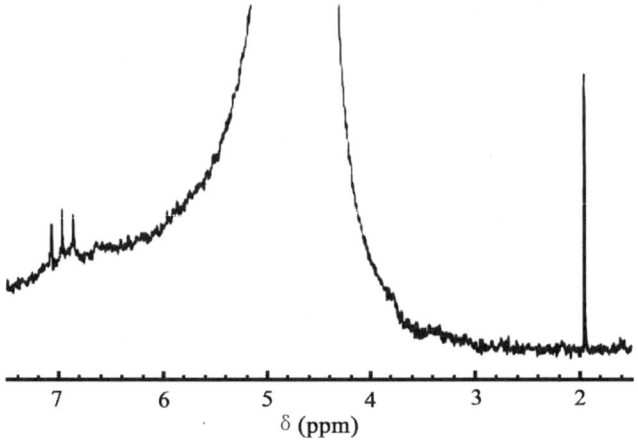

Figure 13.3. ^{1}H-NMR spectrum of the water solution after nitrogen gas discharge.

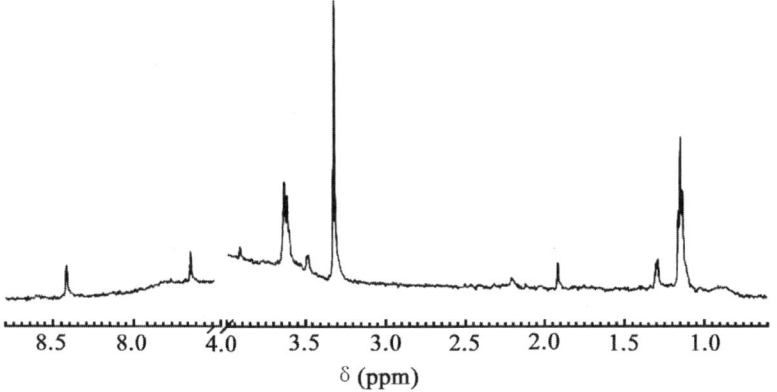

Figure 13.4.   The ^{1}H-NMR spectrum of water after methane ion implantation.

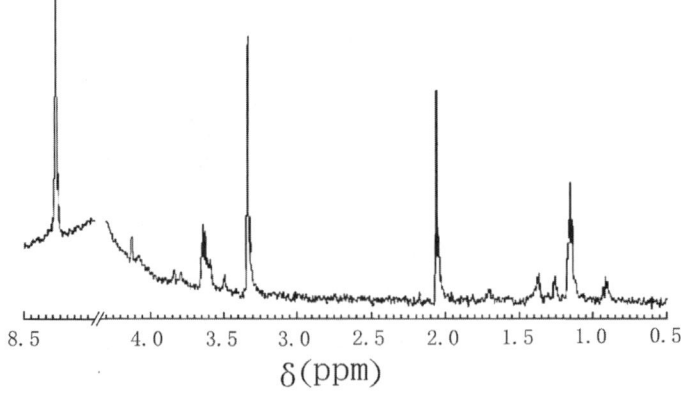

Figure 13.5. The ^{1}H-NMR spectrum of water after carbon dioxide ion implantation.

13.1.3. Gas Discharge Induced Graphite Synthesizing Organic Compounds

In the previous studies, carbon participating in the reactions is in the form of compounds [7], such as CO or CH_4, and some other gases are mixed in to aid the discharge. But very little research has been carried out as to whether and how single-element carbon reacts with other materials to form organic compounds. Carbon, as a main element in organic compounds, is a primary substance in constructing living material. So, providing answers to the question of how single-element carbon is turned into organic compounds and other C-containing compounds is a significant step in studying the bio-molecule origin and evolution processes. But, as pure carbon has very stable chemical properties and does not react with other materials under ordinary conditions, questions concerned have not been answered yet.

In order to look for answers to the questions addressed above, one experiment is performed using the discharge process in the apparatus shown in Figure 13.6. In the figure, B is the graphite anode (made by Hitachi Kasei Co., purity >99.9%), C is a silver wire used as the cathode. To reduce contaminations in the experiment, all the vitreous instruments are heated in an oven at 500°C for 4 hours prior to use.

After discharge, Gas Chromatography/Mass Spectrometry (GC-MS) and ^{1}H-NMR analyses clarify that there are carboxylic acid materials in the water solution. It is seen from the GC-MS spectrum as shown in Figure 13.7 that the products are abundant and the retention time is between 1 min to 2.7 min. Among the peaks there are three relatively noticeable peaks, which have their retention time 1.71 min, 1.86 min and 2.56 min, respectively. By searching and comparing to the standard spectra in the Wiley (The Wiley Register of Mass Spectral Data) and NIST (The NIST Mass Spectral Database from the National Institute of Standards and Technology, United States) spectrum banks, it is concluded that these three peaks are formic acid, acetic acid and propanoic acid.

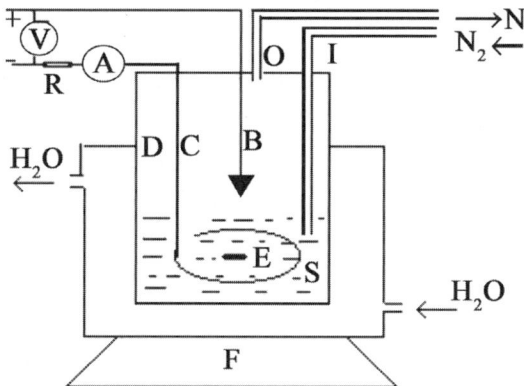

Figure 13.6.   Schematic of the apparatus setup for reaction in aqueous solution induced by low energy N^+ implantation. B: anode, C: cathode, D: glass container, and E: magnet.

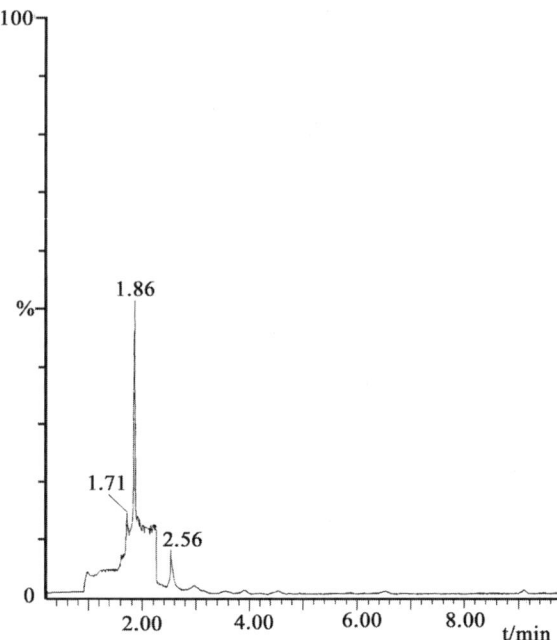

Figure 13.7. GC of the sample from graphite and water after discharge for 6 hours. Peaks at 1.71 min, 1.86 min and 2.56 min are attributed to be formic acid, acetic acid and propanoic acid, respectively.

The ^{1}H-NMR spectrum as shown in Figure 13.8 further demonstrates that some carboxylic acid materials are formed in the process of graphite discharge. The single peak at 8.15 ppm is aldehyde-hydrogen (-C (=O) H) of formic acid, and the single peak at 2.00 ppm belongs to the methyl-hydrogen of acetic acid. The coupling constants of the triple peak at 2.30 ppm and the quadruple peak at 1.00 ppm are the same and the integrated area ratio is 2:3. From the chemical shift value point of view, these two peak groups belong to methylene (-CH$_2$-) in the propanoic acid and hydrogen in the methyl. The high peak at 4.71 ppm is the absorption peak of water. Heavy water (D$_2$O) is used as the solvent and thus the exchanges between the hydrogen in carbanyl and solvent are invisible. In the control, graphite is replaced by a silver thread as the anode for discharge, and the result shows only water absorption peak but without other peaks appearing in the ^{1}H-NMR spectrum of the products. Thereby the carbon source contamination is eliminated in the experiment, namely, graphite is the only carbon source in the reaction system [8]. Graphite, as the anode, vaporizes because of bombardment from the electrons in the anodic potential-drop area, and the gaseous carbon discharges in plasma. When the carbon ions are accelerated in the cathodic potential-drop area and shot into the water solution, they react with water and organic carboxylic acid is produced. Such type of reactions may shed important light on the study of how the single-element carbon is turned into compounds to form organic molecules.

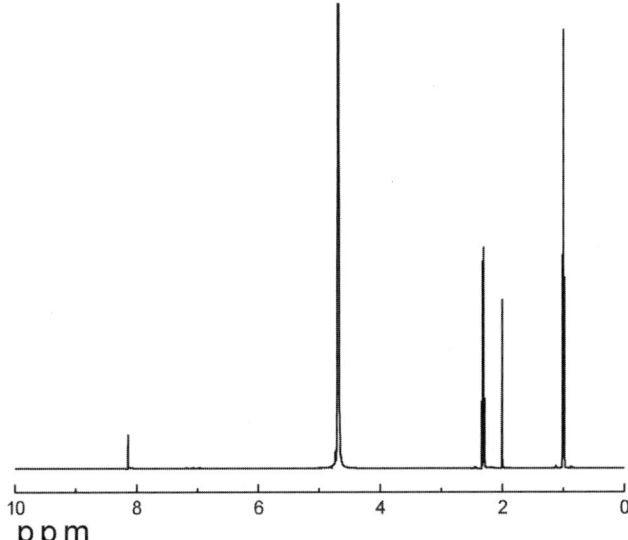

Figure 13.8. ¹H-NMR spectrum of the sample from graphite and water after discharge for 6 hours.

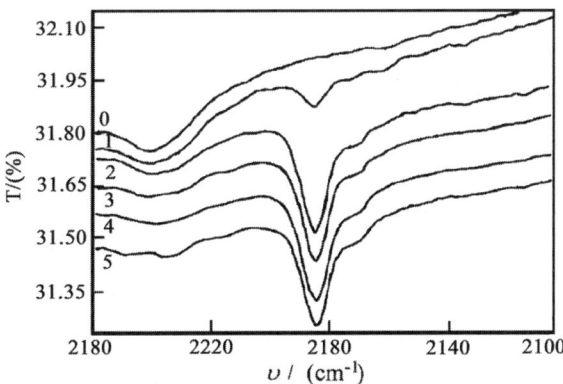

Figure 13.9. IR spectra for the PHCOONa samples with and without N^+ ion implantation. (0) without ion implantation, (1~5) $5D_0$, $15D_0$, $25D_0$, $35D_0$, $45D_0$, with $D_0=10^{15}$ ions/cm².

13. 2. SYNTHESIS OF ORGANIC PRECURSOR MOLECULES

Complex molecules such as ammonia or organic acids can be formed through ion implantation into water, generated by gas discharge at the cathodic potential-drop area. This opens up evolutionary processes in which matter develops from simple single elements to complex molecules and from inorganic to organic molecules. This section discusses solid substances implanted with low-energy ions and analyzes the role of low-energy ions played in the synthesis of bio-organic molecules.

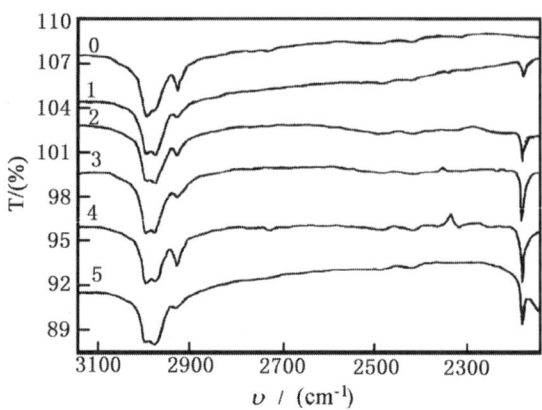

T/(%)

υ / (cm⁻¹)

Figure 13.10. IR spectra for the CH₃COONa samples with and without N⁺ ion implantation. (0) without ion implantation, (1~5) 5 D₀, 15 D₀, 25 D₀, 35 D₀, 45 D₀,   with D₀ = 10¹⁵ ions/cm².

13.2.1. Production of Cyanide

Solid sodium acetate and sodium benzoate samples are irradiated by nitrogen ions generated from gas discharge with different doses. The Fourier Transform Infrared (FT-IR) spectra of the samples, compared with the control, exhibit new peaks at 2190 cm⁻¹ and 2185 cm⁻¹, respectively, as shown in Figure 13.9 and 13.10. The peaks in this range belong to the absorption of the cyano group, i.e. C≡N group. However, this group does not exist in the reactants, so it must be formed by the mass deposition effect of N-ion irradiation [9].

13.2.2. Production of Methylic (–CH₃), Methylene (–CH₂–) and [COO]⁻ Free Radical

Sodium formate is implanted with 20-keV N⁺, H⁺, or Ar⁺ ions to a dose of 15D₀ (D₀ = 10¹⁵ ions/cm²). The IR spectra of the samples before and after implantation are shown in Figure 13.11. A decrease in the intensity of all bands is observed for the implanted samples, indicating that the target material is damaged and spilt. Compared with the control, the FT-IR spectra of the implanted sodium formate samples contain new peaks at 880 cm⁻¹ and 1450 cm⁻¹. The former peak belongs to the C-H out-of-plane vibration of –CH₂- [10], while the latter belongs to the C-H scissoring vibration of –CH₂- or the C-H asymmetrical vibration of –CH₃ [11,12]. These results show that no matter what kinds of ions are used, products containing –CH₂- and –CH₃ groups can be obtained in the implanted sodium formate samples. They may be formed from the rearrangement reactions of the target atoms.

The shapes of the Electron Paramagnetic Resonance (EPR) spectra of the sodium formate implanted by N⁺ ions with different doses have no noticeable difference, as shown in Figure 13.12. The g values of the above spectra are all 1.9997±0.000, indicating that free radicals produced by implantation are the same. The radical concentration increases with the dose when ion energy is the same [13,14]。 The unimplanted sodium formate sample does not contain free radicals, and hence no EPR signal is detected. Previously, Ovenall also measured and obtained similar stable EPR spectra which were formed by the COO⁻ radical ion for γ-ray irradiated sodium formate samples, and found that this radical was very stable [15]. Four lines of the spectra in Figure 13.12 can be attributed to the free electron of the COO⁻ radical ion coupled with a ^{23}Na nucleus with a spin of 3/2, and the value of the ^{23}Na coupling is 8.2258 Gs. The formation process of COO⁻ can be described as following:

$$HCOO^- \xrightarrow{\text{irradiation}} H^+ + COO^-$$

$$H^+ + HCOO^- \longrightarrow H_2 + COO^-.$$

In the above process, H⁺ and [COO]⁻ radicals are produced. As the H⁺ radical is very active and short-life, no EPR signal is detected.

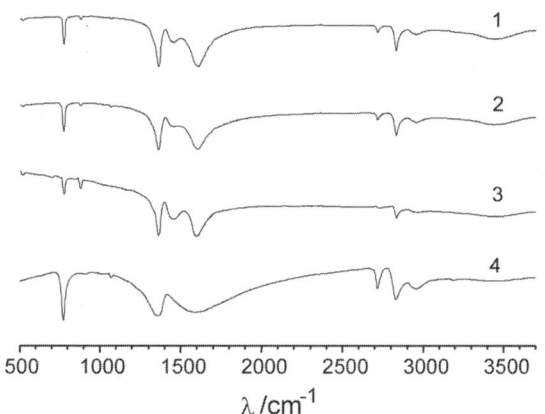

Figure 13.11. IR spectra for the sodium formate samples with and without ion implantation. 1) Implanted by H⁺; 2) implanted by Ar⁺; 3) implanted by N⁺; and 4) without ion implantation.

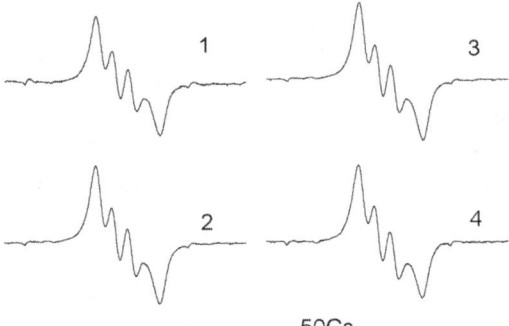

Figure 13.12. EPR spectra for the sodium formate samples implanted by 20-keV N⁺ ion with different doses. 1) $5D_0$; 2) $15D_0$; 3) $25D_0$; and 4) $35D_0$, with $D_0 = 10^{15}$ ions/cm^2.

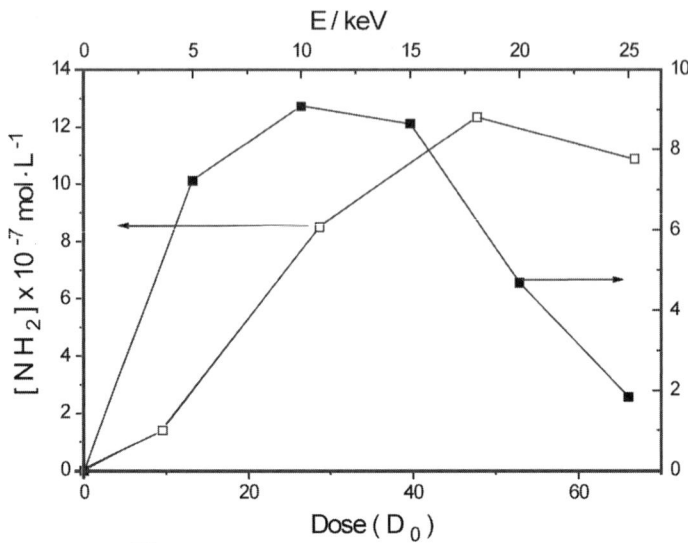

Figure 13.13. Dose (■) and energy (□) dependences of the concentration of NH_2 produced in the N^+-ion-implanted sodium formate samples. The samples were implanted by N^+ ions with different doses at 20-keV energy and with different energy at the dose of 15 D_0 ($D_0 = 10^{15}$ ions/cm^2).

13.2.3. Dose Effect on the Yield of -NH₂ Group

The N^+-ion-implanted sodium formate samples show positive in the ninhydrin test, indicating a new -NH₂ group produced. Figure13.13 demonstrates dependences of the -NH₂ concentration on dose and energy of the N^+ ions. At a N^+-ion energy of 20 keV, the yield of the produced -NH₂ group in the implanted sample increases when the dose is increased from 0 to 50 D_0 ($D_0 = 10^{15}$ ions/cm^2) and then decreased as the dose is higher than 50 D_0. On the other hand, with a N^+-ion dose of 15 D_0, the yield of the produced -NH₂ group decreases when the ion energy is increased from 15 to 25 keV.

As described before, when low energy ions interact with target molecules, some molecular fragments and free radicals are formed through energy transfer. These particles can be recombined to form new complex molecules, such as recombination of $-CH_2-$, $-CN$ and $-NH_2$ groups. It is well known that cyanide and amino acids are important precursors for formation of bioorganic compounds [16]. So, low-energy N^+ ion implantation can not only be a new route for formation of biological molecules, but also play an important role in the production of amino acids for the chemical origin of life.

13.3. SYNTHESIS OF AMINO ACID AND NUCLEOTIDE MOLECULES

As mentioned above, the effect of low energy ions on synthesizing biological molecules is shown in solid organic molecules without nitrogen. However, experimental data have not been enough to demonstrate the synthesis of biological molecules in laboratories. This may be due to the fact that the depth of implanted ions in solid material is limited and the processes of synthesis and decomposition occur at the same time. There are also some new biological molecules which are impossible to detect. In this section, we discuss synthesis of amino acid and nucleotide by using glow discharge.

13.3.1. Synthesis of Amino Acids

Glow-discharge produced N-ion implantation into sodium acetate, sodium malonate and sodium succinate solution is investigated for abiotic syntheses of amino acids. High Performance Liquid Chromatography (HPLC) analyses show that glycine is produced in these processes respectively [17] and aspartic acid is produced in the sodium succinate solution (Figure 13.14).

The experimental result shows that amino acids are formed by low energy N-ion implantation into sodium carboxylic solution. In this experiment, ammonia and amino acids are generated in a nitrogen-free solution. Thus, mass deposition effect plays a clear role. Moreover, carboxylic salt is decomposed as a result of oxidation of the system and the oxidation is enhanced due to nitrogen ion implantation. This oxidation condition is different from the deoxidize condition of prebiotic syntheses of amino acid but more close to the condition of the primitive ocean.

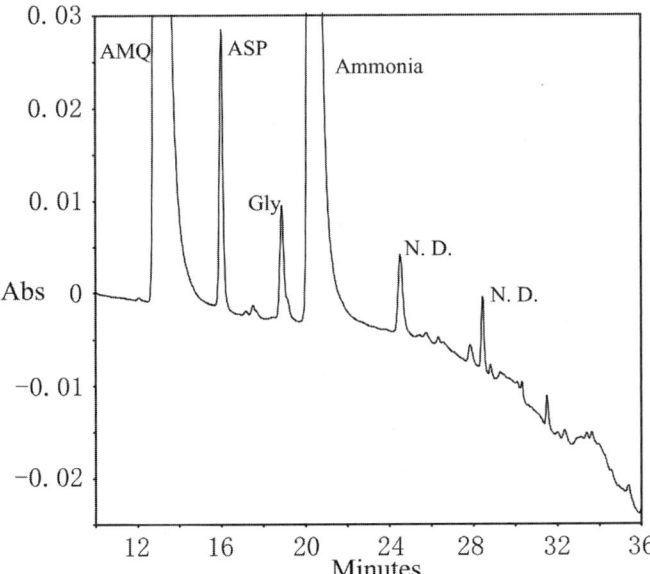

Figure 13.14.   HPLC chromatogram of sodium succinate solution implanted by N ions in glow discharge for 10 hours.

13.3.2. Synthesis of Nucleotide

Abiotic syntheses of nucleotides from their constituents (including bases, ribose or deoxyribose and phosphates) can be studied by low-energy ion implantation into aqueous solution. N-ion implantation into solution containing components of adenyl acid and deoxyadenylic acid can produce adenyl acid and deoxyadenylic acid, respectively, demonstrated by HPLC and ^1H-NMR spectrum, as shown in Figures 13.15 and 13.16.

The synthesis of nucleotide induced by low-energy ion implantation in aqueous solution is complicated. It includes condensation polymer reaction of bases, ribose or deoxyribose and phosphates, and damage of bases, ribose or deoxyribose and phosphates due to decomposition of water by ion implantation. After water is decomposed, various kinds of active groups (free radical, positive or negative ion) can be produced. These active groups can impact the solute and then form new active groups. In addition, ion implantation can induce decomposition of the solute molecules to form free radical and positive or negative ions. These active groups are combined to form new organic molecules. These facts show that low-energy ion implantation can induce synthesis of nucleotide in aqueous solution. This synthesis is not achieved by using conventional chemical methods. Scientists have thought oceans to be the life source, because water existence is critically important for origin and evolution of life. The experimental apparatus used in this study could simulate original conditions of lightening on the ancient ocean. The results showed that the experiment could synthesize nucleotide without any activator and thus provide a new way for the prebiotic synthesis of nucleotide [10,11].

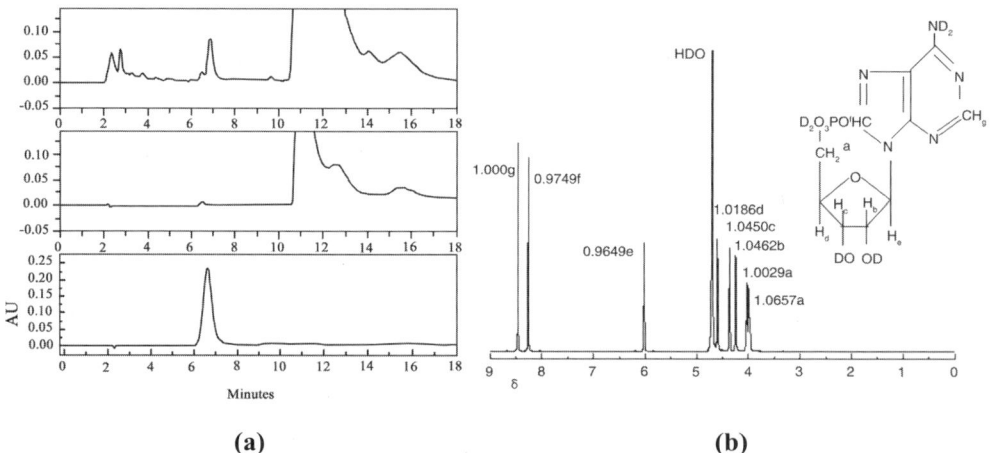

(a)   (b)

Figure 13.15. HPLC (**a**, from up to bottom: the implanted products; pre-implantation sample; 5′-AMP control) and ^1H-NMR spectrum (**b**) of the content detached from N-ion implanted solution containing components of 5′-AMP.

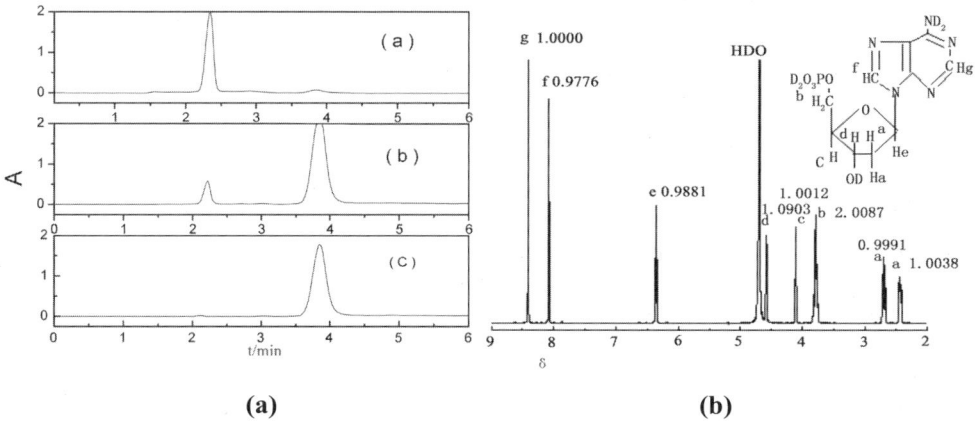

(a)   **(b)**

Figure 13.16. HPLC (**a**, from up: 5′-dAMP control; the implanted products; pre-implantation sample) and ^1H-NMR spectrum (**b**) of the content detached from N^+ implanted solution containing components of 5′-dAMP (D_2O as solvent).

13.4. EFFECT OF LOW-ENERGY ION IMPLANTATION ON CHEMICAL ORIGIN OF BIO-MOLECULES

In 1953, S.L. Miller's experiment in the laboratory of University of Chicago [18] was the first laboratory demonstration of chemistry origins of life and received extensive concerns. Following this, many scholars have imitated the primitive earth environment and synthesized various amino acids. The majority of these studies were generally carried out using highly reduced surroundings, such as the one in Miller's experiment constituted of H_2, CH_4, NH_3, H_2O etc. In this reduced air, with electrical sparks, amino acids as well as other organic substances were generated. However, if reactions were not in reduced surroundings, it was very difficult to get amino acids. What the true components and characteristics of the primitive atmosphere, such as the reductibility and its maintaining time duration, should be is not yet clear. Hence, under the condition of non-reduced surroundings, simulation of the synthesis reaction of bio-molecules in the original earth condition is more significant.

It is commonly believed that life was born in ocean. It is very important for a large quantity of water for vital origins to evolve. In the primitive atmosphere, because of "cluster shoots" of cosmic rays, disintegration of the earth-crust radio chemical elements, volcano sprays, thunder and lightning, air friction with the surface of land or water, ultraviolet rays and so on might make components of atmosphere (H_2, CH_4, CO_2, H_2O, NH3, etc.) ionized and thus generate a lot of low energy ions. These low energy ions could react with water or solution on the surface of ocean and land or in the atmosphere, and new products of the reaction fall into the ocean by the rain. These continuously and repeatedly proceeded.

In experiments of simulating these processes, two points should be noticed. One refers to the non-reduced reaction conditions, and the other is the emphasis of the reaction of primitive atmospheric component ions with the solution. Although the energy of the experiments described in this chapter is supplied by gas discharge, it is essentially different from the experiments done before. The aim of the discharge here is to get ions and provide the energy for the ions to be implanted into the aqueous solution. The temperature is controlled in the macro-reaction system (Fig. 13.6). The implantation is not only to provide energy into the reaction system but also implant reaction substance into the system. This is more close to the situation of the discharges between primitive atmosphere and the water surface.

The results of the experiments above-described in this chapter show that the reaction between nitrogen ions and the water can produce amino and nitrate ions; carbon ions implanted into water are changed into carboxyl; ions of CO_2, CH_4, etc. interacting with water can form abundant organic substances. These experiments display evolution of the molecules from simple to sophisticated formations and from inorganic to organic compounds. N-ion implantation in aqueous solutions of acetic acid, sodium acetate, propyl acid, sodium malonate, and sodium succinate can produce different amino acids including glycin (Gly), Aspartic (Asp), alanine (Ala) etc. N-ion implantation into aqueous solution of nucleotide acid component (base, ribose, deoxidize ribose, phosphate radical etc.) can induce synthesizing reactions of nucleotide acids. These experiments not only show that the technology of low-energy ion implantation is possibly a new method to induce synthesis of organic molecules, but also implies that implantation of component ions of the primitive atmosphere into the sea might be an important mechanism for the origin of chemical biomolecules.

For the origin of chemical bio-molecules, it is important to know how carbon is changed into organic compounds. In the first section of this chapter, it is known that nitrogen ion implantation into water can form NH_3. What will happen if graphite (carbon) reacts with ammonia water through gas discharge using the experimental apparatus as shown in Figure 13.6? Figure 13.17 shows the result of HPLC analysis of the experimented sample. The peak at 20.3 min corresponds to ammonia, and the peak at 13.3 min is the hydrolysate of drivatization reagents AMQ. By comparing the retention time with the standard, it is known that ammonia solution form three amino acids induced by gas discharge, namely, serine (Ser, 17.1 min), glycine (Gly, 18.8 min) and alacine (Ala, 25.5 min). To confirm the three components, standard substances of these three components are added into the solution and analyzed by HPLC in the same condition. The result shows that the areas of the peaks of the three components are increased linearly with increasing adding of the three kinds standard substances, indicating that three amino acids are indeed formed in the sample [17].

This is a new type of reaction in which graphite reacts with ammonia water by gas discharge to induce formation of amino acids. The chemical properties of graphite are very stable. It is very difficult to change carbon into organic carbon by normal chemical reactions, and participation of carbon in the reaction to synthesize amino acids is more difficult. Carbon is even more difficult to react with amino acids and organic acids. Carbon is the basic element of organisms and the basis of life. How carbon is changed into organic compounds is primarily unclear. The reaction described here might be an important way to synthesize bio-micro-molecules in primitive ocean or lake, and could be significant in molecular chemistry to study living creature origins.

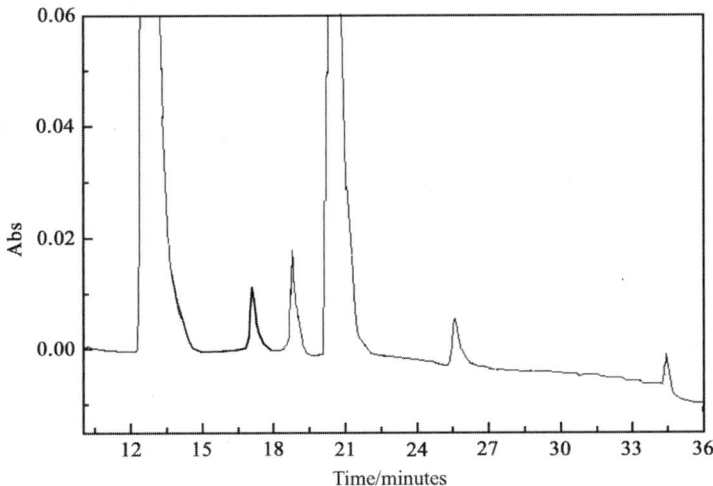

Figure 13.17. HPLC analysis result from the sample of graphite-ammonia water reaction after discharge performed for 10 hours.

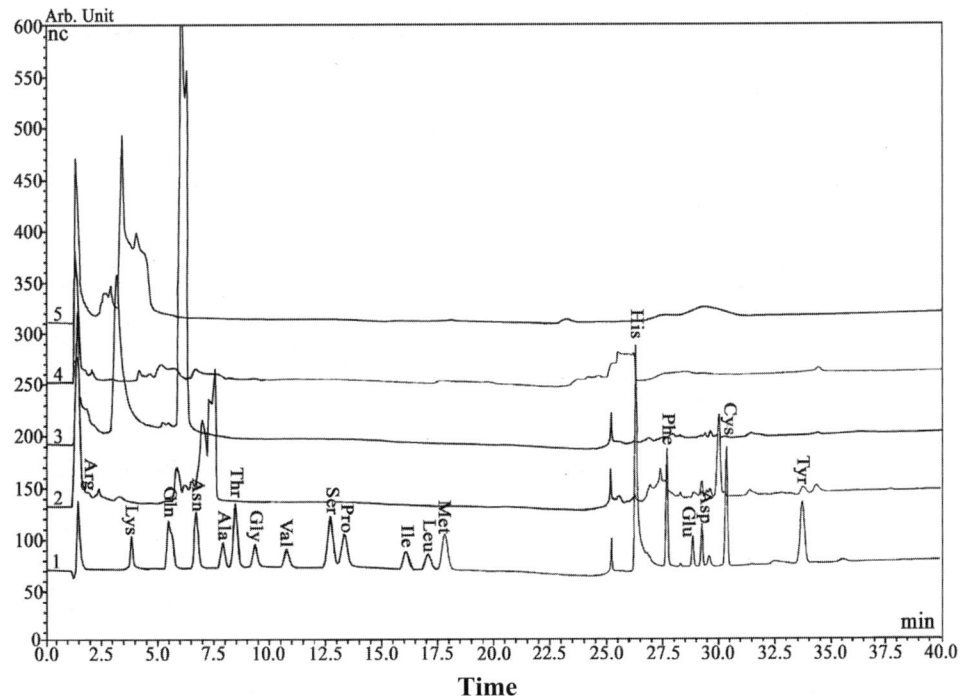

Figure 13.18. Chromatographs of nitrogen-ion-implanted ice and solid sodium acetate and sodium propionate, and amino acids standard. 1: amino acids standard; the peaks here indicate the termed substances; 2: solid sodium acetate; 3: ice sodium acetate; 4: solid sodium propionate; and 5: ice sodium propionate.

In this section, the products of reactions between low energy ions, water and other organic micro-molecules have been analyzed. But, because the contents of these products are insignificant, many new products may not be detected. Figure 13.18 shows the chromatographs of nitrogen-ion-implanted ice sodium acetate and sodium propionate, compared with that of amino acids standard. The ion-implanted sample has some new substances created, which have the peaks not agreeable to the standard peaks of amino acid. This indicates further investigations necessitated for these novel and unknown substances.

It should be realized that the study on effects of low-energy ions on chemical origins of life is still in a beginning stage. The induction of synthesis reactions of organic molecules by low-energy ions may be a route of chemical evolution of biological molecules. But, regarding chemical origins of bio-molecules, an unavoidable question is about the molecular chirality. This is a new research topic for applications of the studies on effects of low energy ion beams on chemical origins of life. Moreover, astronomical observations have shown that minute molecules composed from cyanotype (-CN) and ammonia ($-NH_2$) are important compounds in the interstellar molecule components. It is known that in the solar wind and interstellar molecules, low-energy ions are about 90% of all charged particles. Then, what effects can the low-energy ions have on formation of interstellar molecules? This is also a question worth being investigated. In conclusion, the low-energy-ion-beam technique can be applied to studies not only on the chemical origins of life but also on the interstellar molecule formation. Readers who are interested in these topics may make a trial.

REFERENCES

1. Xu, X.J. and Chu, D.C., *Gas Discharge Physics* (in Chinese, Fudan University Press，Shanghai, 1996)p.160.
2. Han, J.W., *The Study on Damage of Biologic Small Molecules by Low-energy Ion Implantation* (Ph.D. dissertation, Institute of Plasma Physics, Chinese Academy of Sciences, 1998)p.50.
3. Gong, S.H., *The Application of Spectrum Analytical Method in Organic Chemistry* (in Chinese, Science Press，Beijing, 1980).
4. Dyler, J.R., *Organic Spectra*, translated by Chan Xishou (Yinhe Culture Company Limited, Taipei, 1989).
5. Han, J.W., *The Study of Damage of Biologic Small Molecules by Low-energy Ion Implantation* (Ph.D. dissertation, Institute of Plasma Physics, Chinese Academy of Sciences, 1998)p.56.
6. Yu, Z.L., Han, J.W., Wang, X.Q., Huang, W.D. and Shao, C.L., Chemical Synthesis of Life Simulated by Ion Implantation, in *The Role of Radiation in the Origin and Evolution of Life* (Kyoto University Press, Kyoto, 1998)p.175-184.
7. Gettoff, N., Possibilities on the Radiation-Induced Incorporation of CO_2 and CO into Organic Compounds, *Int. J. Hydrogen Energy*, **19**(8)(1994)667-672.
8. Shi, H.B.，*Studies on the Interaction between Low Energy Ions and Biological Micro-molecules and the Role of Such Interaction for the Chemical Evolution of Life* (Ph.D. dissertation, Institute of Plasma Physics, Chinese Academy of Sciences, 2002)p.75.

9. Wang, X.Q., Mass Deposition Effect of Low-Energy Ion Implantation on Solid Sodium Benzoate, *Acta of Physical Chemistry*, **13**(9)(1997)786-789.
10. Shi, H.B., *et al.*, The Synthesis of Adenyl Acid Induced by N^+ Implantation into Solution Containing Components of Adenyl Acid, *Acta of Physical Chemistry*, **17**(5)(2001)412.
11. Shi, H.B., *et al.*, Preliminary Study on the Way of Formation of Amino Acids on the Primitive Earth under Non-reduced Conditions, *Radiat. Phys. Chem.*, **62/5-6**(2001)393-397.
12. Shi, H.B., *et al.*, Synthesis of Amino Acids by Arc-Discharge Experiments, *Nucl. Instr. Meth.*, B**183**(2001)369-373.
13. Wang, X.Q., *et al.*, Synthesis of Amino Acids from Sodium Carboxylic Induce by keV Nitrogen Ions, *Viva Origino*, **26**(2)(1998)106-109.
14. Wang, X.Q., *et al.*, Mass and Energy Deposition Effects of Implanted Ions on Solid Sodium Formate, *Radiat. Phys. Chem*, **59**(2000)67-69.
15. Ovenall, D.W. and Whiffen, D.H., Electron Spin Resonance and Structure of the CO_2^- Radical Ion, *Mol. Phys.*, **4**(1961)135-144.
16. Wang, W.Q., *The Chemical Evolution of Life* (in Chinese, Atomic Energy Press, Beijing, 1994)p.132.
17. Shi, H.B., *et al.*, Preliminary Study on Previous Synthesis of Amino Acid Induced by Low-Energy Ions, *Chemical Research*, **12**(3)(2001)1-5.
18. Miller, S.L., A Production of Amino Acids under Possible Primitive Earth Conditions, *Science*, **117**(1953)528-529.

FURTHER READING

1. Ponnamperuma, C., *The Origins of Life* (Thames Hudson Ltd., London, 1972).
2. Ponnamperuma, C., *Exobiology* (North-Holland Pub. Co., Amsterdam, 1972).
3. Ponnamperuma, C., *Comets and the Origin of Life*: *Proceedings of the 5th College Park Colloquium on Chemical Evolution* (D. Reidel Publishing Co., Dordrecht, 1982).
4. Ponnamperuma, C., *et al.*, *Cosmochemistry and the Origin of Life* (Kluwer Academic Publishers, New York, 1983).
5. Ponnamperuma, C., *et al.*, *Chemical Evolution: The Structure and Model of the First Cell* (Kluwer Academic Publishers, New York, 1995).

14

SINGLE-ION BEAM MUTATION OF CELLS

Health is one of the most important assets of human life and the most significant component of human safety. Environment related health conventionally concerns prevention of harm by microbes and viruses. As the global environment changes, people have increasingly been concerned about effects from exotic environmental health factors. These factors include radiation such as increases in ultraviolet radiation caused by consumption of ozone in the stratosphere, atmospheric cluster radiation of cosmic rays, decay of radioactive elements in the earth's crust, friction between air and the surface of land and sea, radiation of low and medium energy particle radiation produced by lightening, electrostatic induction and electric field action, and so on. Although these types of radiation have very low doses, they directly or indirectly affect human health. For example, in the environment of emission of α particles from radon and its subsequent generations, among every 2,500 human cells, only one receives one α particle per year, and less than 10^{-7} of the total number of human cells receive radiation of more than one α particle. Studies have demonstrated that uranium miners working in high-dose radiation of radon and its subsequent generations of atoms suffer higher rates of lung cancer than other worker groups. However, extrapolation from this high lung cancer incidence caused by the low-dose-radon environment produces considerable uncertainties. Therefore, the exploration of techniques for shooting single ions into cells is very significant in evaluating the harmfulness of exposure to low dose radiation.

Section 4 of Chapter 2 described the single-ion microbeam (SIM) technique which is able to implant a single ion or a precise number of ions into a cell or different locations in a cell, such as the cell nucleus, the cytoplasm, or neighboring cells. This irradiation technique has made many kinds of investigation possible, such as communications inside the cell or between cells, repair of cell damage, and particularly evaluation of environmental harmfulness of low-dose radiation. This chapter introduces methods of localized SIM irradiation of the cell, including experimental preparation, determination of the irradiation location and experimental procedure for studies of biological effect from SIM irradiation of cell nucleus, cytoplasm and neighboring cells.

14.1. METHOD OF LOCALIZED SINGLE-ION BEAM IRRADIATION OF CELLS

The procedure of single-ion beam irradiation of cells includes preparation of the cells, rapid attachment, low-concentration fluorescence staining, computer imaging, cell localization, determination of the origin, automatic irradiation, and biological detection. Details are described here based on the experience in the Institute of Plasma Physics (IPP), Chinese Academy of Sciences.

14.1.1. Preparation of Cells

The sample is a group of live mammalian cells. The cells are trypsinized and counted to be made into a single cell suspension. The cell suspension is diluted and inoculated to the center of a microbeam dish which is specially constructed for the purpose. The dish consists of a ring and a 3.8-mm thick polypropylene film which is attached to the ring on the bottom by means of epoxy resin. The film is coated with Cell-Tak[*] which enables cells to adhere to the film within 5 minutes. For irradiation of nuclei, about 500 to 600 cells are seeded in a 35-mm dish. For irradiation of cytoplasm, only 300 cells are used. The number of cells seeded in the microbeam dish is determined by the SIM irradiation speed and the cell viability under experimental conditions without medium. Thus, the number of the cells irradiated can be increased to accommodate expected speed of irradiation and a corresponding increase in cell death. Before irradiation, the cells in the dish are stained with low-concentration fluorescent dyes. For irradiation of the nucleus, the DNA of the attached cells is stained with a 50-nM solution of Hoechst 33342 dye for 30 minutes, and for irradiation of cytoplasm, the cells are stained with the 50-nM solution of Hoechst 33342 for 30 minutes and then stained with a 100-ng/ml solution of Nile Red for 10 minutes. After the staining, the cells can be used for image analysis and irradiation localization. Experiments have demonstrated that neither Hoechst33342 nor Nile Red, used either alone or in combination under the conditions described here, affects the survival, mutagenesis, or radiosensitivity of the cells.

14.1.2. Cell Image Analysis and Localization

The locations of individual nuclei stained by Hoechst 33342 are clearly visible with a fluorescence microscope under 366-nm UV excitation. The images of Nile Red stained cytoplasm are viewed using green light excitation and red emission. The images may be superimposed into a 24-bit color image using the "merge channel" capacity of Image Pro Plus[**] for image analysis and treatment. For irradiation of nuclei, the image analysis system localizes the center of each nucleus, which is then irradiated with an exact number of particles (Figure 14.1). The aiming points for irradiation of cytoplasm are chosen to be 8 μm away from the ends of the major axis of each nucleus (Figure 14.2).

---

[*] Cell-Tak is a formulation of polyphenolic proteins extracted from Mytilus edulis (marine mussel). These proteins are the key components of the glue secreted by the mussel to anchor itself to solid substrates in its marine environment.

[**] Image Pro Plus is the ultimate image analysis software package for fluorescence imaging.

14.1.3. Irradiation Protocols

Prior to irradiation, the accelerator is adjusted to obtain a monoenergetic ion beam (Figure 2.13). Then, the coordinates of the irradiated point should be determined. The method of determining the coordinates varies with the SIM apparatus. If the number of the particles is counted behind the cells, at the cell dish bottom the light spot of a laser beam through the collimator can be used to determine the coordinates. If the particles are detected before the cells, the light spot excited by the ion-passed scintillator can determine the coordinates (Figure 14.3). After the irradiation coordinates are determined, all cell coordinates are relative to these irradiation coordinates. Computer programs use these coordinate data to move the sample holder to localize the preset positions of the cells at the irradiation coordinates for irradiation of either nucleus or cytoplasm. According to Monte Carlo simulations, the present SIM can achieve the irradiation at the preset position with an irradiation accuracy (the probability of hitting the target) greater than 90%, the number of irradiating single ions with an accuracy greater than 98.4%, and hitting the nucleus from aiming at cytoplasm with an error less than 0.4%.

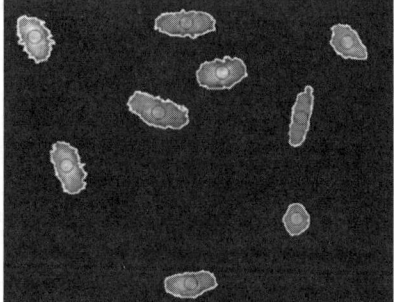

Figure 14.1.   Fluorescent imaging of the A_L cells stained with Hoechst dye viewed by the image analysis system under a 40x objective lens. The nucleus of each cell is outlined in white, and the circles indicate the area where the particle is delivered.

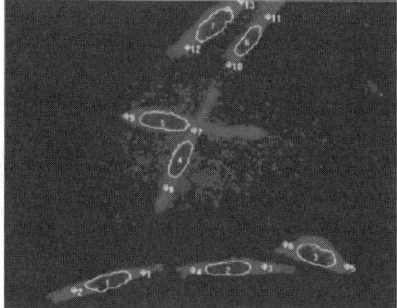

Figure 14.2.   Dual fluorescent imaging of A_L cells stained with Hoechst 33342 (nucleus) and Nile Red (cytoplasm) by the image analysis system under a 40x objective lens. The nucleus of each cell is outlined in white. The image analysis system determines the length of the major axis of each nucleus to calculate the irradiation positions which are chosen to be 8 μm from each end of the nucleus, as shown by the small, numbered circles.

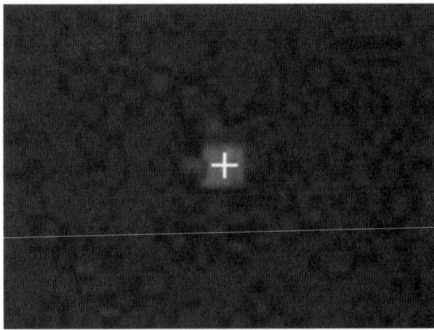

Figure 14.3. A luminous spot produced in the scintillator (Bicron BC400, 18 μm, mounted at the 1.5 μm collimator exit) for positioning the incident microbeam.

14.1.4.   Post Irradiation Treatment

After the cell nucleus or cytoplasm is irradiated, the dish is removed from the stage and the cells are trypsinized at 37°C and inoculated to a cell culture dish to measure the biological effects such as the sensitivity of irradiated cellular spots, mutation and transformation. Or, after irradiation, shortly or after culture for a certain period, detection or measurement can be performed for cell damage including cell wither and death, breaks of single strand and double strands, and production of micro-nucleus as well.

It should be pointed out that the methodology described above only provides a basic reference procedure for researchers, and there can be differences from different microbeam facilities. Researchers should make their own adjustments based on their own facilities for their particular experimental objectives.

14.2.   IRRADIATION OF CELL NUCLEUS

In assessment of the health risk associated with public exposures to ionizing radiations, it has been estimated that more than half of the ionizing radiation dose received by the population originates from α particles produced from radon and its next generation decay [1]. Studies of smoking and non-smoking uranium miners indicate that radon itself is a risk factor for lung cancer at the levels encountered by these miners [2-4]. However, the risk assessment of radon has traditionally been compromised because it is always complicated by concurrent exposure to other chemical and physical environmental contaminants, such as arsenic, ore dust, fluorides, silica and tobacco smoke [2,4,5]. The estimate by the Environmental Protection Agency of 21,600 deaths with confidence limits between 8,400 and 43,200 illustrates the uncertainties inherent in environmental risk assessment using epidemiological data [6].

One of the main uncertainties in estimating risk from the radon exposure based on lung cancer incidence in miners is the need to extrapolate from high to low dose exposure where multiple α-particle traversals are extremely rare [7]. Therefore, in order to extrapolate from the miner data to environmental exposures, it is necessary to be able to extrapolate from the effects of multiple traversals to the effects of single traversals of α particles.

Many investigators have used *in vitro* approaches of broad-field irradiation of cells in culture to quantify the effects of low-dose level exposure. However, due to the stochastic nature of energy deposition by these charged particle tracks and that the positions of tracks of ionization in the cells are random, these attempts to quantify the dose-effect relationship at doses relevant to environmental exposures were unsuccessful [8-10]. For example, if a population of cells in culture is exposed to an average of one ^{210}Po α-particle traversal per cell (equivalent to ∼ 0.4Gy), then the Poisson distributed variation in particle traversals predicts that 37 % of the cells will receive no particles at all, 37 % will receive a single traversal and 26 % of the cells will receive two or more particle traversals. Clearly, the variation in the dose received by different cells will result in an increase in the variation of the biological response observed. Therefore, the variations complicate the determination of the dose-effect relationship.

The availability of the single-ion microbeam irradiation facility where individual targets can be irradiated with either a single or an exact number of particles provides a platform to study the biological effects of a single particle irradiation and the relationship between low dose and high dose ion irradiation. Figure 14.4 shows the dose response clonogenic survival of A_L cells irradiated with defined numbers of α particles through the nucleus. The question of whether traversal of a single α-particle through the nucleus is lethal has been debated for more than three decades. Studies by Barendsen and his colleagues suggested that traversal of the nucleus by a single α particle would in fact be lethal [11]. Moreover, studies based on measurement of induced DNA double strand breaks in C3H10T1/2 cells indicated that virtually 100% of the cells traversed by a single alpha particle would suffer enough damage to be killed by direct action [12]. On the other hand, microdosimetric studies based on particle track structure suggested that the probability of an alpha particle traversal resulting in lethal damage was only 17% in rodent fibroblasts [13]. The measurement of a single particle survival is consistent with low estimates of cell inactivation such that only 20 percent of the irradiated cells are killed. It is amazing that roughly 10 percent of cells irradiated with 8 α-particle nuclear traversals were still viable enough to form colonies. The dose response for survival of A_L cells irradiated with an exact number of α-particles through nucleus was in fact not significantly different from that obtained using average particle traversals based on a random, Poisson distribution [14,15]. These results suggest that, at least for cell lethality, the Poisson estimation gives a fairly accurate projection of the biological effects of broad field exposure compared to that of an exact number of α-particles.

It is widely accepted that one of the early steps in carcinogenesis involves induction of mutations. Figure 14.5 shows the mutant fractions per 10^5 clonogenic survivors at the CD59 locus in A_L cells irradiated with exact number of alpha particles through the nucleus. The fraction of preexisting CD59⁻ mutants in A_L cell population used in this nuclear study averaged 45 per 10^5 survivors. The induced mutant yield by a single alpha particle was ∼3 fold of the background and increased to 8 fold for 8 particles. This mutant fraction of 3 times background was comparable to the frequency induced by an equivalent mean of one particle traversal based on a Poisson distribution [14]. However, a dose of 8 particle traversals per nucleus resulted in an induced frequency that was 8X background level. The incidence was significantly higher than the yield obtained with a mean of 8 particles. It is possible that many cells in the latter group may have either received fewer particle traversals and subsequently fewer mutations, or more than 8 particle traversals and subsequent cell lethality. Thus, it is likely that this difference is due to distortion of the cell population at the time of the assay in the Poisson distributed experiments because of differences in radiation induced division delay. The cells that happened to receive a small number of traversals expand in large population during the expression period than those that randomly got a larger number of hits. Therefore, the single cell irradiation performed here will give a clearer extrapolation from high to low doses. It also suggests a biological basis which supports the validity of the data of high-level radon exposure to uranium miners, who have the most target cells receiving multiple hits, for risk assessment of low-level residential exposures.

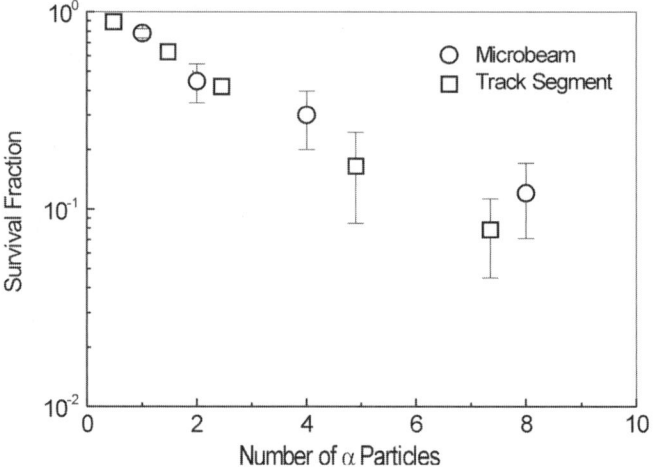

Figure 14.4. Dose dependence of clonogenic survival of A_L cells irradiated with defined numbers of alpha particles through the nucleus.

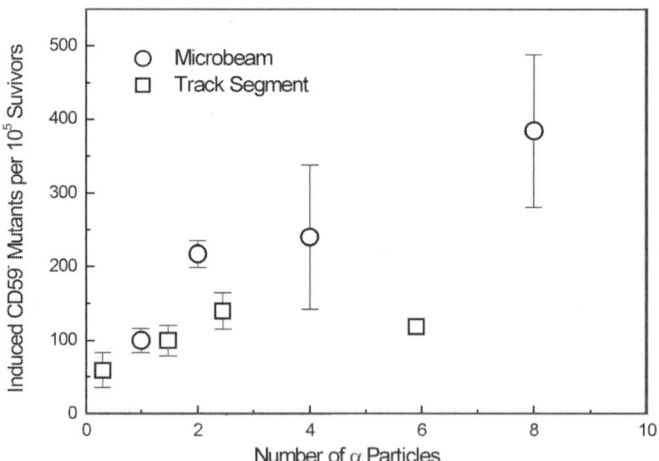

Figure 14.5. Mutant fractions per 10^5 clonogenic survivors at the CD59 locus in A_L cells irradiated with exact number of alpha particles through the nucleus.

Miller et al. further studied differences in α-particle irradiation induced cellular malignant transformation between precise single α particle irradiation and broad α-particle beam irradiation of the C3H10T$_{1/2}$ cell nucleus [16]. For the latter case, the effect of a single α particle irradiation was calculated as the Poisson mean. They found that the nuclear traversal by exact one α particle is not only significantly less effective than traversal by a mean particle (1.2 and 3.1 malignant transformations per 10 living cells, respectively), but also not significantly more effective than a zero-dose (sham) irradiation (1.2 and 0.86 transformations per 10 cells, respectively). They also found that as increasing the number of the irradiating particles, such as 2, 4, and 8 α particles, the yields of oncogenically transformed cells irradiated by microbeam and broad beam were basically similar. These results are very important in estimating the accuracy of extrapolating the data from the risks of underground miners exposed to high-dose α-particle irradiation to domestic radon exposure which, on average, has the dose many times lower than that in the mines. Because, if single α particles only induce very minor cellular malignant transformation, our currently widely used radiation protection models may over-estimate damage effect of single α-particle irradiation. Most of the cellular malignant transformations induced by single mean α-particle irradiation may not come from really single ion irradiation of nucleus, but from multiple α-particle irradiation. Of course, the results were obtained from the study of the C3H10T$_{1/2}$ cell system, and there have not yet been investigations delivered onto normal human cell systems for the malignant transformation. Therefore, further research data should be added to the risk assessment of radon to human health.

14.3. IRRADIATION OF CYTOPLASM

Whether cytoplasm induces nuclear DNA damage is always a very interesting research direction for biologists and geneticists. Early studies from Munro and Puck thought that irradiation of cytoplasm was not harmful, and further pointed out that the target for ionizing radiation to induce biological effects should be nuclear DNA [17,18]. However, increasing evidence has shown that α-particle radiation effects of tissues are not only limited to irradiation of the nucleus [19-21]. Nagasawa et al. found that in CHO cells, α-particles could induce exchange of sister chromosomes in the cell through a mechanism not related to interaction with the nucleus [21]. From this, Deshpande et al. calculated the potential irradiated area, which included cytoplasm and cell outer parts, to induce SCE to be 350 times greater than the CHO cell nucleus [20].

Here is an experiment using a precise number of α particles from SIM to irradiate A$_L$ cells locally and quantitatively at the cytoplasm [22]. The result (Table 14.1) showed that cytoplasmic irradiation induced minimal toxicity in the A$_L$ cells such that the lethal rate was only 12% when 4 α particles irradiated cytoplasm and 76% of the cells could survive and formed colonies after 16 α-particles irradiation of the cytoplasm. The mutation rate of CD59 gene after irradiation of cytoplasm was measured using antibody and complement. The study found that although the lethal rate for irradiation of cytoplasm was light, changes in the gene inside the nucleus were indeed induced; the point mutation rate of CD59 gene of the irradiated cells was as maximum as 3 times as that of spontaneous mutation (Table 14.1). It also indicated that the mutation rate reached saturation when the number of the irradiating particles was more than eight and did not increase as the particle number increased (Table 14.1). It is thought that in this experiment, only two locations of cytoplasm were selected to be irradiated so that the yield of the mutation induction substance could be at a saturated state. A recent study from Narayanan *et al.* provided direct evidence for this suggestion. They found that the production of free radicals O_2^- and H_2O_2 was not linearly proportional to the irradiation dose [23]. Apart from this, the mutation saturation effect is probably also because irradiation of cytoplasm induces repair-protection effect in the cells, and hence increase in cell damage and mutation is restricted. Similar saturation effect has also been reported for low-dose α-particle induction of sister chromosomes [21,22].

PCR (polymerase chain reaction) analysis was carried out for investigation of changes and deletions of WT, CAT, RAS, PTH and APO-A1 genes on chromosome 11 of the spontaneous mutant from the cell that was not irradiated and CD59⁻ gene mutant from the cell whose cytoplasm was irradiated with various numbers of α particles [22]. The results are shown in Table 14.2. From the table it is shown that most of irradiation-induced CD59⁻ mutants only lose the CD 59 gene. The irradiation-induced mutation spectrum is similar to that of the spontaneous CD59⁻ gene mutant, but very different from the case of α-particle irradiation of the nucleus, where the gene deletion increases as increasing the number of irradiating particles. This fact indicates different mutation mechanisms between irradiations of cytoplasm and nucleus.

Table 14.1. Survival fraction and CD59⁻ mutation frequencies (MF) for α particle irradiation of cytoplasm of A_L cells.

Number of α particles	Survival Fraction (%)	Mf (Total)[a]
0	100.0	43±15
1	93.9	78±26
4	88.0	139±50
8	79.0	167.7±73
16	76.0	158±60

Mf[a] = Mutants / 10^5 survivors

Table 14.2. Mutation frequencies of CD59 genotypes.

Number of α particles	Number of mutants analyzed	Mutant class			
		CD59⁻ only		Number of mutants losing at least one marker besides CD59	
		Number	Percentage	Number	Percentage
Control	37	34	92%	3	8%
1	29	28	97%	1	3%
4	32	32	100%	0	0
8	28	26	93%	2	7%

In order to study whether free radicals played a role in induction of CD59 gene mutation when cytoplasm was irradiated, two kinds of chemicals, DMSO and BSO, which have opposite interaction characteristics, were used in the experiment [22]. It was found from the study that treatments of the cells for 10 minutes using 8% DMSO, a free-radical scavenger, before and after irradiation of cytoplasm, respectively, could obviously reduce the CD 59 gene mutation rate after irradiation (Figure 14.6). On the contrary, treatment of the cell using 10 μM BSO, a sulfhydryl binder, for 18 hours before irradiation could increase CD 59 gene mutation rate by 4 – 5 times as the binding could decrease the levels of free radical scavengers such as non-protein sulfhydryl (NPSH) and glutathione (GSH) originally inside the cell (Figure 14.7). These findings directly demonstrate for the first time the important role played by free radicals when cytoplasm is irradiated. The concentrations of DMSO and BSO used in the experiment were examined to have no killing and mutating abilities to the cell.

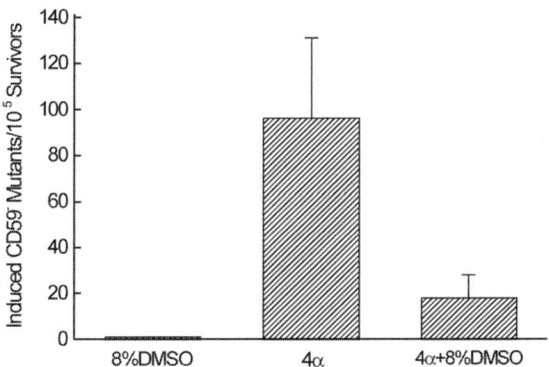

Figure 14.6. Effects of DMSO on induced mutation frequencies in A_L cells treated with alpha particles through the cytoplasm. Data was pooled from 3-6 independent experiments. Bars represent ±SEM.

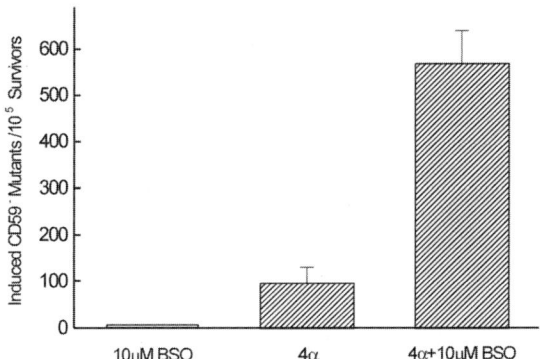

Figure 14.7. Effects of BSO on induced mutation frequencies in A_L cells treated with 4 alpha particles through the cytoplasm. Data was pooled from 3 independent experiments. Bars represent ±SEM.

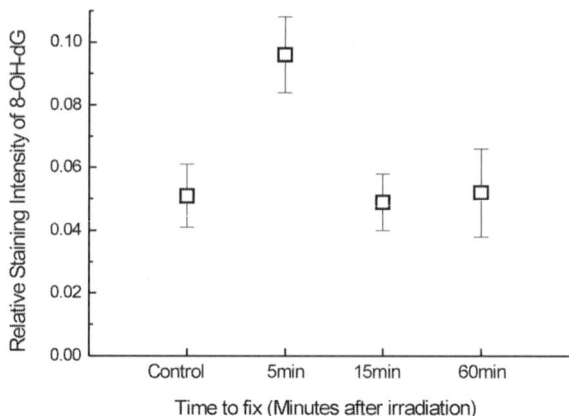

Figure 14.8. Relative immunoperoxidase staining intensity for 8-OH-dG in control and A_L cells irradiated with 8 alpha particles through cytoplasm. Cells were fixed at different time points of 5, 15 or 60 minutes after irradiation. Bars represent ±SEM.

In the same experiment [22], single clone antibody IF7 was used for immunoperoxidase staining reaction. The result showed that the staining relative intensity of the cell fixed stained 5 minutes after irradiation was 0.096 ± 0.012, 1.9 times that of the control (0.051 ± 0.01). However, the relative staining intensity of 8-hydroxy-deoxyguanosine (8-OH-dG) was reduced to the control level when the cell was fixed stained 15 and 60 minutes after irradiation (Figure 14.8). Reasons for this reduction remains unknown. Since 8-OH-dG is the important marker of DNA oxidation damage, this result indicates that the oxidation damage to the nuclear DNA may be one of the ways to induce nuclear gene mutation when cytoplasm is irradiated.

There are currently several conjectures on delivery of damage signals such as free radicals from cytoplasm to an inside nucleus. Based on the fact that DMSO can effectively scavenge free radical $OH^{\bullet}$ in the cell and most free radicals have the extremely short lifetimes and diffusing distance [24,25], one of the possible ways is that free radicals including $OH^{\bullet}$ interact with each other, namely, $OH^{\bullet}$ can induce production of other free radicals via cytoplasm oxidation, like a cascade process continuing, until it is restricted by the reacted substrate. Another possibility for mutation induction by cytoplasmic radiation may be related to mitochondrial damage. It is known that mitochondria are the major intracellular source and target of reactive oxygen species and free radicals [26,27]. It is possible that organic radicals such as peroxynitrite anions generated as a result of mitochondrial damage could also be involved [28,29]. There is recent evidence that mitochodrial DNA damage may also modulate DNA damage, although the exact mechanism of how mitochondrial DNA escapes in the nuclear compartment is not known [30].

It should be especially noted that although irradiation of a nucleus can induce $3 - 4$ times as many mutants as produced by irradiating cytoplasm with the same numbers of irradiating particles, the latter is more important for the occurrence of cancer, as irradiation of cytoplasm causes very few cell deaths. For example, with the doses which have the same killing powers (e.g. 90% survival rate), irradiation of cytoplasm can induce a mutation rate 7 times that of irradiation of nuclei, as shown in Table 14.3 [11]. Therefore, irradiation of cytoplasm should be considered as the major factor harming human health.

14.4. IRRADIATION OF NEIGHBORING CELLS — BYSTANDER EFFECTS

Another important research area for the use of single-ion microbeam is localized irradiation of cell groups (one or several) in tissues to study damage effects produced in cells nearby as well as relationships between this effect and the irradiation dose, time and space. This effect is called the radiation-induced cell bystander effects.

The radiation-induced cell bystander effects were firstly discovered by Nagasawa and Little [31]. Subsequently a number of studies based on statistics have shown that this kind of effect widely exists in various types of cells and radiations, even irradiation of culture medium also able to lead to damage to cultured cells. Employment of conventional broad beams to irradiate cells cannot lead to knowledge about which types of irradiated subcellular organs can induce the bystander effects. Evidence has shown that irradiations of cell nucleus and cytoplasm can both result in lethal and mutation effects for cells. Development of microbeam technology provides conditions for people to irradiate subcellular structure of the cell with a precise number of particles for further understanding mechanisms involved in the bystander effects.

Zhou et al. [32] studied effects on A_L-cell CD 59 gene mutation induced from 5, 10, 20 and 100% of the cell nuclei of single-α-particle irradiated group cells and the result is shown in Figure 14.9. In the figure, the expected mutation rate is under the assumption that there is no bystander-effect interaction between the irradiated and unirradiated cells. Great difference can be seen between the experimental and expected data. For example, The mutation rate of 5% of cells irradiated by single particles reaches 58% of that of total irradiated cells (the mutation rates are 57 and 98 per 10^5 living cells, respectively). The mutation rates for 10% to 100% of irradiated nuclei have no difference. As α-particle emitted secondary electrons have an action range about 0.25 μm [33], the damage to the unirradiated cells is thought to be caused by secondary electrons. When the cells were treated by 1-mM octanol which could restrict communication between cells for 2 hours before irradiation and for 3 days after irradiation, the mutation rate of 20% of irradiated cell nuclei could be greatly decreased (Figure 14.10). Thereby, the cell damage signals are delivered between irradiated cells and unirradiated cells relying on interstitial communication.

Table 14.3. Yields of mutants caused by α-particles striking the cell nucleus or cytoplasm.

Irradiation site	90% survival (particles [a])	M_y[b] / paticle	M_y at 90% survival
Nucleus	0.13	100	13
Cytoplasm	≥ 4	25	≥ 100

(a) Each particle ≈ 12.5 cGy. (b) M_y = mutants/10^5 clonable cells/unit dose = slope of the mutant dose response curve.

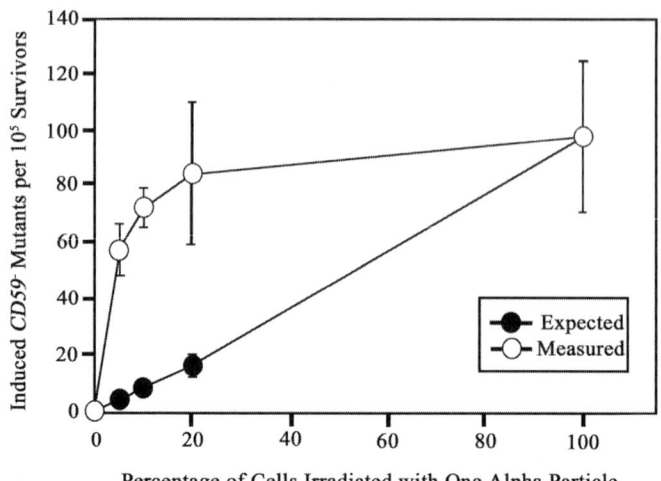

Figure 14.9.   Induced CD59 mutant fractions per 10^5 survivors obtained from populations of A_L cells in which 0, 5, 10, 20, or 100% had been irradiated with exactly one α particle through its nucleus. Induced mutant fraction = total mutant fraction minus background incidence, which was 46 ± 10 mutants per 10^5 clonogenic survivors in A_L cells used in these experiments. Data are pooled from seven independent experiments. Error bars represent $\pm$ SD. The calculated curve deviates slightly from a straight line because of the slight cytotoxic effect of single particle traversal among the irradiated cells.

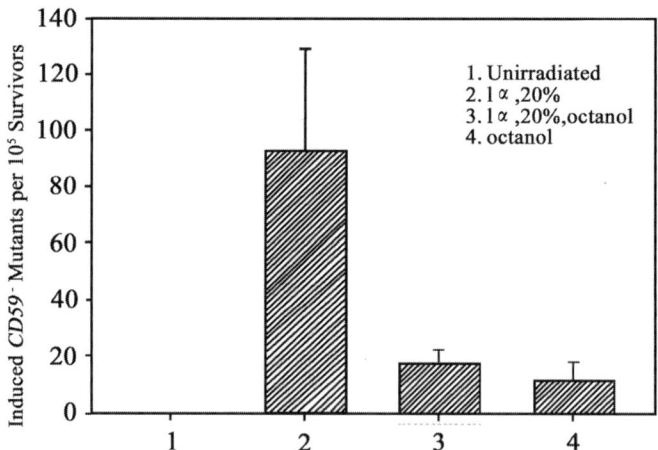

Figure 14.10.   Effect of octanol treatment (1 mM, 2 h before and maintained until 3 days after irradiation) on mutant fractions of A_L cell population of which 20% had been irradiated with a single α particle through the nucleus. Data are from three independent experiments. Error bars represent $\pm$ SD.

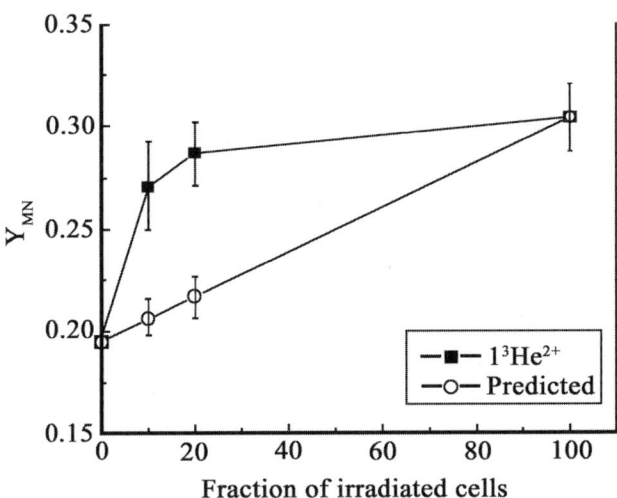

Figure 14.11.   MN yield of the population where a fraction of cells were individually irradiated with one $^3He^{2+}$ ion. The predicted MN yield is calculated by the method that has been described in the text.

Shao *et al.* [34] found that after a single cell nucleus of the human colloid cancer cell, T98G colony, was precisely irradiated with one or five α particles of the microbeam, the micro-nucleus (MN) yield of the cell group was increased by 20% after one-hour further culture (Figure 14.11). This implied that the micro-nuclei could produce several tens of unirradiated other cells. Further study showed that when the microbeam irradiated cell fraction was increased from 0.1% to 20%, the micro-nucleus yield of the whole cell colony rapidly increased and tended to saturation, far greater than the expected value without the bystander effects. When the microbeam irradiated cell fraction was more than 20%, the MN yield was close to the yield when all cells were irradiated. These results directly demonstrate that microbeam irradiation of human cancer cells can also produce the bystander effects.

As most cancer cells do not have interstitial communication, it is speculated that some solvable cellular factors may play an important role in the bystander effects of the cancer cells induced by radiation. In order to investigate what kinds of cell factors play this role, Shao *et al.* used NO free radical scavenger c-PTIO or NO synase depressor, aminoguanidine to treat T98G cells [34]. They found that this treatment could decrease the MN yield induced by microbeam particle irradiation down to the value for only direct irradiation action without the bystander effects (Figure 14.12). This fact indirectly demonstrates that when cancer cells are partly irradiated with microbeam particles, NO is an important cell factor for the bystander effects.

As an important signaling messenger, NO is known to be generated endogenously from L-arginine by inducible NO synthase that can be activated by radiation. However, it is believed that the NO free radical itself does not induce DNA strand breaks. To investigate whether there is any other NO-related signaling factor involved in the bystander response, Shao *et al* [34] performed a medium transfer experiment, in which the medium from the T98G population, where 100% of cells were individually targeted by a precise number of helium ions, was harvested one hour after irradiation and then transferred to another population of nonirradiated cells. Results showed that the MN yield of the nonirradiated population after treatment with the conditioned medium was increased by about 25%, but this increase had no significant relationship to the irradiation dose. Moreover, when c-PTIO was present in the medium during irradiation and subsequent cell culture, the conditioned medium did not show a cytotoxic effect of MN induction in the nonirradiated cells as shown in Figure 14.13, indicating that some NO-downstream long-lived biological active factors contributed to the irradiation-induced bystander effect.

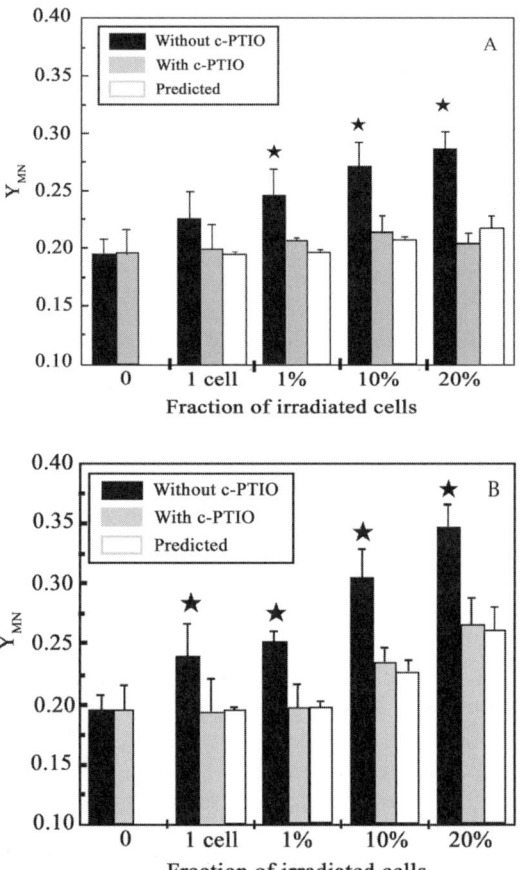

Figure 14.12. MN yields of the population where a fraction of cells were individually irradiated with one ^{3}He^{2+} ion (A) and five ^{3}He^{2+} ions (B), respectively. In some experiments, 20 μM c-PTIO were present in the medium during irradiation and subsequently 1 h after irradiation until MN assay. The predicted MN yield is calculated by the method that has been described in the text. * indicates that the MN yield of the irradiated population is significantly larger than that of unirradiated control, c-PTIO treated population, and predicted values assuming no bystander effect ($P < 0.01$).

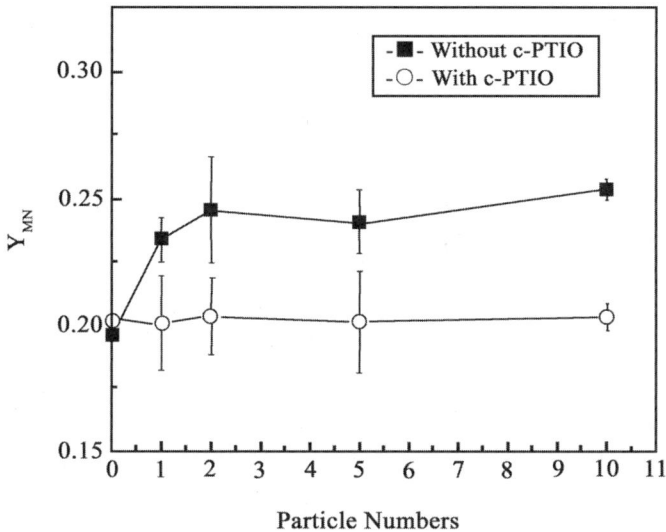

Figure 14.13. MN yields of the conditioned medium-treated cell populations. The conditioned medium was harvested from the cell population where 100% of cells were individually irradiated by a precise number of 3He2+ ions. In some experiments, 20 μM c-PTIO was present in the conditioned medium during and subsequently 1 h after irradiation until MN assay.

The research described above shows that the bystander effects induced by microbeam particle irradiation is significant for radiation therapy. For many years, people have thought that delivering a lethal dose to each cell is the only way to kill cancer cells in radiation therapy. Now the discovery of the bystander effects induced by irradiation of individual cancer cells predicts a new radiation therapy, in which, based on protecting normal cells, a lower dose of radiation is delivered to cancer cells to induce suicide bystander effects resulting in killing more cancer cells. On the other hand, the cancer-cell-produced bystander effect factors imply that damage to normal cells and tissues in radiation therapy processes may be higher than expected. Therefore, how to appropriately control and utilize the bystander effects will determine how people develop new radiation therapies which can more easily kill cancer cells while more effectively protect normal cells.

The primary objective of establishing the single particle microbeam is the study of biological effects of low-dose radiation and relationships between low doses and high doses. Discovery of the bystander effect and unique advantages of the single-ion microbeam in studying the bystander effect further expands application areas of SIM. Currently, researchers are studying hot topics such as instability of gene, developmental biology, combined effects of environmental radiation and toxic substance exposure on health employing the single-ion microbeam with its features of highly localized and quantitative irradiation. The prospects are bright.

REFERENCES

1. Committee on the Biological Effects of Ionizing Radiations, *Health Effects of Exposure to Low Levels of Ionizing Radiation* (BEIR V) (Natl. Acad. Press, Washington D.C., 1990).
2. Samet, J.M., Kutvirt, D.M., Waxweiler, P.J. and Key, C.R., Uranium Mining and Lung Cancer in Navajo Men, *New Engl. J. Med.*, **39**(5)(1984)355-359.
3. Roscoe, R.J., Steenland, K., Halperin, W.E., Beaumont, J.J. and Waxwiler, R.J., Lung Cancer Mortality among Nonsmoking Uranium Miners Exposed to Radon Daughters, *J. Am. Med. Assoc.*, **262**(1989)629-633.
4. Tomasek, L., Swerdlow, A.J., Darby, S.C., Placek, V. and Kunz, E., Mortality in Uranium Miners in Western Bohemia: A Long-Term Cohort Study, *Occup. Environ. Med.*, **51a**(1994)308-315.
5. Kusiak, R.A., Ritchie, A., Muller, J. and Springer, J., Mortality from Lung Cancer in Ontario Uranium Miners, *Br. J. Ind. Med.*, **60**(1993)920-928.
6. NRC Report on Health Effects of Exposure to Radon, *National Academy of Sciences*, 1994.
7. Brenner, D.J., Unfolding the Poisson Distribution: Can Mathematical Tricks Do As Well As a Single-Particle Microbeam? In Michael, B.D., Folkard, M. and Prise, K.M. (Ed), *Microbeam Probes of Cellular Radiation Response* (C.R.C. Gray Laboratory, London, UK, 1993)pp. 1.3.1-1.3.4.
8. Marples, B. and Joiner, M.C., The Response of Chinese Hamster V79 Cells to Low Radiation Doses: Evidence of Enhanced Sensitivity of the Whole Cell Population, *Radiation Research*, **133**(1993)41-51.
9. Lambin, P., Marples, B., Fertil, B., Malaise, E.P. and Joiner, M.C., Hypersensitivity of a Human Tumour Cell Line to Very Low Radiation Doses, *Int. J. Radiat. Biol.*, **63**(1993)639-650.
10. James, S.J., Enger, S.M. and Makinodan, T., DNA Strand Breaks and DNA Repair Response in Lymphocytes after Chronic in vivo Exposure to Very Low Doses of Ionizing Radiation in Mice, *Mutation Research*, **249**(1991)255-263.
11. Barendsen, G.W., Impairment of the Proliferative Capacity of Human Cells in Culture by α-Particles with Differing Linear-Energy Transfer, *International Journal of Radiation Biology*, **8**(1964)453-466.
12. Watt, D.E., An Approach Towards a Unified Theory of Damage to Mammalian Cells by Ionising Radiation for Absolute Dosimetry, *Radiat. Protect. Dosi.*, **27**(1989)73-84.
13. Hei, T.K., Wu, L.J., Liu, L.X., Randers-Pehrson, G. and Hall, E.J., *Proc. Annual Meeting of the Radiation Research Society*, **43**(1996)161.
14. Zhu, L.X., Waldren, C.A., Vannais, D. and Hei, T.K., Cellular and Molecular Analysis of Mutagenesis Induced by Charged Particles of Defined Linear Energy Transfer, *Radiat. Res.*, **145**(1996)251-259.
15. Hei, T.K., Zhu, L.X., Vannais, D. and Waldren, C.A., Molecular Analysis of Mutagenesis by High LET Radiation, *Adv. Space Res.*, **14**(10)(1994) 355-361.
16. Miller, R.C., Randers-Pehrson, G., Geard, C.R., Hall, E.J., and Brenner, D.J., The Oncogenic Transforming Potential of the Passage of Single α-Particles through Mammalian Cell Nuclei, *Proc. Natl. Acad. Sci. USA*, **96**(1999)19–22.
17. Munro, T.R., The Relative Radiosensitivity of the Nucleus and Cytoplasm of

Chinese Hamster Fibroblasts, *Radiation Research*, **42**(1970)451-470.

18. Puck, T.T., in *The Mammalian Cell as a Microorganism* (Holden-Day, San Francisco, 1972).

19. Nagasawa, H., Little, J.B., Inkret, W.C., Carpenter, S., Raju, M.R., Chen, D.J. and Striniste, G.F., Response of X-Ray-Sensitive CHO Mutant Cells (xrs-6c) to Radiation, II. Relationship between Cell Survival and the Induction of Chromosomal Damage with Low Doses of α-Particles, *Radiat. Res.*, **126**(1991)280-288.

20. Deshpande, A., Goodwin, E.H., Bailey, S.M., Marrone, B.L. and Lehnert, B.E., Alpha Particle-Induced Sister Chromid Exchange in Normal Human Lung Fibroblasts: Evidence for an Extranuclear Target, *Radiat. Res.*, **145**(1996)260-267.

21. Nagasawa, H. and Little, J.B., Induction of Sister Chromatid Exchanges by Extremely Low Doses of α-Particles, *Cancer Research*, **52**(1992)6394-6396.

22. Wu, L.J., Randers-Pehrson, G., Xu, A., Waldren, C.A., Geard, C.R., Yu, Z.L., and Hei, T.K., Targeted Cytoplasmic Irradiation with Alpha Particles Induces Mutations in Mammalian Cells, *Proc. Natl. Acad. Sci.* USA, **96**(1999)4959-4964.

23. Narayanan, P.K., Goodwin, E.H. and Lehnert, B.E., Particles Initiate Biological Production of Superoxide Anions and Hydrogen Peroxide in Human Cells, *Cancer Research*, **57**(1997)3963-3971.

24. Roots, R. and Okada, S., Estimation of Life Times and Diffusion Distances of Radicals Involved in X-Ray-Induced DNA Strand Breaks or Killing of Mammalian Cells, *Radiat. Res.*, **64**(1975)306-320.

25. Buettner, G.R. and Jurkiewics, B.A., Ascorbate Free Radical as a Marker of Oxidative Stress: an EPR Study, *Free Radicals in Biol. Med.*, **14**(1993)49-55.

26. Nohl, H. and Hegner, D., Do Mitochondria Produce Oxygen Radicals in vivo? *Eur. L. Biochem.*, **82**(1978)563-567.

27. Chance, B., Sies, H. and Boveris, A., Hydroperoxide Metabolism in Mammalian Organs, *Physiol. Rev.*, **59**(1979)527-603.

28. Wei, Y.H., Oxidative Stress and Mitochondrial DNA Mutations in Human Aging, *Proc. Soc. Exp. Biol. & Med.*, **217**(1998)53-63.

29. Lenaz, G., Role of Mitochondria in Oxidative Stress and Ageing, *Biochem. Biophys. Acta.*, **1366**(1998)53-67.

30. Cavalli, L.R. and Liang, B.C., Mutagenesis, Tumorigenicity, and Apoptosis: Are the Mitochondria Involved? *Mutation Research*, **398**(1998)19-26.

31. Nagasawa, H. and Little, J.B., Induction of Sister Chromatid Exchanges by Extremely Low Doses of Alpha Particles, *Cancer Res.*, **52**(1992)6394-6396.

32. Zhou, H.L., Suzuki, M., Randers-Pehrson, G., Vannais, D., Chen, G., Trosko, J.E., Waldren, C.A., and Hei, T.K., Radiation Risk to Low Fluences of α-Particles May Be Greater Than We Thought, *Proc. Natl. Acad. Sci.* USA, **98**(2001)14410–14415.

33. Magee, J. and Chatterjee, A., Radiation Chemistry of Heavy-Particle Tracks, 1. General Considerations *J. Phys. Chem.*, **84**(1980)3529–3536.

34. Shao, C.L., Stewart, V., Folkard, M., Michael, B.D., and Prise, K.M. Nitric Oxide-Mediated Signaling in the Bystander Response of Individually Targeted Glioma Cells, *Cancer Res.*, **63**(2003)8437-8442.

Appendix

RELEVANT KNOWLEDGE ON RADIOBIOLOGY

A.1. RADIATION EFFECTS AND RADIATION QUALITY

Changes in the physics, chemistry or biology of a target induced by interaction between ionizing radiation and the target material are all called radiation effects. The changes discussed here mean any changes of state – radiation caused heat, ionization, chemical-chain breaking, color change, electrical characteristic change, and biological reaction of the materials. All of these are examples of radiation effects. The measurement of radiation effects depends on these characteristics. For example, the number of unactivated cells, the number of the chromosomal breaks, the expression degree of other biological phenomena (e.g. the change in the biological organism state index), the electrical resistance, optical density, change in atomic valence, and so on, can all be used as indicators of radiation effects and for measuring these effects. No matter what the specific effects, the origin of these effects is the absorbed radiation energy caused by the interaction between incident radiation and the targeted material.

Energy absorption may be inhomogeneous in either space or time. This is extremely important for the manifestation of radiation effects. For example, when a material absorbs energy, the degree of heating (the observed temperature increase) may depend on whether the energy is totally absorbed instantly or progressively absorbed as a function of time. In the latter case, cooling of the material in the interval between two periods of energy absorption can affect the final outcome. Another example is gaseous ionization. If the ion concentration in an irradiated volume is used to indicate the radiation effect, the final effect will be strongly related to the distribution of energy absorption in time and space due to ion recombination.

Material cooling and ion recombination can be considered to be repairing effects, tending toward restoration of the initial state of the material damaged by the ionizing radiation. There are many examples showing this kind of effect. Normally observed radiation effects are not expected, because they usually cause biological cells to die or the cells and their components to be damaged. In these cases, the terms "injury" or "damage" are generally used instead of "radiation effect". However, it must be pointed out that some radiation effects may be beneficial. For example in radiation technology, ionizing radiation is often used to change the properties of materials in order to induce desired chemical processes. Repairing effect is an important reason for observation of radiation effects influenced by the time factor.

When radiation effects damage the sensitive structure of the irradiated material, the effect of the spatial distribution of the absorbed energy is particularly important. For example, for electronic devices composed of semiconductor components, characteristic changes are determined by the absorbed energies at sensitive locations in these components (especially p-n junctions). Genetic effects are also related to specific locations of chromosomes and, therefore, the absorbed energy per volume element containing chromosomes may be the decisive factor for producing radiation genetic effects. Chromosome volume consists of a small part of cell volume. Hence in living cells, even inside sub-cell structures, the spatial distribution of absorbed energy also plays an important role in determining the effects. As the primary reason for radiation-induced changes of materials is absorbed radiation energy, it is very natural to try to quantitatively link the extent of radiation effects with the quantity of energy absorbed. The macroscopic quantity describing the absorbed energy is the radiation dose D, given by

$$D = \Delta E/\Delta m, \tag{A.1}$$

where ΔE is the energy absorbed by the irradiated volume, and Δm is the mass of the irradiated volume. Dose is a macroscopic quantity, the average value of absorbed energy per unit mass. If the experimentally observed radiation effect is designated by η, (which is related to the dose), the dose-effect function can be expressed as

$$\eta = f(D), \tag{A.2}$$

where $f(D)$ is a function of dose.

To what extent is dose a general quantity for describing radiation effects? Is there only one dose high enough to predict the expected radiation effects? In some cases, the effect is really linearly proportional to the dose. In some other cases, this relationship is fairly complicated. In practical applications, equation (A.2) can be more conveniently expressed as

$$\eta = KD, \tag{A.3}$$

where K is a coefficient related to the radiation conditions, indicating those conditions which provide a linear proportionality between effect and dose. The concept of "radiation condition" involves characteristics of radiation types. If the classification of radiation conditions is based on various radiation components, coefficient K is found to be the characteristic parameter of radiation quality.

We know from dosimetry that for radiation safety purposes, the coefficient K (quality coefficient) is related to the linear energy transfer (LET) of the ionized particle. Thus for various types of radiations that have the same LET, the same effect corresponds to the same dose. So a mean LET can be used to describe the radiation quality. But the mean LET is not always a sufficiently characteristic quantity, or in other words, equal mean LETs may not indicate the same effect at the same dose. In this case, a more generally used characteristic for radiation quality may be the spectrum of the LET-dose. It can be expected from this that for the same LET-dose spectrum, the same effect corresponds to the same dose.

Appendix Table 1. LET dependence of some radiation quality factors (Q).

LET in water (keV/μm)	Q	Radiation
≤ 3.5	1	X-rays, γ-rays, electrons
7	2	
23	5	Protons, neutrons
53	10	
≥ 175	20	α particles, heavy recoils

If it is assumed that function (A.2) is continuous and has derivatives everywhere, the relationship between effect and dose can be expressed as an exponential series form such as

$$\eta = \sum_{n=0}^{\infty} K_n D^n, \tag{A.4}$$

where K_n is the expansion coefficient of the nth term.

Since when $D = 0$ there should be no effects produced, it is $K_0 = 0$. Coefficient K_n indicates the radiation quality. A quality factor Q is used to measure the influence of the micro-distribution of the absorbed dose on the effect. It is a function of the collision stopping power at the point concerned in water. Therefore Q is closely related to the LET of the radiation in water. Appendix Table 1 lists relations between LET of various kinds of radiations and Q. In the table, the Q value of X-rays or γ-rays is 1, that of fast neutrons is 10, and thus 10-Gy fast neutrons and 100-Gy X-ray will cause the same damage.

A.2. RELATIVE BIOLOGICAL EFFECTIVENESS (RBE)

A biological effect of ionizing radiation, or more strictly speaking, biological effectiveness, depends not only on the total dose absorbed in a specific time period, but also the energy distribution. The energy distribution along the particle path determines the degree of the biological effect produced at a certain dose. At the same dose, the biological effect of high LET radiation should be greater than for low LET radiation. In radiobiology, relative biological effectiveness (RBE) is normally used to indicate this difference. RBE is also called the relative biological effect coefficient. RBE is normally based on X-rays for comparison. Generally, either of the two following forms is used to express RBE:

RBE = (The dose of a 250-kV X-ray that produces biological effects) /
        (The dose of the actual radiation used that produces the same effect)

or

RBE = (The biological effect produced by radiation actually used) /
        (The biological effect produced by the same dose 250-kV X-ray)

The RBE varies with the dose used for comparison. It is best to compare the different biological effects at the mean inactivation dose or the lethal dose, namely,

$$\text{RBE} = \frac{D_0}{D'_0}, \tag{A. 5}$$

where D_0 is the mean inactivation dose or the mean fatal dose for X-rays, and D'_0 is the corresponding dose for the radiation used. But some literature also uses the medial lethal half dose (LD_{50}) or other indicated dose.

A.3. RADIATION SENSITIVITY

Due to its variety and complexity, living tissue reacts very differently to different kinds of radiation. Even though the radiation is given at the same dose and in the same way, different kinds of tissues or different individuals of the same kind may have great differences in their responses to the radiation. This appears to be a matter of difference between organism sensitivities to the radiation.

Generally speaking, the more highly evolved the organism (the more complicated the organizational structure), the higher the sensitivity to radiation. The radiation sensitivities of animals are higher than for plants, and those of virus and bacteria are still lower than for protozoa. In the vertebrates, radiation sensitivities of mammals are higher than those of lower-order vertebrates.

For the same organism, different varieties, individuals, development stages and physiological states as well as different organs of the same individual have different radiation sensitivities. Generally speaking, young tissue that has exuberant metabolism and fast growth has higher sensitivities than more slowly growing adult and old tissue. Tissues that have fast cell division are more sensitive than tissues with slowly dividing cells; thus the buds and roots of highly evolved plants are more sensitive than stem and leaves, and sprouting buds are more sensitive than dormant buds. Reproducing cells are more sensitive than nutrient cells; the seed embryo is more sensitive than the embryo milk; seeds at the milk-mature period are more sensitive than seeds at the completion of maturity. Note that during the storing period, irradiated seeds may change their response to radiation.

For a single cell, the cell nucleus is generally more sensitive than cytoplasm. The radiation sensitivities are also different when a cell is at different stages of development. A cell at the S and G_2 period is sensitive. In these periods when the cell is radiated, the cell period is much disturbed. Progress of the S and G_2 periods as well as progress from the G_2 period to the M period is thwarted. The radiation sensitivity at the G_2 period is the highest (the sensitivity order: $G_2 > S > G_1 > M$). The closer to the M period the radiation is, the longer the cell period is extended, and giant cells, death and cell non-division may occur.

Radiation sensitivity is a very complicated issue. It is related to many factors such as life morphology, structural characteristics, chemical composition, growth strength, and physiological status. But the essential dependence is on the DNA structure as well as the DNA content in the cell. The sensitivity difference between different varieties of the same species is also dependent on differences in genetic characteristics, self-repair ability and difference in metabolism processes.

A.4. INFLUENCE OF ENVIRONMENTAL FACTORS

A.4.1. Oxygen Effect and Oxygen Enhancement Ratio

Radiation effects in irradiated biological systems or molecules increase with increasing oxygen concentration in the medium. This phenomenon is called the oxygen effect. Normally the oxygen enhancement ratio (OER) is used to characterize the oxygen effect:

OER = (The dose that produces certain effect without oxygen)
/ (The dose that produces the same effect with oxygen)

The OER varies with various radiation types. When the radiation LET increases, OER decreases. For X-rays and γ-rays, OER is about 2.5 to 3.

The oxygen effect is a basic topic in radiobiology. Early interpretations of mechanisms of oxygen effects were limited by the knowledge of biology and biochemistry at the time. With development of science and technology, physicochemical interpretations have become gradually dominant. One of the important theories of the physiochemical mechanisms of the oxygen enhancement ratio is the assumption of oxygen fixing. This assumption considers that ionizing radiation induces free radicals in target molecules. If oxygen exists in the neighborhood of the molecule when it is radiated, the free radicals induced by the radiation will quickly combine with oxygen to form a group which hinders the biological functions of the target molecule:

$$R \rightsquigarrow R^{\circ} \xrightarrow{O_2} ROO.$$

As the lifetimes of many target-molecule free radicals induced by OH°, H° and hydrated electrons are extremely short, there must be oxygen before or during radiation so that the oxygen can effectively react with the free radicals to fix the radiation damage. According to estimation, oxygen fixing usually takes place within $10^{-3} - 10^{-2}$ µs after radiation. If there is no oxygen during this interval, the target molecule free radicals may quickly be transformed into normal biologically active molecules through "chemical repair". There are competitions among oxygen fixing, free radical decay, and intrinsic chemical repair.

Radiation effects at different levels are affected by oxygen to different degrees. For example, in the same bacterial strain, the influence of oxygen on the mutation effect is weaker than on the lethal effect. It has been found using a fast mixing technique that the effect of radiation after fast mixing of a cell suspension and oxygen is biphasic. The fast component may be the result of the radiation interaction with the membrane at cell shallow parts whereas the slow component may be a result due to the interaction with the deeper DNA.

Another interpretation of the mechanism of the oxygen effect involves an electron transfer assumption. Ionizing radiation ionizes the target molecules and produces free electrons, which can return to their original positions in the target molecules to "heal" them. Another possibility is that free electrons are transferred to an electron trap position and cause the target molecule to be damaged. Oxygen reacts with these free electrons and prevents them from returning to their original positions and thus fixes and aggravates target molecule damage.

A.4.2. Temperature Effect

The term "temperature effect" refers to the fact that the radiation sensitivity of many macromolecules and biological systems decreases with decreasing temperature during the radiation. The radiation sensitivity of irradiated saccharomycetin decreases by a factor of 2.5 when the temperature decreases from $0°C$ to $-33°C$. Amazingly, 70% of newly born young white rats survive when irradiated at a temperature below $0°C$ with a usually lethal dose (800R) of radiation.

In plants the temperature effect of radiation is also observed. For example, at very low temperatures of $-80°C$ to $-90°C$, plant seeds usually escape radiation damage. When barley seeds are irradiated at CO_2-ice temperature, the chromosome aberration rate decreases and the plant survival rate increases but the mutation rate is the same as for irradiation at room temperature. When wind-dried seeds of rice are irradiated with ^{60}Co-γ-rays at liquid-nitrogen temperature ($-196°C$), there is an obvious protection effect for the M_1 generation which escapes damage with a protection coefficient of 2. Various mutation rates of the M_2 generation are not decreased because of this treatment and the fruiting rate of the M_2 generation is still protected from radiation damage. Thus we expect to apply high-dose radiation at very low temperatures to increase the total mutation effectiveness of aberration breeding.

The low-temperature protection effect may be due to decreased activity of radiation-induced free radicals by reducing the interaction between them and oxygen. Besides the low-temperature protection effect, heat treatment before or after radiation can also reduce radiation damage of the seeds. Treatment at $75°C$ for $0.25 - 96$ hours before radiation has a protection effect for seeds irradiated with very low to very high dose X-rays. After the seeds are irradiated, treatment at $75°C$ or $85°C$ for 0.25 hours can reduce radiation sensitivity which the seeds exhibit when they germinate under an oxygen condition.

The heat treatment effect is probably due to an increase in free-radical activity, which increases the probability of free radical recombination and thus prevents free radicals that have high damaging abilities arising from interactions with oxygen. Simultaneously, heat treatment before radiation can remove oxygen from the seeds and thereby reduce the oxygen effect.

A.5. DIRECT AND INDIRECT INTERACTIONS

The interaction of ionizing radiation with living tissue includes two mechanisms, direct interaction and indirect interaction. Which interaction is primary and which is secondary have been discussed for many years. Recent studies have corrected the lopsided understanding which takes the radiation of dry systems as direct interaction and the radiation of water-containing systems as indirect interaction, and gives a fairly strict definition of direct and indirect interactions at the molecular level.

Direct interaction with ionizing radiation refers to damage caused by the radiation energy absorbed by some important biological substances (such as DNA molecules) or structures themselves and delivered or released inside them. This means that energy absorption and damage take place inside the same molecule or structure. If one molecule absorbs energy while another molecule is damaged, this is indirect interaction. This means that the primary process of radiation absorption occurs in the "environment" of the damaged biological molecule. This environmental substance can be closely neighboring with other molecules or the medium such as water. After radiation energy is absorbed by the environmental substance, damage is indirectly caused via either energy exchange among molecules or the release and diffusion of highly-active free radicals (including radiation-dissolved water products and biomolecular free radicals) that attack biomolecules.

How to identify direct interaction and indirect interaction and their respective contributions when radiation interacts with the cells?

When living cells are killed by ionizing radiation, death can be caused by either direct interaction or indirect interaction, or both. It is generally thought that both interactions are important inside living cells. In order to estimate quantitatively the direct interaction effect, three methods can be used. They are freezing, drying and adding protectants into the system. It can be assumed that in these cases cell death is mainly the result of direct interaction. Cell death under humid circumstance without protectants and freezing is caused by both direct and indirect interactions. The difference between these two cases of cell death is thought to be the result of indirect interaction. As freezing does not cause damage to many kinds of cells (such as bacteria, yeast, cultured mammal cells), most researchers use this method of freezing. Drying is only used for some special types of organisms that can survive in extremely dry conditions, such as bacterial bud-cells, seeds, eggs of sea shrimps and virus. The method of adding protectants is not often used.

In most cases, radiation biology effects (such as cell death) produced by direct and indirect interactions are both very important and each contribution relates to radiation conditions. For example, RNA enzyme irradiated by γ-rays in the dry state and in solution (5 mg/ml) produces inactivation from the latter condition that is 100 times more sensitive than from the former (D_{37} = 420 kGy and 4 kGy, respectively). This indicates that at the usual solution concentration, about 99% of the RNA enzyme molecules are inactivated due to the action of water free radicals, but only 1% become inactive due to the enzyme molecules themselves absorbing radiation energy. Another example is that the radiation required for tobacco flower-leaf virus particles to lose their activity in the almost dry state is 10^6R, but at a solution state diluted 1 million times, the required radiation drops down to 10^3R. Thus, in diluted solution, radiation indirect interaction is dominant. However there has not yet been any evidence or example to show radiation effects determined solely by indirect interaction.

The above mentioned "environmental substance" is not only limited to water; there exist indirect interactions also for the dry state. For cells and organs, "dry" does not mean absolutely no water. Even though a non-water enzyme is irradiated, indirect interactions still occur. The effect of producing atomic hydrogen (i.e. the hydrogen free radical H°) from biomolecules (MH) may be the most important indirect interaction mechanism: MH $\rightarrow M^+ + H^+$. Miller (1968) discovered that when dry DNA and bacteriophages were irradiated, hydrogen atoms were released as the characteristic spectral line of hydrogen atoms was shown in the electron spin resonance wave spectrum. Other evidence is that when hydrocarbon compounds, amino acids, proteins and nucleic acid components are irradiated, the gas which is produced is mainly composed of molecular hydrogen. The reaction between the released H^+ in radiation and undamaged molecules is mainly the addition at double bonds and thus another hydrogen atom can be removed to form a hydrogen molecule.

A.6. TARGET THEORY

A.6.1 Basis and Application Conditions of Target Theory

Target theory estimates the existence and size of the "target" in the radiated cell from a biophysics point of view. In 1924, Croether proposed that there exist "target" structures sensitive to radiation reaction within the cell. Since then, by continuous improvement, the American radiobiologist Lee extended this simple model to various LET rays and more extensive biological systems. His "Actions of Radiation on Living Cells" is the representative work on target theory.

Generally speaking, the basic points of target theory are:

(1) There exist plant and animal parts sensitive to radiation. Damage to these parts will result in production of biological effects. This sensitive structure is called the "target".

(2) Light and ionizing radiation impact the target region in the form of photons and ion clusters, and the probability of a target hit follows a Poisson distribution.

(3) One or more than one of the hits can produce radiation biological effects.

The one-hit effect is the simplest assumption in target theory. If we consider that the interaction of the ray with the biological materials follows the "one-hit" principle, the interaction of ionizing particles with the target follows a Poisson distribution. If $vD_0 = 1$, then $v = 1/D_0$, where D_0 is called as the inactivation dose or mean lethal dose. The target volume can be deduced from this formula. The smaller the target volume is and the lower the ionizing density of the ray is, the higher the accuracy of the ray. The larger the target is and the higher the ionizing density of the ray, the higher the error.

Target theory introduces the quantum viewpoint into biology for the first time, and has played a great role in promoting the study of radiobiology from the whole-body level to the cell and molecular level. There are many examples successfully showing that target theory explains "one-hit"-caused radiobiological phenomena and measures the biological macromolecules having molecular weights of $10^4 - 10^8$ d. However, the radiobiological phenomena that the theory can explain are limited. Its application must be limited within the organism radiated without various biochemical reactions or biophysical processes. These limitation conditions are normally:

(1) All of the studied effects must be caused by interaction between ionizing radiation and biological macromolecules and without the influence of indirect interactions.

(2) The effects studied must be completely dependent on the entire macromolecule, namely, the macromolecule should not have extra accessory structures insensitive to the radiation.

(3) It must be guaranteed that before the radiation produces active radical reactions, there is no occurrence of repeating reactions recapturing electrons, i.e. no chemical repairing of damage.

(4) All of the measured macromolecules are completely identical and reactions are not affected by the dose rate.

Therefore, applications of target theory to explain damage to biological macromolecules and calculations of target size must be very carefully undertaken.

A.6.2. Cell Survival Curve Models

On the base of the target theory, many theoretical models of cell survival curves have been developed, such as the one-hit model for a single target, the model of one-hit of many targets, and the linear-square model using DNA double strands as the target.

A.6.2.1. One-Hit Model for a Single Target

This model assumes that each cell has only one target and if the target is hit the cell is inactivated. If all individual cells in a cell group are completely identical, V as the cell volume and v as the target volume, then when a cell is hit once the probability of hitting the target is $\rho = v/V$. When the cell is hit d times, the probability of hitting the target h times is

$$P(\rho, h, d) = \frac{(\rho d)^h}{h!} e^{-\rho d}. \tag{A.6}$$

The cell survival probability when the cell target is hit 0 times is

$$S = P(0, 0, d) = e^{-\rho d} = e^{-(v/V)d} = e^{-vD}, \tag{A.7}$$

where $D = d/V$ is the hitting times per unit volume of the cell (related to dose). This formula is the same as the one-hit inactivation formula in target theory. Normally it is written as $S = e^{-\lambda D}$, where λ is called the inactivation constant.

A.6.2.2. One-Hit Model for Many Targets

This model, proposed by Elknid, assumes that a cell has N identical targets, that only when all of the targets are inactivated the cell dies, and that the inactivation of each target follows the one-hit inactivation principle. Therefore, for dose D, the probability of each target being inactivated is $1 - e^{-\lambda D}$ and the probability for N targets being inactivated is $(1 - e^{-\lambda D})^N$, and hence the survival rate of the cell group is

$$S = 1 - (1 - e^{-\lambda D})^N. \tag{A.8}$$

This model is frequently used, but the biological meaning of N is not clear. The theory assumes N to be the target number of the cell. However, in experiments the N value is generally not an integer, but varies widely for the same cell under different experimental conditions. Thus, N is now not called the target number but the "extrapolation number".

This model has zero initial slope. But in reality, many cell survival curves do not zero initial slopes; instead, the survival rates start to decrease at very low doses. Thus, Bender et al modified this model as

Appendix Table 2. Some cell survival curve models in radiobiology.

Category	Model name	Mathematic expressions
Bio-physical Models	Single target one-hit	$S = \exp(-\lambda D)$
	Many-targets one-hit	$S = 1-[1-\exp(-\lambda D)]^N$
	Modified many-target one-hit	$S = \exp(-D/D_1)\{1-[1-\exp(-\lambda D)]^N\}$
	Many-points many-targets one-hit	$S = \{1-[1-\exp(-\lambda D)]^\tau\}^u$
	Single-target many-hits	$S = \displaystyle\sum_{k=0}^{N-1} \frac{(\lambda D)^k}{K!} \exp(-\lambda D)$
	Double-unit interaction	$S = \exp(-\alpha D-\beta D^2)$
	Double-strand breaking	$S = \exp[-p(\alpha D+\beta D^2)]$
Repair Models	Repair–missense-repair	$S = \exp(-\beta D)(1-\beta\theta TD/\mu)^u$
	LPL repair	$S = \exp[-(\eta L-\eta PL)D\{1+\eta PLD[1-\exp(-\varepsilon PLtT]/\varepsilon\}^\varepsilon$
	Saturated repair	$S = \exp\{-\rho(\alpha D-Co)/[1-(Co/\alpha D)\exp[kT(Co-\alpha D)]]\}$

$$S = e^{-D/D1}[1-(1-e^{-D/Dn})^N].\qquad (A.9)$$

A.6.2.3. Linear-Squares Model

This model takes the DNA strand as the target and considers that for dose D the average breaking number of double strands in each cell is $N = \alpha D + \beta D^2$. If the probability of cell death caused by an initial double-strand breaking is P, the cell survival rate is

$$S = \exp[-P(\alpha D + \beta D^2)].\qquad (A.10)$$

The above models are developed based on either the basic content of target theory or micro-dosimetry and the theory of double-strand breaking. They are in the category of biophysics or physics. There is another kind of models, the so-called repair models. These models assume that radiation damage can be repaired by repair systems inside the cell, and this sort of the repair ability is dependent on the radiation dose or can be saturated by increasing the dose.

Both kinds of models have similarities. Some of their expressions can be deduced from either model. Nevertheless, these biophysical models have broader applications. Appendix Table 2 lists some radiobiological models of cell survival curves.

INDEX